Digital Peacebuilding?

Digital Peacebuilding?

The Socio-Technicality of Transforming Armed Conflict

Andreas T. Hirblinger

OXFORD
UNIVERSITY PRESS

OXFORD
UNIVERSITY PRESS

Great Clarendon Street, Oxford, OX2 6DP,
United Kingdom

Oxford University Press is a department of the University of Oxford.
It furthers the University's objective of excellence in research, scholarship,
and education by publishing worldwide. Oxford is a registered trade mark of
Oxford University Press in the UK and in certain other countries.

Published in the United States of America by Oxford University Press
546 Fifth Avenue, New York, NY 10036, United States of America

British Library Cataloguing in Publication Data
Data available

Library of Congress Control Number: 2026935457

ISBN 9780198986522

DOI: 10.1093/9780198986553.001.0001

Printed and bound by
CPI Group (UK) Ltd., Croydon, CR0 4YY

The manufacturer's authorized representative in the EU for product safety is
Oxford University Press España S.A. of Parque Empresarial San Fernando de Henares,
Avenida de Castilla, 2 – 28830 Madrid (www.oup.es/en or product.safety@oup.com).
OUP España S.A. also acts as importer into Spain of products made by the manufacturer.

MIX
Paper | Supporting responsible forestry
FSC® C013604

"To Laila, my love
and to the survivors of war and violence."

Acknowledgements

This book is not the product of my own making. While I authored every single word of it, the knowledge, ideas, and observations it contains were co-produced by many people and things. A description and acknowledgement of these socio-technical relations in all their complexity, accounting for the distributed agency that made me produce this book, would certainly go beyond what the reader can reasonably be asked to digest. It would also require a method that seems rather impossible to produce, as I explain in Chapter 2. Therefore, I will rely on established conventions but not before having acknowledged all that can't be named.

To start, I would like to express my gratitude to the Swiss National Science Foundation (SNSF) for the generous financial support that made the research for this book possible, particularly through the Ambizione Grant for the project 'An Apomediate Peace: The Role of Digital Technologies in International Peacebuilding'. The SNSF also funded the Open Access publication of this book. I owe special thanks to the Centre on Conflict, Development and Peacebuilding (CCDP) at the Geneva Graduate Institute, especially Keith Krause and the CCDP team, whose encouragement and multifaceted support provided more than just an institutional 'birthplace' for my research on digital peacebuilding. They played a pivotal role in the years before and during the 'An Apomediated Peace' project—offering critical advice, introducing me to the 'Geneva ecosystem', and helping me establish the many contacts and insights that shaped the design of this research. The Peace Research Institute Oslo (PRIO) became my second intellectual home during the research and writing of this book. My heartfelt thanks go to my many colleagues at PRIO for the stimulating discussions on technology, peace, and conflict, and for their warm welcome during my time there as research associate. PRIO's Strategic Initiative on Digital Peacebuilding offered an invaluable platform to develop and refine some of the central concepts of this book, and I am thankful for the support the Initiative received. One of its outcomes was the co-authored article 'Digital Peacebuilding: A Framework for Critical–Reflexive Engagement', published in volume 24, issue 3 of *International Studies Perspectives*, which informs parts of Chapter 2. Kalypso Nicolaïdis and Michele Giovardi at the European University Institute (EUI)'s School for Transnational Studies and the Global PeaceTech Hub provided critical support in the final stretches of the project, by co-organizing an international research workshop relevant to the book's core themes, held in Florence in autumn 2024. Finally, I am grateful for the support of many colleagues at the Norwegian Institute of International Affairs (NUPI), where I started a new role as I finalized the book in early 2026, and where I hope to carry this book's research agenda forward.

Equally critical were the many exchanges with colleagues working on related questions. Roger Mac Ginty challenged my idea of apomediation by waving two Northern Irish newspapers at me; Bruno Oliveira Martins inspired deeper reflections on the socio-technicality of digital peacebuilding; and Kristofer Lidén persistently pushed my understanding of, and approach to, the study of peacebuilding—sometimes through late-night comparisons to witchcraft. Martin Wählisch grounded my work with his practical insights and pragmatic optimism; and Lisa Schirch helped sharpen my thinking through feedback at several stages of the project. I also benefitted enormously from collaborations and exchanges with friends and colleagues in policy and practice, including Ville Brummer and Felix Kufus at the CMI-Martti Ahtisaari Peace Foundation who collaborated with me for the research of the article 'Leveraging Digital Methods in the Quest for Peaceful Futures: The Interplay of Sincere and Subjunctive Technology Affordances in Peace Mediation' published in volume 27, issue 11 of *Information, Communication & Society*, which informs parts of Chapter 5. The book was also inspired by the thoughts and observations of many other practitioners, including Guy Patrick Banim, Jonathan Cohen, Véronique Dudouet, Ahmed Eleiba, Enrico Formica, Julie Hawke, Sanjana Hattotuwa, Asif Khan, Helena Puig Larrauri, David Lany, Johanna Poutanen, Juan-Díaz Prinz, and Lena Slachmuijlder.

This book would not have been possible without the dedicated work of several research assistants—Zineb El Ouassini, Swetha Ramachandran, Sara Kallis, and Fabian Hofmann—who helped me build the empirical and conceptual basis of this book, through desk studies, data collection, and data analysis. Benedicte Engesland at NUPI supported with the final touches and a human-made book index. I would especially like to mention Sara's desk research and data analysis for the article 'When the Digits Don't Add up: Research Strategies for Post-Digital Peacebuilding", published in volume 59, issue 3 of *Cooperation and Conflict*, which informs parts of Chapter 2. Likewise, Fabian's contribution to the empirical research and concept development of Chapter 6 deserves a special thanks, as well as his feedback on the early drafts of Chapters 1–4. Sara and Fabian's dedicated support for the project went far beyond my expectations. Working with them also taught me the value of leading research projects with humility, and the merits of collaboration that transcends the conventional hierarchies of academia.

I am also indebted to the several country-based researchers in Sri Lanka and South Sudan, without whom the empirical research for several chapters would not have been possible. They did far more than support fieldwork, interviews, and focus group discussions by generously sharing their experiences, contextual knowledge, and professional networks. I am especially grateful to Hasini Haputhanthri, who taught me much about Sri Lankan politics and society. Together with Sara Kallis, Hasini co-authored the article 'DIY-Online Reconciliation? The Role of Memes in Navigating Inter-Group Boundaries in the Context of Sri Lanka's 2022 Political Crisis", published in volume 10, issue 4 of *Social Media + Society*, which informed parts of Chapter 6. I would also like to thank two anonymous South Sudanese research assistants who helped me navigate the intricacies of the country's peacebuilding landscape and undertook field trips I could not carry out myself. Not least, I would like to acknowledge the important role of the many research participants and interlocutors who willingly shared their insights and knowledge during interviews and discussions, at times at great risk to themselves. I appreciate their courage and resilience in the face of conflict, violence, and insecurity and acknowledge my own privilege in having the means to learn from them while not sharing the same burdens.

I am also grateful for the curiosity, openness, and willingness to collaborate shown by individuals working in peacebuilding organizations. They granted access to their projects and activities, shared critical insights, and supported my empirical work logistically. While not all their initiatives are covered in this book, learning from them was an important part of my research journey. My sincere thanks go to the staff at the CMI–Martti Ahtisaari Peace Foundation, the Search for Common Ground offices in South Sudan and Sri Lanka, the United Nations Mission in South Sudan, the United Nations Support Mission in Libya, DefyHateNow South Sudan, the Community Empowerment for Progress Organization (CEPO), the Sentinel Project, Hashtag Generation, Sri Lanka Unites, the International Centre for Ethnic Studies (ICES), the Sarvodaya Institute of Higher Learning, the National Peace Council of Sri Lanka, FHI360, Groundviews, the Gesellschaft für Internationale Zusammenarbeit Sri Lanka Office, and the United Nations Resident Coordinator's Office in Sri Lanka.

This book is dedicated to my wife Laila Matar, who helped ground my research in deeper meaning. She was also the first person to agree to read the book from start to finish without financial compensation. I'm deeply grateful for everything that she provided along the way, for her genuine excitement and encouragement, and her intellectual curiosity and incisive thinking. And for always lifting me up.

Contents

List of Abbreviations x

Introduction **1**

1. **Digitalization, Technology, and the Voids of Peacebuilding** **25**
2. **The Socio-Technicality of Digital Peacebuilding** **50**
3. **Preventing Conflict: Countering Harmful Social Media Speech** **82**
4. **Mitigating Conflict: Localized Early Warning and Response** **112**
5. **Making Peace: AI-Enhanced Mediation and Dialogues** **153**
6. **Building Peace: Everyday Online Interactions for Political Change** **189**

Conclusion **224**

References 236
Index 263

List of Abbreviations

AI	Artificial intelligence
ANC	African National Congress
API	Application Programming Interface
APSA	African Union Peace and Security Architecture
ASRSG	Acting Special Representative of the Secretary General
AU	African Union
AUSC	African Union Security Council
CAIN	Conflict Archive on the Internet
CCC	Conflict Carrying Capacity
CDC	Community Development Centre
CEPO	Community Empowerment for Progress Organization
CEWERU	Conflict Early Warning and Early Response Unit
CEWS	Continental Early Warning System
CMI	Finnish Crisis Management Initiative
CPA	Center for Policy Alternatives
CPRMS	Community Peace Response Mechanisms
CRD	Crisis Risk Dashboard
CSOs	Civil society organizations
DD&R	Digital divide-and-rule tactics
DIY	Do-It-Yourself
DPPA	Department of Political and Peacebuilding Affairs
DSM	Diagnostic and Statistical Manual
DUP	Democratic Unionist Party
ESM	European Single Market
EU	European Union
EUI	European University Institute
EWER	Early Warning and Early Response
EWOs	Early Warning Officers
FEWER	Forum for Early Warning and Early Action
GEDS	Global Event Data Survey
GIS	Geographic Information Systems
GNC	General National Congress
GNU	Government of National Unity
HoR	House of Representatives
ICES	International Centre for Ethnic Studies
ICT	Information and Communication Technologies
IDP	Internally Displaced Persons
IGAD	Intergovernmental Authority for Development
IPI	International Peace Institute
IPIE	International Panel on the Information Environment
IR	International Relations

ITU	International Telecommunications Union
LLMs	Large Language Models
LPDF	Libya Political Dialogue Forum
LTTE	Liberation Tigers of Tamil Eelam
M&E	Monitoring and evaluation
MDGs	Millennium Development Goals
MENA	Middle East and North Africa
NGOs	Non-governmental organizations
NLP	Natural Language Processing
NPP	National People's Power
NSD	National Strategic Dialogue
NSS	National Security Service
OFAI	Austrian Research Institute for Artificial Intelligence
ONUR	Office of National Unity and Reconciliation
PA	Palestinian Authority
PANDA	Protocol for the Assessment of Nonviolent Direct Action
PCCA	People's Coalition for Civic Action
PoC	Protection of Civilians
PRG	Peace and Reconciliation Group
PRIO	Peace Research Institute Oslo
PTA	Prevention of Terrorism Act
PVE	Preventing Violent Extremism
RTLM	Radio Télévision Libre des Mille Collines
SDGs	Sustainable Development Goals
SfCG	Search for Common Ground
SNSF	Swiss National Science Foundation
SRSGs	Special Representatives of the Secretary-General
SSPRC	South Sudan Peace and Reconciliation Commission
STS	Science and Technology Studies
TF-IDF	Term Frequency-Inverse Document Frequency
UK	United Kingdom
UN	United Nations
UNDP	United Nations Development Programme
UNGA	United Nations General Assembly
UNPBF	UN Peacebuilding Fund
UNSC	UN Security Council
UNSC	United Nations Security Council
UNSG	United Nations Secretary-General
UNSMIL	United Nations Support Mission in Libya
US	United States of America
USIP	United States Institute of Peace
VPNs	Virtual private networks
VRA	Virtual Research Associates
WEIS	World Events Interaction Survey
WSIS	World Summit on the Information Society

Introduction

Moments of crisis can be very revealing. When the COVID-19 pandemic swept the world in early 2020 and interrupted almost all spheres of human activity, peacebuilding efforts also seemed in peril. Diplomats, and many experts involved in international efforts to maintain peace and security, expressed concern that lockdowns and travel bans would create more violence and suffering. Among them was United Nations (UN) Secretary-General António Guterres, who called for a global ceasefire in April 2020, warning that the pandemic would exacerbate existing conflicts and create new ones (UNSCOL 2020). Many also expressed the worry that the lack of in-person contact would erode trust among population groups and society and the state, create more inequality, marginalization, and grievances, and ultimately lead to more tensions and armed conflict (United Nations 2020a). At the beginning of the pandemic, the UN and many other international, national, and local peacebuilding organizations struggled to maintain their activities, as the need for social distancing massively reduced opportunities for the in-person interactions that many believe are at the heart of almost all efforts aimed at preventing, ending, or transforming conflict: from monitoring ceasefires to keep hostilities at bay, to facilitating dialogues to discuss the content of peace agreements, and organizing and mobilizing for peaceful protests against injustice and violence (see, for instance, Van Der Lijn 2022; Mercy Corps 2020).

However, the need to avoid human-to-human contact also unveiled a thus-far little-acknowledged reality. Decades of digitalization and technical innovation had led to the global spread of digital Information and Communication Technologies (in short, digital ICTs or digital technologies) which had inserted themselves in the relations between us humans in the most intricate ways, shaping how we wage wars and unleash armed violence, as well as how we prevent such conflicts and build peace. This became evident when, in weeks, if not days, most professional peacebuilding organizations moved almost all of their in-person activities online—testing, expanding, and improving their capabilities to facilitate dialogue and virtual meetings, collecting data and evidence remotely, and supporting social and political change processes meant to reduce the risks of conflict and violence through social media campaigns, to give just a few examples. This embracing of digital peacebuilding approaches appeared rapid to many. Yet, it wouldn't have been possible without a decades-long prelude of innovation and experimentation. Nonetheless, during much of the training given to boost digital skills and competencies, many practitioners

Digital Peacebuilding?. Andreas T. Hirblinger, Oxford University Press. © Andreas T. Hirblinger (2026).
DOI: 10.1093/9780198986553.003.0001

and researchers raised concerns about how 'going digital' would affect peacebuilding dynamics. For instance, some argued that the reliance on digital technologies would reduce opportunities for 'meaningful' participation (Rahmaty and Jaghab, n.d.) and the building of trust and relationships (Clark and Alberti 2021), or indeed that it could lead to a 'missing sense of peace' (Bramsen and Hagemann 2021).

Despite decades of experimentation with digital technology, there thus remained a widespread sense that building peace was and should remain primarily an effort between humans. In contrast, making war always involved technology—from Acheulean hand axes thrown at enemies in the Stone Age to crewless aerial vehicles and artificial intelligence (AI)-guided targeting systems that not only make today's wars seemingly less human but also less and less humane. Yet, peacebuilding appears to be about and for humans who meet, negotiate, and change how they interact and relate to one another. Maintaining this view, though, requires a wilful blindness to how ICTs have long begun to shape the practices that we today subsume under the term 'peacebuilding', contributing to more effective approaches and, likewise, new challenges and, at times, failures. However, some have begun to reckon with the long durée of technological innovation (Diplo 2021), citing, for instance, late nineteenth- and early twentieth-century examples of how the manipulation of information transmitted via telegraphs, sometimes wilfully and sometimes accidentally, jeopardized peacemaking efforts and favoured war. For instance, in 1870, German Chancellor Otto von Bismarck intentionally truncated a telegram that he was asked to forward to his French counterparts to provoke a war between France and Prussia. In the months leading up to the First World War, European diplomats struggled to understand how telegraphy affected their communication. Quickly exchanged telegrams triggered boiling tempers for reasons that also help to explain why contemporary mobile messaging services are rather unsuitable for deescalating conflicts, such as the speed of communication, the deliberate manipulation of information in ways that wouldn't be possible in in-person interactions, and a failure to interpret the meaning of messages due to lack of context (Kern 2003, 274–5; Frankenberger 2024; García-Gómez 2020).

While electric telegraphy, such as that patented by Samuel Morse in 1838, is not commonly viewed as a 'digital technology', its way of operation and effects resemble those of contemporary digital messaging services: the breaking down of human natural language into text transmitted via a digital signal enables remote communication (Reeve 2003). This vice versa means that the technology 'inserts' itself between the humans who exchange information. Through such features, tools such as the telegraph began to influence not only how information critical for maintaining or building peaceful relations was exchanged, but also how those who used it made sense of war and violence, and how they understood and embarked on the task of preserving or building peace. For instance, the brevity of the telegraph, Bismarck's ability to truncate it, and the receiver's inability to verify the information through other means, must sound oddly familiar to those who aim to fight disinformation and misinformation on social media, which thrives on the shortness of social media posts, the users' ability to manipulate content, and challenges to obtain facts about what really has happened. The same holds for many other contemporary digital technologies. Whether from a century ago or today, many consequences of digitally mediated human interaction are comparable, for instance, the fact that digital data often lack the complexity and contextual richness of the information that humans can obtain through physical presence and sensory encounters with the world.

Nonetheless, in the face of the widespread conviction that building peace is and should be a human endeavour, the term 'digital peacebuilding' entered the field. In academic research, it has come to delineate the sum of peacebuilding activities that rely on the processing of digital information (Hirblinger et al. 2022, 268). Practitioners associate it mainly with the harnessing of digital tools to enhance peacebuilding outcomes, including those focused on the online sphere (see, for instance, Alliance for Peacebuilding 2021). These understandings prompt a delicate and vital question that permeates the research, policy, and practice of digital peacebuilding: how should we comport ourselves vis-à-vis the many 'tools' and 'applications' that fill the arsenal of digital peacebuilders? As in many other spheres of life, the digitalization of peacebuilding is primarily viewed as something that affords us all sorts of software and hardware—things we can touch and use. Yet, in many ways, tools are only the visible 'tip of the iceberg'. Underpinning, feeding off, and nourishing these material artefacts of the digital is a tremendous number of additional factors—some material and some not—that shape the prospects of building peace today. This book sets out to explore them.

Digital Escapism

You may know what digital escapism is and what it feels like. You find yourself slouched on the sofa after a long day of work, your hands loosely clasped around your precious smartphone, your fingers semi-automatically scrolling through a social media feed to stimulate your brain's reward system. Or perhaps, defying your exhaustion, you download the latest mobile application to start a new meditation routine, work out faster and more efficiently, buy more goods cheaper, or optimize your household budget. All the while, what you really need is a good night's sleep to have the headspace to tackle the more fundamental, underlying challenges that have piled up to make your life less manageable and joyful, such as the combination of a taxing job market due to cuts in academic funding and consumption habits conditioned by personalized ads that make you spend more money and less time with friends and family. At least, that's how digital escapism can feel for me. The finger on a smartphone screen is a good metaphor for the relationship with the world that digital tools encourage. It is often a superficial one. However, connected to that screen, lying below and beyond it, is a much complex constellation of factors determining how you meet your interests and needs. Some of these factors are digital, such as recommender systems that serve you a mix of content and ads based on your social media profile. Others are social, such as the capitalist market and regulatory environment that conditions the design of applications and platforms. However, in your digital escape, you won't see how these factors shape your behaviour, experiences, and the long-term effects on your life.

Something similar can happen when we study digital peacebuilding. Our work often remains too focused on the newest and most exciting tools that promise to make a difference while paying too little attention to the context that conditions their design, use, and impact on peacebuilding. Consequently, our engagement with (digital) peacebuilding remains somewhat superficial. Let's explore this by taking a look at the recent scholarship. I embarked on the research for this book at a time when (information and communication) technologies for peacebuilding and PeaceTech had already been the topic of such a large number of academic articles, forums, and book chapters that it's impossible to do justice to all of them here (for instance, Kahl and Larrauri 2013; Welch, Halford, and Weal 2015; Tellidis and Kappler 2016; Firchow et al. 2016; Tellidis 2020; Duursma and Karlsrud 2021). On the other hand, I came across only one monograph that exclusively discusses digital peacebuilding and PeaceTech. The book, written by Christine Bell (2024), describes the emerging global PeaceTech 'ecosystem' and focuses on the practical and normative challenges of 'doing' PeaceTech. Notably, Bell (2024, 44) cautions against the appeal of 'shiny new things', illustrating how the plethora of current PeaceTech projects is fuelled by the 'conviction that new forms of digital innovation offer immediate improvement of peacebuilding practices'. I very much agree with her.

There is by now a considerable body of published research focused on specific digital applications relevant for a broad range of activities falling under the umbrella of peacebuilding, such as conflict forecasting, peacekeeping, or peace mediation. Discussing this diverse body of work in detail would require much more space than is available in this Introduction. What's important to highlight, however, is that many of these contributions tend to take a tool- and solution-centred perspective when discussing the role of technologies in peacebuilding. A recent example is Branca Panic and Paige Arthur's joint volume, *AI for Peace*, which provides a practice-oriented introduction to emerging AI technologies in peacebuilding and conflict prevention. The book makes the case that, if used in the right way, machine learning, Natural Language Processing, and image processing harbour the potential to support the collection and processing of vast amounts of data, help uncover hidden patterns, and enhance the productivity of peacebuilders. However, the authors also acknowledge that employing AI for peace requires balancing technological benefits 'with an awareness of its malicious use and unintended consequences' (Panic and Arthur 2024, 4). This view is not unique; the discourse on digital technology and peace is commonly Janus-faced, emphasizing both the benevolent and the malevolent uses of tech (Hirblinger et al. 2022). However, the 'negative effects' of well-intended applications are commonly reduced to an afterthought or a section at the end of the project template. Yet, there's a more fundamental mechanism at work: we tend to be concerned first and foremost with the utility of a tool, while the social and political context of its use becomes a matter of concern mainly when considering the tool's impact.

The risk with focusing on tools first and on their impact on society and politics second is that we pay too little attention to how the social embeddedness of technology conditions the design and use of technology in the first place. With that, we might also lose focus on the bigger picture of global politics and how it relates to digital peacebuilding. Many discussions of individual digital applications do briefly address the fact that the context in which these tools are designed and employed matters, including the role of national politics, international bureaucracies, or a market of demand and supply. However, what's missing is a research perspective that does so comprehensively. Indeed, signs for the need to look beyond the immediate utility of tools were arguably already visible in the earliest explorations of what we conceive of as digital peacebuilding today, which focused on the potential of data-driven conflict early warning but also discussed the risks that such systems would be dominated by staff based in the headquarters of organizations in the Global North (Rupesinghe and Kuroda, 1992). Early warning continues to be a vibrant research field today, mainly concerned with the scientific exploration of the opportunities and limits of statistical and machine-learning models (see, for instance, Cederman and Weidmann 2017; Hegre et al. 2019), yet operational challenges and political sensitivities seem likewise crucial to consider (Wählisch 2020).

In addition, there is considerable research on how digital technology can enhance international peacekeeping, including conflict analysis, intelligence generation, and reconnaissance (see, for instance, Karlsrud 2014; Duursma and Karlsrud 2019); ceasefire monitoring (see, for instance, Hug and Mason 2022; Verjee 2022), or strategic communications (see, for instance, Williams 2018; Leib 2023). Such contributions commonly raise concerns about how the effectiveness of digital solutions is conditioned by aspects such as bureaucratic routines, operational mandates, and the political will of conflict parties. Furthermore, a growing body of research focuses on the use of digital technologies in peace mediation and dialogues (see, for instance, Varela 2021; Hirblinger 2022), much of which points to the need to align and reconcile digital approaches with the methodological preferences of professional mediators. The use of digital technologies for reconciliation, dealing with the past, and longer-term conflict transformation is comparably less studied (see, for instance, Ataci 2022; Silva 2018), yet emerging research points to the need to consider the cultural and psychological dimensions of building peace. In addition, some research has focused on how digital technologies can enable empowerment, participation, and crowdsourcing approaches to peacebuilding, which highlights the relevance of digital innovations for civil society mobilization and social movements (see, for instance, Kahl, McConnell, and Tsuma 2012; Abubakar and Dasuki 2018).

This book builds on such earlier research with one central goal: to advance an approach to the study of digital peacebuilding without using the digital as an escape from society and politics. Suppose we were to occupy ourselves only with individual applications and their effects. In that case, there's a risk that we would be carried away by some sort of tech-solutionist optimism that is powered by the sentiment of a 'digital revolution', while losing sight of the bigger picture of global politics that conditions peacebuilding as a professional field, as well as the prospects of peace. Indeed, news about groundbreaking innovations in digital technology, and currently especially AI, floods our social media threads, podcasts, and newsletters almost daily, promising to make our work faster, easier, and more impactful. At the same time, the number of armed conflicts drastically increased over the past fifteen years, and the number of battle deaths reached the highest point for the post-Cold War period in 2022 (Rustad 2024). Many dedicated to the work of preventing, terminating, and overcoming armed conflicts find themselves confronted with a sense of doom and powerlessness in the face of the breakdown of the canon of liberal policy that underpinned most international peacebuilding efforts for two decades, a crisis of multilateral collaboration, dwindling resources, and emboldened authoritarian and populist governments that not only manage but also instigate and exploit conflicts for their own ends. I will look at this next.

Peacebuilding in Turmoil

Contemporary digital peacebuilding takes place in a global environment that has changed considerably since the end of the Cold War. In the past two decades, international efforts to terminate armed conflict, build peace, and construct political, social, and economic order after war have become increasingly characterized by tense frictions between competing agendas and interests. Peacebuilding and post-conflict reconstruction were conventionally associated with processes of political transformation (Darby and Mac Ginty 2008). In the period following the Cold War, internationally supported peacebuilding efforts were underpinned by the assumption that a single-minded international community of states and organizations could build peace by supporting the establishment of liberal, political, and economic orders in the aftermath of war. In fact, since the early 1990s, international organizations have been given unprecedented authority to intervene in conflict-affected countries and have taken on the responsibility not only to govern but to reshape states, economies, and societies (Caplan 2005). However, the outlook of building a 'liberal peace' has since then been significantly challenged by a changing geopolitical landscape, global democratic backsliding (Bermeo 2016; Niblett 2017), and the resurgence of geopolitical rivalries that prioritize questions of territorial control and military dominance over questions of global governance (Mead 2014).

Unsurprisingly, liberal-style international interventions, which were the primary object of scholarly debates after the end of the Cold War, have appeared increasingly 'toothless' over the past two decades (Debiel, Held, and Schneckener 2016; Sending and Neumann 2006). Rather than giving way to the linear, steady, and comprehensive processes that one would expect when reading many of the international policies that defined peacebuilding in the 1990s, many of the internationally supported peacebuilding efforts we have witnessed were shaped by political struggles between competing approaches and agendas. These struggles may be traced in policy debates taking place in intergovernmental bodies such as the United Nations Security Council (UNSC) or the African Union Security Council (AUSC), but they play out concretely in the domestic arenas where peacebuilding actors operate. The fantasy that efforts to end wars and build peace would simply unfold on some sort of empty *tabula rasa* and could follow a blueprint of measures is, of course, a rather colonial one (Ashcroft 2001, 39–40). Foreign actors that enter peacebuilding arenas must factually attune their efforts to national or local counterparts that tend to have their own objectives. This is also reflected in the normative principles that are broadly endorsed in peacebuilding research, not only by scholars of decolonial thought, but also by those working from the vantage point of liberal peace theory or pragmatism. De facto, this had led to peacebuilding outcomes that, compared to the ideal of liberal peacebuilding, seem 'compromised' (Barnett, Fang, and Zürcher 2014), 'hybrid' (Mac Ginty 2011), 'frictional', 'precarious' (Björkdahl and Höglund 2013), 'relational', and 'non-linear' (Chandler 2013; Chadwick et al. 2013; Hunt 2017).

In the face of disillusionment with liberal peacebuilding and growing contestations about the right approach and the right values and principles that should underpin it, international peacebuilding measures have also increasingly been influenced by a new sense of pragmatism, encouraging a context-dependent, step-by-step approach to peacebuilding. Rather than rushing to implement pre-defined reform agendas with fixed objectives, there is an increased emphasis on slow-paced implementation, adaptive approaches, and learning-by-doing that anticipates failure (Moe and Stepputat 2018; de Coning 2018; Bargués 2020). Scholars often portray this trend in positive terms. Yet, its digital sibling is one in which technological experimentation, ill-fitting prototypes, and innovations that may cause more disruption than positive impact are rolled out in the most vulnerable contexts. The turn towards pragmatism not only represents a retreat of liberal idealism and the crumbling of the hubris that underpinned liberal peacebuilding (Moe and Stepputat 2018), but also a more resounding failure of peacebuilding that manifests in more violence and suffering. The hope that frictions between international and local peacebuilding actors would simply result in something new or 'emancipatory' (Leonardsson and Rudd 2015) did not materialize in most contexts. Instead, in an increasing number of cases, the struggle has been won by actors that prefer authoritarian approaches to conflict management, relying on a combination of state coercion, use of force, and selective distribution of economic resources to exercise authority and control (Wolff 2009; Smith 2014; Piccolino 2015; Lewis, Heathershaw, and Megoran 2018; Piccolino 2015). Moreover, conflict and armed violence simply rage on because conflict parties successfully recreate the preconditions for violence, using the same strategies and techniques that have worked for centuries, such as producing and exploiting group-based grievances and animosity.

These dynamics also affect the use of digital technologies for peacebuilding, but we know little about how exactly. In parallel to these global developments and the impact they had on peacebuilding dynamics and outcomes, we have seen considerable investments in digitally enhanced peacebuilding approaches, which have led to thousands of proofs-of-concept, prototypes, and fully developed applications that ultimately prove challenging to scale up, maintain sustainably, and—most importantly—apply in ways that are impactful. Admittedly, measuring the impact of digital peacebuilding approaches on all efforts to prevent and end conflict is a difficult, if not impossible, task. However, it's also clear that, despite the at times braggy sales pitches of digital peacebuilding initiatives, none of them have claimed to make *the* difference. Indeed, many involved in the field will agree that digital applications must be more than mere gadgets that are 'nice to have'.[1] However, even if they meet what peacebuilders perceive to be real needs, their impact tends to be bound to a particular context, often limited and temporary, and clearly not enough to balance out what are perceived as more fundamental challenges in the way of building peace.

What, then, is the limited impact, or even failure, of this digital revolution in the face of the world's current political turmoil supposed to tell us? On the most fundamental level, we shouldn't think of technological innovation as a remedy for the shortcomings of peacebuilding, but as an integral manifestation of it. The dichotomy between (digital) technology on the one hand and politics and society on the other is a false one. The problem with this focus on tools is that it can make us blind to the larger social and political dimensions that underpin the current crisis of peacebuilding and thus lead us to avoid a more difficult question, namely how technological innovation and use drive, and are driven by, (global) politics and how both may contribute to, deepen, or ameliorate the current crisis of peacebuilding. A core assumption underpinning this book is that this question can only be meaningfully answered if we study the digitalization of peacebuilding not as something 'virtual', in the sense of being different from something physical or real, but as a phenomenon with material and tangible dimensions that are closely related to many other factors that condition the prospects of peace and conflict—ranging from narratives and discourse, to policy and practice. This means that we need to make sense of the reality of digital peacebuilding not only by accounting for digital tools and their effects, but by studying the socio-technicality of peacebuilding—and, as I will suggest, of conflict transformation.

The Socio-Technicality of Conflict Transformation

In many other sub-fields of International Relations (IR), scholars have begun to explore the nexus between digital technology and a changing global environment, characterized by geopolitical competition and a retreat of liberal institutionalism, but also a resurgence of social and political activism. For instance, the role of digital technologies has been studied in regard to efforts to enable new forms of political participation (Sæbø, Rose, and Skiftenes Flak 2008; Steibel and Estevez 2015); facilitate social movements that ask for social justice and equality (Freelon, McIlwain, and Clark 2018; Bastos, Mercea, and Charpentier 2015; De Choudhury et al. 2016); help political mobilization for protest movements (González-Bailón et al. 2011; Bennett and Segerberg 2012; Halverson, Ruston, and Trethewey 2013; Bastos, Mercea, and Charpentier 2015); and facilitate the promotion of democratic values and mechanisms that increase government transparency, openness, and accountability (Avgerou 2008, 132; Thompson and Walsham 2010, 117–18). On the other

[1] I learned about the differentiation between technologies that are 'nice to have' and that one 'needs to have' from Martin Wählisch.

hand, just as much attention has been paid to how digital technologies have been used to undermine liberal orders and promote populist and nationalist political agendas, for instance, through the automated spreading of political propaganda (Christensen 2019) and misinformation (Badawy, Ferrara, and Lerman 2018). As has been prominently pointed out by Shoshana Zuboff (2015), digital technologies commonly enable new forms of surveillance, and they allow governments to assert more robust political control by filtering and blocking content (Heeks 2018, 263).

We can rightly infer from this research that the political contexts in which armed conflicts unfold and are transformed, and relatedly, the struggle between different peacebuilding approaches in such contexts, shape the objectives for which digital technologies are used in peace processes. Therefore, digital technology and peacebuilding should not be treated as separate objects of research but as closely entwined. This is not a unique position, but one that reflects a stance that now informs research in many other fields of IR. Indeed, a growing number of researchers convincingly make the point against isolating research on 'technology' and its agency and impact on international practices from research on 'society' more broadly conceived, including factors such as the international organizations and institutions or private sector interests and the market. The heightened interest in the implications of technology in political processes and outcomes has probably been expressed most vocally by the sub-discipline of Critical Security Studies. Drawing on insights from the study of the relationship between science, technology, and society (STS), this body of work explores the intricate entanglements between technology, knowledge production, and social practices and politics in 'socio-technical systems' (Hoijtink and Leese 2019). Others have argued in a compatible manner that the study of digital IR should not limit itself to a preoccupation with the digitalization of previously analogue entities and processes but should study the 'disruptive interplay of digital technologies and power structures in global politics' (Bjola and Kornprobst 2023, 2).

Inherent in such calls is a reckoning with the fact that the forces that shape outcomes in global politics today necessarily form relations that bridge the social and the technical. Translated to the topic of this book, instead of thinking about the digitalization of peacebuilding as a process of technological innovation that supplies 'tools' to the sector, we may want to investigate how peacebuilding dynamics and outcomes result from socio-technical relations. The two opening chapters of this book are exclusively dedicated to this effort. Chapter 1 reconstructs how pioneers in the field have, to a varying degree, acknowledged how digital peacebuilding is shaped by both the social embeddedness of technology *and* the technological embeddedness of society. Chapter 2 introduces a 'post-digital' perspective on peacebuilding through which we can better grapple with its socio-technical character and develop a framework for applying it to a diversity of digi-political contexts.

However, before I get into the nitty-gritty of what I mean by socio-technicality (which I will do in Chapter 2), it's essential to clearly state what we want this socio-technicality to achieve. This book explores how these socio-technical relations that underpin digital peacebuilding may help or hinder the effort of *conflict transformation*. I'm intentionally putting the emphasis on transforming conflict instead of merely managing, resolving, or ending it because this directs the focus towards changing the relations that make an outbreak or continuation of armed conflict more likely. Conflict transformation is sometimes understood in terms of activities that take place after the active phase of an armed conflict is concluded, such as hearings, conferences, or dialogues organized by dedicated peace and reconciliation commissions. However, I suggest thinking of transforming conflict more as an approach and objective that can inform peacebuilding activities at different stages of a conflict, including conflict prevention, conflict mitigation, peace mediation, and efforts to promote long-term social and political change. This choice is grounded in several important considerations, as I will explain below.

As a topic of research, (digital) peacebuilding constitutes a problem that is ontological as much as theoretical and methodological (Jackson 2011). It poses questions about what past, contemporary, and future peacebuilding is made of (ontology), how we should conceptualize it (theory), and how we can study it meaningfully (methodology). Focusing on the socio-technicality of conflict transformation puts an emphasis on the relations between the variety of elements that cause violent conflict. Focusing 'simply' on such relations is ontologically parsimonious because it avoids prescribing the worldly elements that make up peacebuilding. Theories of peacebuilding that are prescriptive often struggle to be relevant across various contexts because putting ontology first tends to create a bias in perspectives. As Jackson (2011, 30) argues, research guided by strong theoretical predispositions tends to retreat behind specific scientific ontologies, and this risks that the 'partisans of each worldview [will] simply reassert their central postulates and go on reading the world in their own way'. In fact, many academic engagements with digital peacebuilding are strongly guided by specific theoretical assumptions (see Hirblinger et al. 2022 for a detailed discussion). Most prominently, liberal peacebuilding theories have been rightly accused of applying a 'blueprint' or 'template' of entities and actions that are supposed to build peace, such as democratic governments elected to office or transitional justice mechanisms that demonstrate the value of accountability and human rights. This is what has made liberal peacebuilding theory prone to allegations of Eurocentric bias, with critics arguing that it problematically assumes that Western or European experiences with ending conflict and building peace are universally applicable. However, many countertheories inspired by critical or decolonial thought likewise operate with theoretical concepts that suggest that the world is made up of certain entities, such as 'local actors' or 'bottom-up' and 'top-down' approaches (Sabaratnam 2013).

Putting ontology first is particularly unsuitable for research on digital peacebuilding for several reasons. First, for the simple reason that digital peacebuilding is a research field that is just emerging, it seems apt to take a largely inductive approach that enables us to discover new phenomena. The empirical chapters in this book all contain detailed, close-up qualitative accounts of specific attempts to leverage digital technology in support of peace, and they often follow specific tech pioneers, initiatives, or activists to make sense of these efforts. Applying a deductive approach that would limit what we could look at seems unsuitable for an emerging field of research that isn't characterized by hardened theoretical propositions (at least very few that are intellectually convincing) and where we frankly often don't know what we will discover. Second, the specific concern of this book, namely, to understand the effects of digitalization on peacebuilding, means we might discover, and must make sense of, an even larger diversity of relations, actors, objects, and practices than is the case with earlier studies of peacebuilding. What's particularly challenging is that the 'things' that digitalization creates are not on the same 'ontological plain'. For instance, peacebuilding scholars would discuss the relevance of 'international actors' vs 'local actors', compare the efficiency of international vs regional peace operations or mediation initiatives, or whether retributive or restorative justice mechanisms are more effective. However, when it comes to the topic of digitalization, we are suddenly confronted with the difficulty of having to make sense of how a lack of digital infrastructure or literacy affects peace activism, the impact of dashboards and data visualizations on conflict analysis, or how clustering algorithms affect the ability of participants in online consultations to relate to one another.

Put differently, the dynamics I have discovered during the research for this book often do not fit into one neat worldview or theoretical frame. Therefore, I'm positioning the focus on conflict transformation as a methodological choice that enables studying a diversity of *relations* between the many heterogeneous elements relevant to building peace while also directing the focus of inquiry towards what is needed in terms of a *change* to these relations to terminate conflicts in a sustainable and lasting manner. The individual empirical chapters are not written in a way that builds a theory of conflict transformation through digital means nor as a scientific exercise of arguing for or against its validity. Instead, I am using the theoretical angle of conflict transformation to provide a horizon and heuristic 'compass' for the book's journey through digital peacebuilding that ultimately won't reach its goal. Yet, while we keep our bearings in the direction of conflict transformation, we'll discover many stumbling blocks. Along the way, I will point to the various issues in the socio-technical relations that make up digital peacebuilding and demonstrate how they stand in the way of transforming conflict.

Let's take a closer look at how we can understand the objective of transforming conflict to which digital peacebuilding may or may not contribute. One of the most influential understandings of conflict transformation was advanced by John Paul Lederach, who argued that what makes the difference between armed conflict and peace is a change in relationships between the conflict parties. In his view, building peace requires the transformation of 'negative' or 'destructive' relationships and interactions into 'positive' or 'constructive' ones. This is the parsimonious part of his theory of conflict transformation. Yet Lederach adds ontologically more loaded and demanding elements to it when arguing, for instance, that this change can be encouraged through the pursuit of truth, justice, and forgiveness (Lederach 1995). In so doing, Lederach's approach to conflict transformation echoes earlier influential works, such as those of Robert Bush and Joseph Folger, who argued since the early 1990s that conflict transformation requires a change in relationships through a dual focus on 'empowerment' and 'recognition', which requires the parties to understand their own situation and needs, and those of their foes (Folger and Bush 2007). More recent takes on conflict transformation, such as that by Oliver Ramsbotham, propose thinking of conflicts as 'socially constructed relationships' and declare that the aim should be 'to transform actually or potentially violent conflict into non-violent forms of social struggle and social change' (Ramsbotham 2010, 53). Yet his work likewise points to goals such as working towards justice and the fulfilment of material needs.

To be clear, I don't see a problem with these more prescriptive or directive aspects of conflict transformation if they are reflective of what the actors we study have signed up for and pursued. The empirical chapters in this book highlight in many places how pioneers of digital peacebuilding thought about the task at hand and their often quite prescriptive suggestions, such as the need to address the underlying causes of hate speech. However, there is a risk that frontloading such prescriptive elements into a theory of conflict transformation, instead of following a situated and interpretative approach, will lead to research that is less attuned to the field. This seems particularly problematic given the current crisis of peacebuilding, the state of global politics discussed above, and especially the widespread disenchantment with liberal peacebuilding approaches. All this means that we can't take for granted that (digital) peacebuilding unfolds along a set of clearly defined activities.

Therefore, I suggest thinking of conflict transformation as an objective that can inform a large variety of peacebuilding practices at different stages of a conflict, while providing a precise understanding of the properties of peace-supportive socio-technical relations. To this end, it's useful to think of conflict transformation as enabling an agonistic peace. Agonistic peace theory, which has evolved over the past decade or so, assumes that the transformation of violent conflict requires a change of antagonistic relationships between enemies into peaceful, agonistic relationships between adversaries (Shinko 2008; Maddison 2015). In such networks of adversarial relations, conflicting needs and interests can be acted out without the use of armed violence aimed at destroying the opponent or the elimination of difference, which means that conflict parties can 'dare' to differ (Hirblinger and Landau 2020). This take on conflict transformation builds on Chantal Mouffe's (2002, 2005) conception of agonistic democracy, which is grounded in a critique of liberalism as unduly ignoring the constitutive role of difference in politics, which in turn gives way to variants of technocratic governance that are less legitimate and thus more prone to conflict. Agonistic peace theory recognizes that the 'creation and maintenance of peace is itself an inherently political act' through which contemporary political structures and their underlying values are constantly contested (Shinko 2022). At first sight, an agonistic peace seems to resemble a negative peace that is characterized by the absence of direct physical violence between the conflict parties, yet agonistic peace theory is also concerned with the context conditions that are necessary for conflicts to be 'fought out' without violence. While mainly concerned with political relations between conflict parties, agonistic peace theory is also cognizant of the tangible and structural dimensions of violence. It focuses on the conflict parties' ability to engage with them in communicative, cognitive, and emotional processes that ultimately humanize opponents.

Let's take these considerations back to the idea that digital peacebuilding is made of socio-technical relations to see how they can guide the research presented in this book. As should have become clear above, there are many ways in which digital technologies can be employed to respond to armed conflict, and likewise, many ways in which digital technologies shape how we think about armed conflict and the response to it. When exploring the socio-technicality of transforming armed conflict, my aim is to initiate a conversation about what's required to enable digital peacebuilding dynamics that have the power to transform conflict because they further socio-technical relations that enable conflict parties to develop constructive relations between them. What exactly these 'constructive relations' entail and how they can be achieved will differ from case to case. However, at a minimum, they require that the parties acknowledge the difference between them in terms of identity, interests, needs, etc., and that they aim to resolve conflicts between them that emerge because of these differences without resorting to armed violence.

This approach to the socio-technicality of conflict transformation is theoretically parsimonious because it doesn't prescribe the approaches, tools, or methods through which the transformative work is carried out. Put differently, it doesn't seem 'too much to ask' that digital peacebuilding should enable conflict parties to fight out their differences without attempting to kill each other, while leaving the path to get there open. However, the empirical chapters in the book point out that socio-technical relations can hinder the prospects of an agonistic peace in a large variety of ways, for instance, when efforts to counter hate speech on social media focus primarily on deleting hate while failing with its deeper causes, or when AI-enhanced dialogue efforts cause rigid designs and reduce the participants' ability to express their needs and interests freely.

Some readers may object that the objective of conflict transformation that enables an agonistic peace may set the bar too high, because many digital peacebuilding activities pursue more modest goals. While this is indeed often the case, I suggest thinking of conflict transformation as something that can and should be mainstreamed across the diversity of initiatives and activities that contribute to digital peacebuilding. High-level negotiations and dialogue processes, such as those covered in Chapter 5, are only the most prominent examples that can help groups in conflict transition from violent antagonism to political adversity. Efforts to humanize 'the Other' and to recognize their legitimate identities, interests, and needs can likewise be supported by many other interactions in which digital technologies play an increasing role. This includes the efforts to counter online hate in Chapter 3, efforts to mitigate incipient conflict through mobile early warning and response mechanisms in Chapter 4, or everyday online interactions in Chapter 6 that challenge divisive political regimes. Taking this broader, and arguably more holistic, view on the socio-technicality of transforming armed conflict, the book explores how antagonisms that are conventionally understood as a human relationship are conditioned by the socio-technical.

The Main Contributions of the Book

It has already become evident that digitalization is an open-ended and fast-moving process, which means that its impact on global politics and its professional areas of engagement, such as peacebuilding, is impossible to determine in any conclusive manner. Quick cycles of innovation and the central role of emergent technologies such as AI mean that empirical studies about specific applications have a relatively short shelf life. The nascent body of research that looks at the impact of digitalization on peacebuilding has nonetheless often discussed particular tools or applications, such as crowdsourcing systems, virtual reality simulations, or online dialogue platforms, usually considering their beneficial and detrimental effects on peacebuilding dynamics and outcomes. In contrast, this book's central message is that we need to grapple with the socio-technicality of contemporary peacebuilding instead of merely approaching digital peacebuilding as something that happens when peacebuilding actors utilize digital technologies. While the existing research tends to take note of the operational contexts in which digital technologies are employed, it provides very little in terms of theories, concepts, and methods that can help study the social embeddedness of digital technology. Therefore, the book aims to introduce a research perspective that can be employed independently of individual digital applications and helps us to make sense of, and critically engage with, the digital peacebuilding as a socio-technical dynamic. To this end, the book draws on and furthers existing scholarly work on the role of technology in global politics, situated in social constructivist, material-semiotic, and critical-reflexive traditions, which share a strong interest in practices of knowledge production *with* and *about* technology and their entanglements with politics and society.

The book's chapters represent my effort to develop this research perspective into a comprehensive whole by integrating suitable theoretical concepts and methods, and employing them in empirical study. Notably, the research for this book was organized around one central idea, namely that due to the increasing integration of digital technologies into the peacebuilding sectors, the ability of conventional peacebuilding actors to suggest and shape approaches to ending armed conflict has further eroded. This phenomenon has been described through the term 'apomediation', which was initially coined to make sense of the increasing impact of the internet on personal healthcare decisions. It captured how individual users rely on decentralized human–machine networks, that is, 'apomediaries', to obtain relevant information about their health conditions and possible cures and preventative measures, instead of relying on a single authoritative source, such as a doctor. The comparison of armed conflict to a disease can, of course, be somewhat misleading. There is a risk, for instance, that we might think of armed conflict as a deviation from a 'healthy' normal. This ignores the fact that globally, 'healthy' zones of peace and 'unhealthy' zones of war are co-constitutive and much of the relative peace that populations have enjoyed in Europe and North America over the past seventy years has come at the cost of conflict and violence perpetuated in other parts of the world (Barkawi and Laffey 1999). However, we know from Michel Foucault's seminal work that the 'gaze' of medical discourse rarely reflects only biological reality but contributes to the social construction of normality and pathology, and thus of what is viewed as healthy and what is not. These constructions determine what is worthwhile treating, while leaving important contextual factors affecting our health out of sight (Foucault and Foucault 1994). Peacebuilding, just as healthcare, often focuses on the symptoms of antagonistic conflict through security and stabilization measures that 'pathologize' certain actors and practices, instead of doing the deeper work of tackling institutional, governance, economic, or historical drivers of armed violence.

That said, I will employ the term apomediation *not* to draw parallels between armed conflicts and diseases, but because the research in this book has been inspired by similarities between healthcare and peacebuilding in terms of the impact of digitalization specifically. These similarities mainly concern the enabling of the decentred production, sharing, and consumption of information in human–machine networks, and how this redistributes authority of knowledge and the agency to act on it. This is the main conceptual intervention, which is also contained in the Greek and Latin meaning of the preposition *apo*, which denotes separation and a motion away from an origin or source. Apomediated peacebuilding is thus peacebuilding that unfolds in decentred socio-technical networks, instead of something that's carried out by specific peacebuilding actors, such as UN agencies, civil society organizations, or community-based peace committees or 'traditional'[2] leaders.

More concretely, the book suggests studying digital peacebuilding as a relational dynamic in which understandings of conflict and approaches to building peace are apomediated. Apomediaries have been defined as 'tools and peers standing by to guide consumers to trustworthy information or adding credibility to information' (Eysenbach 2007, 166), by forming decentralized networks of humans and machines that guide users in obtaining relevant information about a specific topic (Eysenbach 2008, 129). The digitalization of peacebuilding has led to strikingly similar dynamics. The rise and spread of social media platforms and mobile and smart technologies have enabled what has been referred to as the 'post-internet life', characterized by the ubiquitous utilization of digital technologies in our everyday social practices and their intricate and often tacit agency in shaping bodies, minds, our relationships, and more broadly, our societies and political systems (Mosco 2017). This phenomenon also affects armed conflicts and efforts to end them. For instance, most, if not all, parties, stakeholders, and ordinary citizens involved in or affected by contemporary armed conflict use social media and online resources daily, which informs their view on the conflict. Many of them will also use the same set of technologies to learn about and advance ideas for how the conflict could be ended.

More importantly, professional peacebuilding actors, such as mediation teams or local civil society organizations, aim to make sense of, and respond to, armed conflict via digital approaches that establish decentred relations between humans and

[2] I'm putting 'traditional' in inverted commas to remind ourselves that traditions are often invented and re-invented, and so were and are 'traditional' leaders in many colonial and post-colonial settings (Hobsbawm and Ranger 1983; Mamdani 1996).

technical devices. Beyond their everyday and ubiquitous uses, such as obtaining news from social media, digital technologies are now commonly employed for specialized functions, such as conflict early warning or promoting dialogue. New technologies such as machine-learning-based Natural Language Processing increase the ability of their users to gather, process, and analyse large amounts of text data, such as is produced on social media. International organizations increasingly employ such tools to monitor changes in the political atmosphere and understand who and what drives public opinion. They enable the remote monitoring of political developments and targeted interventions by peacebuilding organizations based on a fine-grained, machine-supported analysis of stakeholders' views and perceptions, instead of relying solely on some insider experts that may have privileged access to information. Moreover, digital technologies and platforms have enabled new forms of human interaction, for instance, in large, decentralized, and heterogeneous networks of volunteers that crowdsource information, such as on security incidents, the adherence to ceasefire agreements, or rumours that could trigger conflict outbreaks. Instead of helicoptering a team of international observers into a local conflict zone or relying on the authority of an external monitoring force by the UN Security Council, crowdsourcing relies on the buy-in, acceptance, and participation of the local population to collect the data. Rather than verifying the information through a team of independent experts, such systems increasingly rely on computational data verification and triangulation systems.

All this begs the question of how the digitalization of peacebuilding changes the role, agency, and authority of conventional peacebuilding actors, such as UN Special Envoys, religious leaders, expert commissions, or pools of consultants. Speaking to this challenge, the concept of apomediation inspires us to explore how authoritative knowledge that guides decisions and actions is no longer controlled by conventional authorities, such as professional peacebuilding actors and their staff, but emerges in decentralized, peer-to-peer networks of information exchange that involve humans and machines. Based on this perspective, the book asks how apomediation dynamics affect what peacebuilding is, how it is done, and what kind of peace results from it. More specifically, I suggest approaching apomediated peacebuilding as an interplay of relation-making and meaning-making in socio-technical systems and asking how this leads to specific peacebuilding dynamics and outcomes. This straightforward operationalization comes with the benefit of enabling us to shed light on the socio-technicality of peacebuilding in a flexible manner and at various scales, ranging from the effects of software interfaces on human cognition and emotions, to the role of data collection and visualization methods in shaping group dynamics in dialogue settings, and the instrumental role of digital infrastructures in the maintenance of violent political regimes. Moreover, it also invites us to reflect on how assumptions and claims about digitalization and technology affect peacebuilding approaches. This latter part of the inquiry is essential to do justice to the idea that technology and society are co-constitutive. For instance, Chapters 4 and 6 explore variously how tech-solutionist idealizations of remote connectivity and AI-data processing power have been presented as answers to the common challenge that peace processes often lack inclusivity, and how this has paved the way for experiments with mobile applications and AI-enhanced remote consultations in highly volatile settings. With that, I understand apomediated peacebuilding as more than peacebuilding that is mediated by (digital) technology; it also entails the shaping of digital innovation and the use of technology through discourse and practice.

By putting the socio-technicality of peacebuilding upfront, the book situates itself in a research agenda that aims to be both critical and reflexive, which, simply put, requires asking how the specific ways we talk about technologies and go about using them benefit somebody and some purpose (inspired by Cox 1981; and the more recent work of Hamati-Ataya 2013). As vast human–machine networks produce and circulate more or less explicit claims about the causes of armed conflicts, how they should be resolved, and the role of technology in these efforts, it's undoubtedly the case that some bits of the circulating information will have more impact than others. The book thus explores how claims about society and technology lead to specific digital peacebuilding agendas and outcomes, and how digital technologies are always useful for some purpose, while other potential uses of technologies and their outcomes remain much more obscure and wanting.

To direct the focus on these dynamics, I draw on Martin Heidegger's (1977) reflections on 'modern technology', which, while written well before what is commonly viewed as digitalization, remain highly relevant. Heidegger noted that technologies are often instrumental, 'ready-to-hand' objects that 'enframe' human existence. In simpler and arguably more reductionist terms, when humans use technology, their relationship with it shapes who they are and how they relate to the world. Translating this to the study of digital peacebuilding, we can ask how the use of technology changes how peacebuilding is *done*, as well as what peacebuilding *is*. Notably, the utility of technology and the enframings that come with it are not simply pre-given, but they are, of course, at least partly a consequence of human design. As Chapter 1 makes clear, the question of whether technologies are mere neutral 'tools' or whether they are 'socially embedded' has also informed the thinking and approaches of some of the pioneers in the field of digital peacebuilding. What I'm mainly interested in, however, are the kinds of enframings that do not materialize when peacebuilders use technology, and thus the socio-technical constellations that remain missing-from-hand for peacebuilding. To scrutinize these 'voids' of digital peacebuilding, each chapter thus explores how specific approaches that leverage digital technologies fall short of contributing to transformative efforts.

Finally, and importantly, I wanted my research to engage with the fact that the digitization of peacebuilding occurs in the changing and heterogeneous global environment I've attempted to describe further above. Omitting this aspect would leave the book's ambition of providing a socio-technical perspective somewhat incomplete. Thus, the research for this book was also driven by an interest in how global variations in political context and degree of digitalization will affect digital peacebuilding approaches and outcomes. More specifically, I've aimed to better understand how digital peacebuilding efforts unfold and what their chances of transforming conflict are in places that are characterized by a relative absence of digital connectivity and skills, as well as resurgent or resilient authoritarian rule. To some degree, these conditions characterize the vast majority of peacebuilding cases. The empirical research presented in this book, conducted in a variety of places, ranging from Northern Ireland to Sri Lanka and South Sudan, demonstrates that factors such as social media censorship and digital surveillance shape meaning-making and relation-making in apomediation networks. Likewise, forms of 'offline' political oppression also shape how professional peacebuilders and conflict-affected populations employ digital technologies in the pursuit of peace because they affect how users can connect online and what kind of information they can exchange. However, just as important are factors such as digital literacy and connectivity because they influence the degree to which peacebuilding initiatives can counter and undermine authoritarian measures.

Therefore, the empirical chapters of this book compare variations in the degree of digitalization and their political context. To this end, I suggest thinking of apomediated peacebuilding as taking place on a 'digi-political playing field', which enables or curtails apomediation dynamics depending on its degree of 'slantedness'. The slantedness of the field results from two interdependent dimensions. On the one hand, there is the degree to which conflict parties, especially those in government, can slant the field, by stifling the possibility of articulating, sharing, and utilizing diverse understandings of conflict and peace. On the other hand, the digital adaptability of the field refers to the degree to which ordinary people can employ digital technologies to cope with and challenge such authoritarian practices. I argue that variations in these two dimensions help us to understand core aspects of the sociotechnicality of digital peacebuilding because they condition both meaning-making and relation-making practices that unfold in apomediation networks. The empirical studies presented in Chapters 3–6 of the book highlight concretely how various factors influencing the political slantedness and digital adaptability of the field shape apomediation dynamics—and importantly, how this shapes the prospects for conflict transformation.

Research Approach and Ambition of the Book

The book draws on six years of qualitative research, conducted mainly under the framework of the 'An Apomediated Peace?' project, which I led as Principal Investigator.[3] However, the book also benefitted from insights generated through several other research projects that I led in the years before. The results presented across the chapters would not have been possible without the dedicated support of several student research assistants and junior researchers employed in these projects, and country-based researchers in Sri Lanka and South Sudan who actively contributed to data collection and production of written research outputs. The research for the

[3] The project's full title was 'An Apomediated Peace? The Role of Digital Technologies in International Peacebuilding'. It was funded by an Ambizione Grant of the Swiss National Science Foundation (SNF) from January 2021 until August 2025.

book entailed a large diversity of methods and tasks, ranging from comprehensive desk research, including regular reviews of academic, policy, and practice-relevant literature, to the analysis of documents such as manuals and guidelines on PeaceTech, and social media analysis and ethnography involving regular study of social media content such as written posts and memes. However, I obtained the most important insights through the many interviews and focus group discussions with project staff and project participants, as well as participatory mapping and visualization exercises that helped me to understand the socio-technical relations underpinning the individual projects we studied, and especially how information about conflict and peace was produced, shared, and employed. I also regularly participated in practice- and policy-relevant meetings relevant to the book's topic and engaged with professional audiences during conferences, workshops, roundtables, and training sessions.

The case studies presented in Chapters 3–6 are based on remote empirical research and fieldwork in Northern Ireland, South Sudan, and Sri Lanka. Studying concrete digital peacebuilding initiatives requires taking a close-up look at how projects operate and thus repeated interactions with the projects' staff, volunteers, and populations engaged with and affected by them. Therefore, I aimed to establish trusted relations with the staff of several peacebuilding organizations, including missions and agencies of the United Nations, international and national civil society organizations, and governmental entities. Formalized partnerships or research agreements were the exception. In most cases, my proposition was simply to conduct research that would help advance knowledge on digital peacebuilding overall while ensuring that all empirical insights I obtained would be used confidentially and safely. I also provided assurances that I would not attribute what I observed singularly to individuals, projects, or organizations, and that I would not publish any compromising information about partners and research participants. This was a promise I could make without curtailing my academic freedom since my research aimed to produce knowledge about the distributed effects of digitalization as they unfolded in apomediation networks, not the agency of individual actors. Indeed, every single chapter of the book has something critical to say about digital peacebuilding dynamics and outcomes while demonstrating that they result from socio-technical relations in which no single actor enjoys independent agency.

The organizations and individuals I engaged with received no material benefits from the collaboration. However, I offered regular exchanges, the sharing of insights based on interim results, and a space for reflection that might have informed their thinking and work even during the project. While some researchers may think that this risks introducing a 'response bias', I felt that sharing my opinion and insights was a moral duty, not only because it often proves challenging to translate academic knowledge back into practice but also because research on peacebuilding stands in a long trajectory of a global economy of knowledge production that has usually benefitted the Global North. At times, I also saw immediate risks with some approaches and developments, and I felt it was my moral duty to raise them upfront and without delay. Through this approach, the research team gained access to ongoing or recent peacebuilding initiatives that rarely share such information because of the potential operational risks. Not all empirical research in this book is produced in the framework of these partnerships, and not all projects with which I partnered are represented in this book, and certainly not to an equal extent. However, even projects that are not explicitly mentioned, or only in passing, have informed the results presented in this book.

The book develops a theoretically informed and empirically grounded account of digital peacebuilding. It uses qualitative and interpretive research methods and empirical case studies to represent the point of view of the research participants involved in the observed digital peacebuilding activities while generating theoretical insights and a comparative perspective that can be applied to make sense of future digital peacebuilding activities. Given the fast-paced change in the field and the many digital innovations that have the potential to alter peacebuilding practices almost on a daily basis, the book's contribution must necessarily be mainly theoretical. Importantly, the theoretical perspective, concepts, and claims that I make do not stand in the tradition of positivist social science research but in a research tradition that thinks of theory as containing a set of guiding insights and experiences that help us to understand and engage with the world prudently, and of context-specific case studies as material for reflection and learning (Flyvbjerg 2001, 2006).

The specific observations I share across the various chapters are ultimately my personal observations alone, which have emerged through context-specific, time-bound, and situated research. However, the content of the book has also been inspired and shaped by many colleagues, through their previous research and engagement with the project. The observations and considerations of my research team also shaped my own to a considerable degree. It has been my ambition to represent the experiences, views, and factual details that were shared with me as accurately as possible. The book often takes an interpretivist approach that represents experiences of digital peacebuilding from a situated point of view, which has either been documented in the artefacts I studied, such as documents or social media posts, or through interviews or focus group discussions. This means that my own thoughts, views, and questions shaped how I interpreted specific observations and especially what the research participants had to say. Where the book attempts to recount facts, such as statistical data or events, I have done my utmost to ensure that they are accurate and reproducible by choosing trusted sources, triangulating information that is not publicly documented, and providing as much detail as I could obtain.

Importantly, I do not describe the conclusions I drew based on these observations as objective, scientific 'facts'. Another objective strikes me as more relevant. The empirical material helped me to develop a research perspective, which, alongside corresponding theoretical concepts and concrete insights, has value as heuristic knowledge that may matter for the readers of this book as they engage with future cases of digital peacebuilding. The book's empirical findings, read against the comparative framework and the theoretical apparatus developed in the opening chapters, are thus intended as learning examples—nothing more and nothing less. A key strategy for realizing this objective was the oscillation between in-depth and context-specific empirical case discussions on the one hand and structured comparisons between cases (in terms of peacebuilding contexts and uses of technology) on the other, enabled through the employment of the same theoretical research perspective across use cases and peacebuilding contexts.

The Structure of the Book

The book starts with two introductory chapters that present its core theoretical considerations against the backdrop of a short historical reconstruction of the contemporary origins of digital peacebuilding.

Chapter 1 reconstructs the discourse around terms that shaped the emergence of digital peacebuilding in the past two decades, namely 'ICT4Peace' and 'PeaceTech'. It argues that we can identify two competing visions that characterized the effort to professionalize digital peacebuilding. The first vision was one of digital technologies as tools independent of society and politics, yet able to influence them. The second thought of digital peacebuilding as deeply entwined with society and politics. Based on a reconstruction of this tension, the chapter argues for a research perspective that captures the social embeddedness of technology. Drawing on Martin Heidegger's discussion of how technology enframes our existence, the chapter argues that digital peacebuilding tends to be caught up in a utilitarian discourse that circles around the question of how technology can be 'ready-to-hand'. In contrast, the chapter suggests that critical engagement with the adverse effects of socially embedded technologies on peacebuilding benefits from asking how, in digital peacebuilding, something has become, or remains, 'missing-from-hand' for those who aim to transform conflict.

Chapter 2 operationalizes this research perspective by further exploring how we can think of digital objects, such as smartphone applications and dashboards, as socially embedded. To this end, it first suggests taking a 'post-digital' posture, which helps us ponder questions related to the long durée and the globality of digital peacebuilding, its tangible and material manifestations, and how digital peacebuilding is shaped by distributed agency that emerges in human–machine networks. The chapter then discusses how we can make sense of this socio-technicality of peacebuilding. This entails asking how technologies for peacebuilding and peacebuilding with technology are co-produced. The chapter translates this insight to the study of peacebuilding through two manoeuvres. First, it introduces the concept of apomediated peacebuilding, which suggests that peer-to-peer networks of humans and machines produce understandings of conflict and peace that underpin digital peacebuilding approaches. Second, drawing on critical-reflexive thought, the chapter directs attention to how claims about technology and society shape apomediation agendas and outcomes. Finally, the chapter shows how we can account for global variations in digitalization and political context by introducing the conceptual framework of the digi-political playing field. The subsequent chapters all make the point that digital peacebuilding tends to be strongly conditioned by variations in the political slanting of this playing field, shaped by factors such as internet censorship, blackouts, and online harassment, as well as the ability of the field to adapt to these challenges, shaped by factors such as connectivity and digital literacy.

Having laid the conceptual groundwork, the book's remaining chapters explore digital peacebuilding across all peacebuilding activities, ranging from conflict prevention to early warning, peacemaking and dialogues, and long-term reconciliation.

Chapter 3 discusses international efforts to fight harmful speech (e.g. hate speech and disinformation) as a new approach to conflict prevention. It starts by examining how harmful speech in contemporary practice and policy discourse is primarily portrayed as a problem of digitalization through social media and less as a manifestation of deeper, underlying causes of conflict and violence. This position is contrasted with historically evolved problematizations of harmful speech, which are inter alia rooted in the experience of Nazism and racism in Europe and North America. The chapter then shows that the digitalization of harmful speech renders the 'deep work' on the underlying causes of conflict more difficult. Digital peacebuilding initiatives focussed on harmful social media speech 'datafy' hate, which enables the efficient identification, collection, and analysis of relevant speech acts. They also support efforts to delete and falsify content. However, this work rarely tackles the drivers of hate, including marginalization, elite manipulation, or long-standing grievances and narratives. This tendency is examined through empirical research with several social media initiatives in South Sudan and Sri Lanka. The chapter shows that this is especially the case when such initiatives operate in conflict-affected contexts characterized by a slanted digi-political playing field where digital literacy and digital access are low, and civil society risks being punished for promoting evidence-based information.

Chapter 4 discusses how digital technologies are employed for conflict mitigation in the context of localized early warning and response initiatives. It examines how such efforts have been underpinned by a call to integrate vulnerable and conflict-affected populations into such mechanisms. To this end, the chapter revisits these considerations of some of the pioneers of early warning, identifying with the Global South, as well as the assumptions of the crowdsourcing community, demonstrating their hope that digitalization could enable horizontal, spontaneous, and organic forms of social organization, instead of hierarchical, 'top-down' approaches. Comparing several localized early warning initiatives in Northern Ireland and South Sudan, the chapter finds that digital approaches often struggle to do so. Concretely, the chapter argues that datafying early warning information tends to produce 'skimmed' understandings of conflict. It shows that many early warning initiatives have a clear preference for quantified and aggregable data that are machine-readable and suitable for aggregated analysis. This leads to a reduction in situated, context-specific knowledge and creates a 'gap' between local understandings and responses to imminent conflict, which makes a comprehensive and lasting peacebuilding response difficult. Moreover, in settings with a slanted digi-political playing field, remote systems increase the risk that data are obtained and used by actors that prefer top-down, hard security responses to violence, instead of sustainable de-escalation.

Chapter 5 examines the use of digital technologies in peace mediation and dialogue efforts. Discussing the uptake of AI in peacemaking, it suggests that such efforts are shaped by the interdependent collaboration of human and machine that generates 'hybrid intelligence'. It argues that the adaptation of digital methods has, amongst other factors, been driven by the normative commitment to increase broad-based participation in peace processes and make them more inclusive. Applications leveraging AI in particular promise to enable the large-scale and instant analysis of data and are increasingly embedded in participatory approaches, such as online dialogues and consultations involving hundreds of participants at a time. Furthermore, digital technologies have been idealized as vehicles that could make peacemaking more effective by transcending the track logic of peace processes that usually privileges high-level negotiations. Against this background, the chapter discusses the international support to the Libya Political Dialogue Forum (LPDF) during the COVID-19 pandemic to argue that digital approaches will have little impact on the design and dynamics of the peace process when organizational cultures and mandates continue to privilege formalized, high-level negotiations. Moreover, the chapter discusses some of the algorithmic processes employed in AI-enhanced large-scale dialogues to argue that they may curtail opportunities for conflict transformation if they structure participant interactions in ways that privilege the generation of representative data over the recognition and empowerment of the multiplicity of diverging voices that are in, and affected by, conflict. The chapter concludes by discussing how digital approaches could be used less to satisfy a sincere desire to generate evidence about the present conflict. Presenting a case study of intra-Palestinian dialogue efforts, it highlights how digital peacemaking can foster a subjunctive, future-oriented attitude concerned with how conflicts could be transformed.

Finally, Chapter 6 explores the role of apomediation in long-term, societal efforts to reconcile antagonistic groups, asking whether and how it can change how groups and communities relate to one another. Many contemporary armed conflicts and their aftermath are characterized by what the chapter describes as 'digital divide and rule' (DD&R) tactics, because conflict parties with authoritarian ambitions exploit social media to maintain societal divisions. In these contexts, organized online reconciliation projects tend to have limited reach and traction, which makes it worthwhile asking if everyday online interactions of ordinary social media users could provide a valuable alternative. To this end, the chapter identifies three distinct apomediaries prominent on social media, which advance different understandings of conflict and peace. First, programmatic apomediaries that build relations around specific issues related to the canon of formal peacebuilding policy, which professional peacebuilding organizations and policymakers usually dominate. Second, reformative apomediaries that build vertical relations to communicate popular grievances, often led by civil society groups and the political opposition. Third, integrative apomediaries that build horizontal relations to counter divisive political designs and aim to connect population groups that identify with different identity categories. The chapter then compares everyday online interactions in Northern Ireland, South Sudan, and Sri Lanka to show variations in the digi-political playing field that shape the likelihood that these three apomediaries can jointly contribute to transformative apomediation dynamics. It finds that when DD&R plays out on playing fields that are strongly slanted and little adaptive, these conditions are rarely met. However, a close look at short periods of political upheaval in South Sudan and Sri Lanka suggests that a 'digital critical mass' of social media users can make a substantial difference by helping to initiate processes of political change that have the potential to contribute to long-term conflict transformation.

Building on the empirical work presented in this book, the Conclusion provides a summary of the book's main findings and a brief outlook on how current trends in technological innovation may affect the apomediation of peacebuilding. It discusses how the increasing employment of generative AI could lead to the 'death' of socio-technicality as I have come to think about it in this book, since the foundational models that power generative AI often lack transparency, are difficult to modify by end-users, and may lack training data from conflict-affected contexts. This brings the risk of 'hyperreal' peacebuilding, through which the world behind machine-generated understandings of peace begins to vanish. In a future increasingly shaped by generative AI, experiences of marginalization, deprivation, and insecurity would be reduced to mere input data for the probabilistic calculations of foundational models, which means that the human voice needed to transform conflict could virtually get lost in the 'black box'. This trend would be difficult to reconcile with transformative ambitions because machine-generated data are replacing the voice of those affected by violent antagonism. Against this backdrop, apomediation—as a process that keeps the human 'in the loop' in increasingly automated peacebuilding efforts—is not only worthwhile studying. It may become something worthwhile to defend.

1
Digitalization, Technology, and the Voids of Peacebuilding

Whether we use an artificial intelligence (AI) chatbot to generate suggestions for resolving an armed conflict or a mobile application to order food from the hottest new restaurant, digital technologies promise to revolutionize our lives—but often, they fail to deliver. How many applications have you downloaded on your phone and deleted them soon after because they turned out to be dysfunctional or had effects you did not foresee? For instance, have you ever waited too long for the food you ordered online, only to receive the wrong item, or end up with a less healthy meal than you wanted? Did you ever have the sense that a shiny mobile application did not deliver, in the literal or symbolic meaning of the word, what it promised? Of course, much more is at stake when digital technologies promise to 'deliver' peace. Still, the fundamental dynamics are the same: digitalization changes how we obtain information, how we communicate, and how we relate to one another, and while it often promises improvements to those affected by violence, the outcomes are usually more mixed. Capturing what can be lost when digitalization is invoked to find solutions to the complex problems underpinning war and armed conflict is a much harder task still, because it requires going beyond individual user experiences and shedding light on much vaster and more complex processes.

First, however, the digital revolution needs to be called off. We tend to associate anything digital with an arriving future, thus maintaining a sense that digitalization is becoming or unfolding, even though the first innovations in digital technology now date back many decades. Five years from now, could high-level peace mediators employ a generative AI model to draft the most intricate and disputed peace agreement provisions—and could they succeed? Questions like this, and the sentiment they evoke, are grounded in the everyday tangible reality of digital innovation, ranging from the endless stream of software updates sent to our mobile devices to disruptive moments of groundbreaking technological advancements, such as the Large Language Models (LLMs) that power the current advancements in generative AI. These dynamics also affect the practices of preventing conflict and building peace, where staff at the headquarters of international agencies, just as much as employees of local community-based organizations, are confronted with a constant flow of innovations, ranging from minor adaptations in data security protocols or the functionality of social media platforms to the more remarkable disruptions caused by breakthrough advancements in remote sensing, big data analytics, or virtual reality, among others.

Digital Peacebuilding?. Andreas T. Hirblinger, Oxford University Press. © Andreas T. Hirblinger (2026).
DOI: 10.1093/9780198986553.003.0002

It is therefore not surprising that an increasing number of observers speak and write about the 'digital revolution' in peacebuilding or a 'turn' towards digital peacebuilding. This turn is proclaimed in policy- and practice-oriented contributions that equate 'digital technologies' with 'new technologies' (Wählisch 2019), mobilize the 'data revolution' for the Sustainable Development Goals (SDGs) (Independent Expert Advisory Group 2014), or aim to guide the 'digital transformation' of the sector (United Nations Peacekeeping 2011). The emerging academic scholarship on the topic likewise takes note, albeit with more scepticism of an apparent 'digital revolution' (Firchow et al. 2017, 5). However, it also discusses the possibility of a 'digital renewal' (Richmond and Tellidis 2020), and indeed proclaims a 'potential digital turn in peace and conflict studies' caused by a 'shift from "analogue" to "digital" approaches in international relations' that affects the 'international peacebuilding architecture' (Richmond, Visoka, and Tellidis 2023).

Such invocations of a 'revolution' or 'turn' suggest that we are facing a moment of radical change driven by digital innovation, altering how peacebuilding is done and the odds for peace. However, such notions come with a risk of 'technology hubris': excessive confidence that new or digital technologies can be a 'game changer' (Jobarteh and Natalina 2020; Hirblinger et al. 2023, 12; CMI 2024), 'bring revolutionary change to the practice of mediation' (UNDPPA and HD Centre 2019, 7), and more generally contribute to 'maintaining international peace and security' (UNSC 2022). If digital innovation is so rapid and powerful, why does it have such limited leverage in current efforts to end wars and violence? Why, despite decades of technical innovation and digitalization, is today's world ravaged by armed conflicts and devastating violence?

Part of the problem may be a naïve belief that digital innovation for peace is straightforward and impactful. For instance, in 2018, a comment in the prestigious journal *Nature* suggested that machine learning could make predicting wars as straightforward as forecasting the weather. It would simply require us to 'retool AI' so that 'conflict prediction and prevention' could be based on a 'global data-driven system' comparable to those used for forecasting the weather or epidemics (Guo, Gleditsch, and Wilson 2018). Yet, the authors seem to omit that efforts to develop conflict forecasting methodologies date back more than four decades, that machine learning models continue to struggle with obtaining sufficient amounts of relevant data for their training and operation, and that conflicts are often 'black swan events' that cannot be anticipated based on knowledge about prior cases (Hegre et al. 2017). Moreover, while they mention the need to build a 'global consortium' for innovation, the political challenges of forming such an alliance comprised of international organizations, states, academia, and civil society do not receive their attention.

Put differently, it isn't immediately clear who or what those 'rulers' are, figuratively speaking, that have been 'dethroned' by the digital 'revolution', what the 'game' is that must be 'changed', or what distinct aspects of the analogue past we are turning away from when we are shifting towards digital peacebuilding. What is more, the transformation that the digitalization of the field brings doesn't always seem very profound. After all, the approaches and practices of peacebuilding and the international architecture that is supposed to support them have not fundamentally changed since the end of the Cold War (Hearn, Bujones, and Kugel 2014), despite a deepening crisis of multilateralism and the resurgence of great power rivalry. International peacebuilding may be 'in crisis' (Chandler 2017a) due to its limited effectiveness and impact; it may be based on ill-founded assumptions, theories, and often on empty promises; and it may contribute to, rather than change, global, national, and local power dynamics that ultimately play a role in the reproduction of conflict and violence. Yet, none of these challenges are new, and more importantly, they prevail despite the increasing presence of digital technologies in the day-to-day efforts of building peace.

Thus, maybe there are other motivations behind the invocations of the digital revolution or the digital turn. We know that policy and practice often rally around 'buzzwords' that create increased attention for a specific topic. Academic peace research does so similarly, for instance, when hyping up terms such as 'inclusion' (Hellmüller 2019). There has also been a tendency to organize scholarly inquiry, and especially the debates among self-described 'critical' scholars, around various 'turns'. The most prominent example might be the 'local turn', which is prominent in conference panels, research proposals, and journal special issues. However, these 'turns' tend to stand less for a shift in the empirical world these disciplines study than in the academic discourse itself. This was well visible, for instance, in the context of the cultural turn in the political and social sciences, which introduced a whole new set of post-structuralist methodologies into what was otherwise a rather positivist and empiricist scholarship, which brought significant changes, for instance, to the study of collective identities such as nationalities, ethnicity, and gender. However, turns can also be the result of scholarly attention-seeking. Indeed, declaring a turn tends to enhance scholars' position and build their social capital in the field (Baele and Bettiza 2021).

We have learned from past turns in peacebuilding research that they come with the risk of ignoring developments in other fields and disciplines and investing too little into the exchange of ideas, tools, and concepts with them (Schierenbeck 2015). This makes it more likely that intellectual mistakes that have already been grappled with elsewhere are repeated. Turns may be, as Hunt (2023) has argued, a means to mobilize an academic audience and fellowship to jointly scrutinize a particular empirical phenomenon and bring corrective change to it. Yet, they commonly fail to do so because they tend to have a limited impact on the policy and practice of peacebuilding. Some readers of this book may sympathize with such peculiarities and quirks of the academic profession. However, it is essential to note that the invocation of a digital turn in peacebuilding mystifies, rather than clarifies, the role of digital technology because it cements a distinction between a 'digital' and a 'pre-digital' practice of peacebuilding. As I argue in the next chapter, this distinction has little analytical value.

First, however, this chapter will attempt to describe what we mean by the digitalization of peacebuilding. To this end, the chapter first reconstructs the emergence of ICT4Peace and PeaceTech as two central ideas that organized the growing professional field of digital peacebuilding. Moreover, I describe how the discourse around technology for peacebuilding was characterized by two competing views: one concerned with how technological tools can have an impact on society, politics, and, by extension, peace, and the other concerned with how technologies for peacebuilding are always conditioned by, and mirror, politics and society. The chapter then demonstrates. the need for a research perspective that can grasp how digital technologies employed in peacebuilding are always socially embedded. I demonstrate this by drawing on Martin Heidegger's observations that 'modern' technologies 'enframe' the world through their utility, that is, their state of being 'ready-to-hand'. This helps to explain why the practice and research on digital peacebuilding often struggles to move beyond an obsession with the usefulness of technology. The final section explores how these insights can be translated into a critical research perspective on digital peacebuilding interested in how digitalization, while creating utility for some peacebuilding agendas, produces shortcomings, and a sense of void for others.

The Digitalization of Peacebuilding

There is little doubt that contemporary peacebuilding would not have been possible without digital technologies. Ranging from standard programs and applications that enable everyday communication, such as email, text messaging, and video-conferencing software, to highly specialized technologies for specific tasks, such as earth observations to remotely gather information about ongoing armed violence or Natural Language Processing (NLP) of social media data to identify hate speech and misinformation, digital technologies often form a necessary and irreplaceable element in contemporary peacebuilding. Only a couple of decades ago, anything involving the gathering and processing of digital data was viewed as avant-garde. For instance, efforts to systematically collect and analyse large amounts of conflict early warning data constituted a highly specialized field of research pioneered by a select number of academic institutions and think tanks. In the 1980s, a small group of computer scientists around Kumar Rupesinghe at the Peace Research Institute Oslo began pioneering the use of data-driven early warning systems, which became a 'growing concern for a variety of international organizations' during that time (Rupesinghe 1987, 217). His contemporaries explored options for 'highly sophisticated information system[s]', arguing that due to 'the astounding development of computers during the 1970s and 1980s, extraordinarily sophisticated handling of information has become possible' (Onishi 1987, 271). As will become apparent in Chapter 4, such systems are widely employed today, including in some of the least digitally connected contexts, such as South Sudan.

On the other hand, it is also still widely assumed that the core functions of peacebuilding require direct interaction between humans. Preventing the recurrence of conflict in places such as Liberia or Sierra Leone involved lengthy political reform processes implemented through in-person meetings of experts participating in dedicated commissions or hands-on reconstruction and development projects that rebuilt and improved public infrastructures and service delivery. In Rwanda, efforts to reconcile former warring factions and overcome inter-group grievances were implemented through local transitional justice institutions that gathered in rural churches and often under trees. Even the more technical tasks, such as the monitoring and enforcement of ceasefires or demilitarization zones, first and foremost involved the presence of human peacekeepers wearing the widely recognized blue helmets of the United Nations (UN). Undoubtedly, all these tasks usually required *some* technical equipment, such as radios to communicate, cameras and microphones to record events, vehicles to move from place to place, and so on—and these tools increasingly relied on digital transmission, storage, and processing of data. However, all this did not fundamentally change the common perception that (digital) technology ultimately mattered little in the efforts of keeping, making, or building peace.

Arguably, the notion that peacebuilding is a human activity continues to prevail today. Even during the COVID-19 pandemic, when most organizations had to resort to online conferencing and remote collaboration, many in the peace mediation community continued to stress that, while online meetings were necessary and tools that enabled remote conflict analysis or stakeholder engagements could come in handy, facilitating dialogue and negotiations between conflict parties remained an inherently 'human-centred' task (Eleiba 2020). At the launch meeting of the 'Cybermediation Network', a community of practice comprising experts from many mediation support organizations, which took place at one of the wood-panelled rooms of the Palais des Nations in Geneva, a UN representative recounted that parts of the Iran nuclear deal had been negotiated in the very same room. In preparation for the talks, all non-essential technical equipment was removed to avoid sensitive information being leaked (Hirblinger 2022). While an aversion to technology may be most pronounced among professionals supporting peace processes and negotiations, who first and foremost perceive the presence of microphones, cameras, and instant messaging applications as a confidentiality risk, in many other activities that fall under the umbrella concept of peacebuilding, (digital) technologies have until recently played at best a minor role. Often, they were seen to be a mere *tool* to enable their efforts: the plane or conference platform that would get dialogue facilitators to meet with conflict parties, the audio recorders used to collect testimonies of ceasefire violations, or radio and television used to broadcast important events to the broader populations.

Yet, even while playing a mere support role, digital technologies have become increasingly indispensable for the practices of building peace. This is partly because recent decades have seen the increased digitalization of conflict-affected contexts. Consequently, the peacebuilding efforts in these places have become increasingly digitalized. In the three countries that are at the core of the research for this book—the United Kingdom, Sri Lanka, and South Sudan—the number of internet users grew steadily in the decade before the onset of conflict and continued to climb during periods of political upheaval or armed violence. It likewise grew during the various peace processes in those countries, as well as during the many less formalized efforts to build peace and overcome conflict. However, as will become apparent in the following, only looking at the aggregate picture through household surveys and statistics capturing the increasing spread of digital connectivity will not suffice for a nuanced understanding of what causes and conditions the trend towards digital peacebuilding.

The Emergence of ICT4Peace and PeaceTech

The past two decades have seen the emergence of several practice- and policy-focused initiatives that aimed to promote the use of digital technologies for peace. However, as will become clear in Chapter 2, these initiatives should not be read as the 'beginning' of digital peacebuilding as we understand it today, because they are in many ways embedded in a much longer trajectory of technological innovation. It is also important to note that these initiatives did not simply unfold in the winding corridors or air-conditioned meeting rooms of the UN or other international institutions, insulated from those affected by armed conflict and violence. Rather, they were noticeably conditioned by them, including by experts and tech enthusiasts who describe themselves as being part of the Global South. That said, the visions of technology and its role in peacebuilding that emerged in those contexts often competed with the ideas and influence of bureaucrats, diplomats, and tech entrepreneurs located in countries that had been at the forefront of global digital innovation, especially the United States, but also smaller players such as Switzerland or Finland. Several aspects of these conflicts over how technology should be understood and used also characterize digital peacebuilding today, and they are therefore worthwhile recounting in greater detail.

It is difficult to pin down what the outcomes of these initiatives were concretely, because they consisted partly of ideas, visions, and knowledge of technology use, partly of declarations, regulations, and norms, and partly of heterogeneous and fluid communities of practice and intersecting professional networks. Nonetheless, we can identify two competing visions that characterized the effort to professionalize digital peacebuilding: on the one hand, a vision of digital peacebuilding as employing tools that are independent from society and politics yet able to influence them; on the other hand, a vision of digital peacebuilding as deeply entwined with society and politics and of digital technologies as their outcome. In the first vision, digital peacebuilding simply requires designing and employing the right tools to effect change. In contrast, in the second vision, designing and employing digital technologies helps with identifying, gauging, and facilitating what needs to change. We can call the first position a tool-centred perspective and the second a society-centred perspective on digital peacebuilding. The first position came with a tendency to promote packaged, off-the-shelf solutions that were often also proprietary, whereas the second tended to encourage a Do-It-Yourself (DIY) culture that preferred the bottom-up development of open-source applications. This vision of open-source digital peacebuilding implied that large parts of the population, and not just a highly specialized community of tech designers and professional digital peacebuilding experts, would employ digital tools to effect societal change. These perspectives were not mutually exclusive but nonetheless contained tendencies and priorities capable of shifting attention, resources, and capacities in either direction. The battle between these perspectives primarily unfolded around two key terms: 'ICT4Peace' and 'PeaceTech'.

ICT4Peace

In November 2005, the second phase of the World Summit on the Information Society (WSIS) concluded in Tunis, and heads of state and government declared that they 'value the potential of ICTs to promote peace and to prevent conflict'. The declaration noted that 'ICTs can be used for identifying conflict situations through early warning systems preventing conflicts, promoting their peaceful resolution, supporting humanitarian action, including protection of civilians in armed conflicts, facilitating peacekeeping missions, and assisting post conflict peace-building and reconstruction' (ITU and WSIS 2005). After several years of intensive preparations, paragraph 36, which contained the declaration above, summarized the emerging thinking of a small group of diplomats around the Swiss Ambassador Daniel Stauffacher, who was involved in organizing the Summit. Inspired by the ICT4Development movement that began to emerge at the same time, Stauffacher was keen to create a policy momentum to leverage digital technologies not only in service of the then recently formulated Millennium Development Goals (MDGs), which was the focus of the first WSIS phase, but also more specifically for peacebuilding and conflict prevention—a thought that was also advanced by people close to the then UN Secretary-General, Kofi Annan.

The central idea evolving in these circles was that Information and Communication Technologies (ICTs) shouldn't merely be looked at as a possible threat to peace and security but as tools that can help maintain them. The declaration remained ambivalent about how exactly ICTs would contribute to peace. However, it emphasized their utility as *tools* with specific affordances of relevance for professionalized peacebuilding efforts. This tool focus was also echoed in the accompanying report published the same year, which discussed several examples of 'ICT4Peace in Action', ranging from information dissemination to early warning, field operations support, electoral assistance, and government capacity building to journalism and reconciliation efforts. The report defined ICT4Peace as encompassing:

> the varying types of activity that are carried out in relation to armed conflict, including conflict prevention and management, peace operations, humanitarian relief and disaster assistance, and post-conflict peace building and reconstruction. (Stauffacher et al. 2005, 6)

In the following years, the outcome document of the WSIS summit turned into a programmatic vision for what is now known as 'digital peacebuilding'. More concretely, it formed the substantial basis for the 'ICT4Peace project' and, later, the creation of the ICT4Peace Foundation based in Geneva, Switzerland. The Foundation was meant to serve as the custodian of paragraph 36 of the WSIS outcome document and to support the development of ICTs for democracy promotion, crisis management, humanitarian operations, peacebuilding, and, later, the promotion of cybersecurity and peaceful cyberspace.

Undoubtedly, the ICT4Peace movement was driven by a range of influential figures who thought of technology primarily as a *tool* that could be *used*, such as former President of Finland, Martti Ahtisaari, who was at the time involved in several high-level peace efforts, including in Northern Ireland and later Kosovo. Ahtisaari was also at the time the Chairperson of the International Crisis Group and the founder of the Finnish Crisis Management Initiative (CMI), which later spearheaded the use of digital facilitation methods in Track 2 processes (see Chapter 5). Importantly, Ahtisaari's involvement came at a time when Finland was at the helm of enabling mobile connectivity, with Finnish Nokia phones dominating the European market. After the decline of Nokia's popularity, observers of the Finnish efforts to spearhead the use of digital tools for peacemaking would jokingly speak of the 'Nokia Hangover', triggering a sense that Finnish digital innovations should seek other outlets. Interestingly, technologists and large tech companies not only shaped the founding moment of the ICT4Peace movement but also the conception of the US-based variant of PeaceTech, discussed in the next section. Nonetheless, the discussions around ICT4Peace were, from the onset, characterized by some concerns with the broader social, political, and infrastructural context that would determine the effects of digitalization on peace. Indeed, the conclusion of the report referred to certain 'basic problems' that should be 'kept in mind at all times', including issues related to the digital divide, such as lack of infrastructure or poor regulatory frameworks, aspects such as confidentiality and security, or concerns with the neutrality or impartiality of ICT4Peace initiatives (Stauffacher et al. 2005, 52).

However, the idea of using digital technologies in peace support did not originate somewhere between Switzerland and Finland, nor necessarily in the Global North. A couple of years before the WSIS summits, the Swiss Foreign Ministry supported the Norwegian-facilitated peace process in Sri Lanka, in which a think tank called the Center for Policy Alternatives (CPA) played a critical role. Part of the team was a largely self-taught tech enthusiast named Sanjana Hattotuwa, who supported the process design for Track 2 and Track 3 activities. Through his initiative, the first practical experiences of employing digital technologies to support peace negotiations were made, documented, and shared with a global audience. Having completed his undergraduate degree at the University of Delhi, where he witnessed the fast-paced adoption of digital technologies in India, Hattotutwa was keen to employ computer software to support the process, including through early variants of cloud computing software developed by Groove Networks (later acquired by Microsoft), which enabled asynchronous data exchange while maintaining a high level of data security. Hattotuwa collaborated with mediation support experts to create the first 'virtual' negotiation table by developing what came to be known as the 'One Text Method'. It allowed dialogue facilitators to document and compare the statements and positions of dialogue participants and to work on a joint, single text, taking all perspectives into account. The results of these exercises were made available to senior (and often older and less digitally literate) participants in the Track 1 negotiations. In addition, the method also helped the negotiation parties to experiment with different positions and negotiation scenarios, as will be discussed in Chapter 5.

Following Hattotuwa's work with the CPA, the Swiss Ambassador to Sri Lanka at the time recommended Hattotuwa to Stauffacher because he had left 'an impression'.[1] Interestingly, Hattotuwa had garnered attention not only through his work on the One Text Method but also because of his public criticism of the ICT4Peace process, flagging its socio-political context. Hattotuwa had called out the 2005 ICT4Peace report for its 'Northern Gaze', lamented the 'top-down approach to research', and demanded that participants 'engage with Southern actors and give voice to their ideas and ICT solutions that answer the challenges of conflict transformation' (Hattotuwa 2006). His critical comments caught the attention of Stauffacher and Hattotuwa was hired as Special Advisor to the ICT4Peace Foundation. According to Stauffacher, Hattotuwa added a critical component to the organization's work, including criticizing technologies when these were 'thrown at situations by white people' without being 'contextual' and without 'addressing the root causes of conflict'.[2]

Hattotuwa expressed a degree of tech-scepticism and formulated a radical vision of a networked, bottom-up, and community-centred use of technology. With that came the understanding that digital approaches would naturally have to grapple with the deeper, underlying causes of conflict and violence. As he wrote in his first blog post on this matter, published in July 2005:

> Where crumbling frameworks of governance exist cheek in jowl with advanced telecoms infrastructure and national ICT agencies and the spectre of terrorism hangs in a Damoclean balance atop sustainable development, where corruption and zero sum partisan politics undermine the development of legitimate peace processes, where hope itself is strangled by the Hydra of inflation, human rights violations, cultures of impunity and violence, one could be forgiven at laughing uncontrollably at the mere prospect of using ICT for peacebuilding and by extension, sustainable development and social empowerment. (Hattotuwa 2005)

Speaking from the vantage point of the Sri Lankan context, he argued that 'ICT must not be seen as a panacea that is the magical harbinger of a new social order overnight' (Hattotuwa 2005). Instead, they should be employed as part of a 'holistic framework' that fashions technologies that are 'resonant to demands from the grassroots' and 'empower communities by taking ICT to the people', which would also require the eschewal of 'all technological frameworks written by those in the West that are ill-equipped to grasp the complexity of the vigorous dynamic of inter-locking interests that define politics and communal interactions in our region' (Hattotuwa 2005).

[1] Interview 124, Daniel Stauffacher, Founder of the ICT4Peace Foundation, via videotelephony software, 25 April 2024.
[2] Interview 124.

Hattotuwa's version of ICT4Peace was thus one of context-specific technology development and use. In the case of Sri Lanka's decade-long armed conflicts and ethnic violence, it was the long durée of colonial legacies and global inequality that had to be taken into consideration in any effort to implement the idea of ICT4Peace. This position was not too different from that of many other Southern activists and scholars, as well as discourses and conversations that had at that time begun to emerge not only in development studies but also in the research and practice of peacebuilding, namely a concern with Eurocentrism (Sabaratnam 2013).

Notably, Hattotuwa's understanding of the utility of technology was inspired by exchanges with various civil society activists, some of whom had been involved in the struggle against Apartheid in South Africa. Among the learning examples that inspired his work was 'Operation Vula' by the African National Congress (ANC), during which self-taught activists built an encrypted communications system to enable secure communication between the exiled ANC leadership and the military wing inside South Africa. For Hattotuwa, this initiative demonstrated how tech could be employed for 'subverting power relations' and 'fostering social change' (Hattotuwa 2008). Unsurprisingly, his vision of technology wasn't one of 'neutral' tools. Instead, their design and use were always for a particular purpose, and they were embedded in society and politics. This also meant that the very meaning of peace and how it was to be achieved couldn't be separated from the technology and the technological approach employed towards it.

However, putting a society-centred vision of digital peacebuilding into practice was another matter altogether. Much of the work that the ICT4Peace Foundation conducted in the years after the WSIS summits was of an applied nature, focused on operationalizing the idea of ICT4Peace. Importantly, this effort was underpinned by an elastic understanding of what it meant to do digital peacebuilding. As the 2005 report put it, 'ICT4Peace activities are very broadly defined at present, and rightfully so; it is up to the actors themselves to further develop the definitions involved' (Stauffacher et al. 2005, 18). In the years after, numerous international and civil society organizations began to roll out additional initiatives with the support of the Foundation, and donors introduced specific themes and objectives. The further professionalization of the field and the adaptation of the ICT4Peace agenda by a growing number of organizations, many of which were located in the Global North, meant that it became more challenging to implement the vision of ICT4Peace as a means for grassroots empowerment. The design and use of digital tools became written into programmes and project proposals with formal theories of change, implementable within allocated resources and timelines. This projectification meant that initiatives emerging from within conflict-affected contexts could easily be co-opted by international initiatives that promised funding or well-paid jobs. It also meant that the meaning of 'peace' in the ICT4Peace became increasingly elastic, capable of serving a broad range of objectives and activities, ranging from supporting humanitarians and promoting economic development to strengthening social cohesion, empowering women, promoting gender equality, and fighting violent extremism and terrorism.

PeaceTech

In the two decades since the WSIS summits, many peacebuilding and peace mediation organizations established or supported a large number of initiatives, programmes, labs, and new civil society organizations and consultancies that specialized in the design of technologies for peacebuilding (Dajer 2018; Larrauri and Jung 2017). International organizations, including the UN, began to systematically consider how to respond to, and make use of, the increasing use of new technologies in support of their development, peace, and security agenda (Wählisch 2019) and created dedicated units such as the UN Innovation Cell and Global Pulse. Working in conflict-affected contexts, UN peace operations increasingly started exploring remote sensing and satellite-based applications, as well as big data, for surveillance and reconnaissance (Karlsrud 2014), and many regional organizations developed conflict early warning and response capacities by establishing data-gathering and analysis systems capable of assessing the likelihood of conflict outbreaks (Rohwerder 2015). In this rush of activities, a second term emerged: PeaceTech. Unsurprisingly, obtaining a comprehensive overview of this rapidly evolving, increasingly heterogeneous, and dynamic field is challenging. For instance, Lisa Schirch (2020) identified twenty-five different spheres of PeaceTech, ranging from digital citizen journalism, election and conflict monitoring, and social media fact-checking, to promoting peaceful narratives, intergroup dialogue, and peace negotiations. However, this list is neither complete nor a neat typology of mutually distinct activities. Maybe it suffices to say at this point that these technological innovations have given rise to a broad range of 'evidence'- and 'data'-driven approaches to peacebuilding that have also enabled new forms of communication, exchange, and engagement (Panic 2020).

One of the most influential actors that shaped the digital peacebuilding space formed around a group of tech enthusiasts under the name of 'Build Up'. The organization was co-founded by Helena Puig Larrauri, who had previously worked for the United Nations Development Programme (UNDP) Sudan, where she formed part of the so-called Standby Task Force of the Crisis Mapping Movement, a network of volunteers that experimented with geographic information systems (GIS), open source, and crowdsourcing approaches to document humanitarian needs and conflict dynamics in acute crisis situations. This community was driven as much by a fascination with the potential of networked, 'bottom-up' approaches that could generate relevant information with the help of actors 'on the ground' as it was in specific technological solutions. In collaboration with Anne Kahl, who worked at the UNDP at the time, Larrauri wrote the first academic article on the subject, entitled 'Technologies for Peacebuilding', published in the *Stability Journal* in 2013, in which they argued that through new technologies 'new avenues of participation, engagement, and accountability [were] emerging', and that the 'empowerment of people to participate in localized conflict management efforts [was] one of the most significant innovations and opportunities created by new technologies'—yet, this wouldn't happen automatically but required 'designing for empowerment' (Larrauri and Kahl 2013).

Created during the Arab Spring, Build Up's vision was inspired by the then much-discussed potential of social media to catalyse political participation and democratic social movements. 'Designing for empowerment' meant that the social and political aspects of conflict prevention should remain central to all technology-focused interventions and that technology should be viewed as a catalyst for social and political change, but not as its replacement. In the following years, Build Up's declared mission became to 'transform conflict in the digital age' and was carried by the ambition to enable peacebuilding organizations and their staff to employ technology in purposeful ways for their declared objectives. Build Up, first and foremost, acted as a platform for discussions surrounding the use of technologies for peacebuilding, and its members provided training and advice to many peacebuilding organizations. These activities, importantly, aimed to leverage technologies for peacebuilding by starting with the question of what was needed for peace instead of asking what specific tools could achieve. This approach was underpinned by a view on technology that was comparable to that of Hattotuwa, namely that technologies were not neutral because they couldn't be separated from the social and political contexts in which they emerged and in which they were used. As Larrauri described it:

> I always ask people, 'Are clocks neutral'? . . . If your answer is well, you know, time can be used for good or for bad, then, you know, sure. But that's a very superficial conversation to have. If your answer is the way that time is constructed, and measured, is inherently entangled with the structures of oppression, of capitalism, then maybe we can have a conversation now. And so, it's very similar with this. So, what that means is that in our programmes, we never begin from the tool; we always begin from the change.

In her view, technology itself was part of the conditions that needed to change. Change couldn't simply be achieved by employing technology as a tool, given the many factors that contributed to hatred and violence, including histories of oppression and exploitation, failing public institutions, violent regimes, business interests, and profit-making models. These could not simply be fixed through a new mapping application or text data analysis tool. Build Up attempted to implement this perspective by promoting a version of 'human-centred' design that should ensure that digital peacebuilding approaches would be context-appropriate. However, developing technologies to meet the needs of end-users operating in conflict-affected contexts often does not help with resolving the underlying, and especially the structural, causes of conflict, as I will demonstrate in Chapter 4.

Build Up emerged in a community that found itself at the interface of tech enthusiasts, digital activists, social change-makers, and development consultants. However, the newly emerging field of digital peacebuilding was also shaped by peacebuilding actors with much closer linkages to large technology companies. In March 2011, Richard Solomon, the then President of the United States Institute of Peace (USIP), sent out invitation letters to a meeting to discuss an 'ambitious initiative', namely the 'creation of the first facility dedicated to the development and deployment for conflict management and peacebuilding, the PeaceTech Lab'. Besides the USIP's letterhead, the letter also contained the logo of the US National Academy of Engineering, and was co-signed by known tech personalities. This included Vinton Cerf, recognized as one of the 'fathers of the internet' and self-described as the 'Chief Internet Evangelist', and Irwin Jacobs, the CEO of Qualcomm, an American multinational corporation building wireless technology semiconductors and software and involved in the creation of the 2G mobile internet standard. The USIP-based PeaceTech Lab was initiated by Sheldon Himmelfarb, who had pioneered the use of radio and television for peacebuilding at Search for Common Ground (SfCG) in the 1990s and returned to professional peacebuilding work after a period in the private sector. The well-resourced initiative displayed a somewhat different spirit than its counterparts. Coining the term 'PeaceTech', trademarking 'PeaceTech Lab',[3] and securing the backing of some major internet technology industrialists, the US-based initiative was comparably more tool- and market-centred. It promised, according to the invitation letter, to provide 'virtually unlimited opportunities to try new approaches to age-old problems of corruption, ethnic tension, boundary disputes, water and food shortages, election fraud, [or] gender violence'. Unlike Hattotuwa's notion of ICT4Peace as socially and politically embedded, the foundational vision of the PeaceTech Lab was that technology could simply provide a solution to a pre-given problem. In this thinking, technology appeared as an independent variable, a 'treatment' that could be entered into the equation that makes the difference between war and peace.

However, while 'new' technological innovations were meant to resolve 'old' problems such as ethnic hatred, it wasn't always clear what difference *exactly* technology was really supposed to make. The wording in the letter suggests that technology could make a qualitative difference by fundamentally altering how peacebuilding and conflict prevention are done. As mentioned above, this notion of PeaceTech as a 'game changer' has also been prominent among other parts of the PeaceTech community. It was also justifiable in certain cases, for instance, when digital platforms enabled crowdsourcing approaches that simply weren't possible before (see, for instance, Martin-Shields 2015, 33). On the other hand, Himmelfarb also described technology as a 'force multiplier', which would suggest the much less extraordinary role of a catalyst of already existing approaches and efforts. Nonetheless, what these two different visions of technology shared was the underpinning assumption that technology would be neutral. As Himmelfarb put it:

> I'm a big believer that the tech itself is neutral. It's not good or evil. But you do have to focus on the kinds of resources, one elevating its power for good. Otherwise, . . . history has shown, the resources will be deployed in other ways.[4]

[3] USIP stopped short of trademarking 'PeaceTech' so as not to inhibit the replication of the concept by other peacebuilding organizations, but it did secure 'PeaceTech Lab', which would provide a minimum protection against actors who would divert PeaceTech too much from the main idea.

[4] Interview 123, Sheldon Himmelfarb, Founder of the PeaceTechLab at the United States Institute of Peace, via videotelephony software.

Since technology could be employed for both good and bad purposes, Himmelfarb also cautioned against the possible adverse effects of PeaceTech applications. Citing the impact of digitalization on democracy and voting, he argued that the same digital infrastructures could provide 'the means for creating inter-religious, inter-ethnic violence'.[5] However, the answer to the challenge of addressing malevolent technology use, according to Himmelfarb, would be simply to scale up PeaceTech: 'We need more PeaceTech, we need more users, in more places, more often'. While many PeaceTech initiatives struggled to run sustainably, Himmelfarb argued that going beyond the 'pilot phase' would require building a PeaceTech industry that would employ government resources only initially and then run on a sustainable business model.

This idea became inscribed in the PeaceTech Lab Accelerator, which the Lab launched after its formal incorporation in 2014 in partnership with major technology companies, including Amazon Web Services and SAP NS2, which provided cloud technologies to support PeaceTech development. The accelerator's idea was to prepare PeaceTech ideas for the 'market', not only in the sense of professional peacebuilding organizations, but indeed the private sector that could productize them.[6] To this end, the programme recruited entrepreneurs who promised to generate social impact on a global level, supported them in incubating their idea and in building a business case, identified customers, partners and investors, and helped to promote their product. For Himmelfarb, the intention behind this comprehensive effort was no less than building a 'PeaceTech industry' that would be 'uniquely positioned to not only alter the world's current trajectory, but to lead the planet into a new era of conflict prevention'. The PeaceTech Industry would 'beat technological swords into plowshares for the good of humanity', as he argued jointly with Vinton Cerf on CNN Opinion in 2018.

Examples of PeaceTech start-ups that managed to secure start-up funding from technology companies or foundations, and monetize their services, provided encouraging proof of concept and suggested that a 'reboot' of peacebuilding would indeed be possible (Himmelfarb 2015). Thus, Himmelfarb's vision of PeaceTech was distinct in that it suggested understanding it as a product that could be developed and sold and, more importantly, that it should be able to attract a big enough client base that it would be able to do so sustainably. While tool-centred at the core, the need to consider the social embeddedness of technology is acknowledged in this vision, at least in terms of the need to create solutions that could be sold on the 'market'. Indeed, pairing this tool-focused perspective on PeaceTech with assumptions about tech-neutrality is somewhat paradoxical, as I will make clear in the following chapters. How can tools be neutral if they have to be bought by a specific consumer base that holds strong views on what's right and what's wrong? Innovating for a large enough market and at scale ultimately means designing for and responding to specific contexts and demands, and this means that a tool-centred perspective ultimately does little more than conceal the

[5] Interview 123.
[6] Interview 123.

social, political (and, indeed, economic) context conditions that shape digital peacebuilding. Vice versa, many actors in the peacebuilding sector and elsewhere may use technology and claim that it is 'neutral', 'scientific', or 'objective' exactly to portray the sense that their work is apolitical. Instead, as will become clear throughout this book, the 'tools' they employ are closely entwined with the assumptions about conflict and peace, their peacebuilding agendas, and the political context in which they operate.

The PeaceTech Lab was closed by the USIP in 2024. One of the reasons for this decision was a lack of interest by potential investors; another was that digital peacebuilding and PeaceTech had become increasingly mainstreamed, making specialized initiatives such as the PeaceTech Accelerator redundant. Many organizations with a peacebuilding mandate have over the past years created dedicated 'digital peacebuilding' teams, focal points, or programmes, including the UN DPPA's Innovation Cell, the Alliance for Peacebuilding's Digital Peacebuilding Community of Practice, the HD Centre's Digital Conflict team, to mention only a few. Besides this proliferation of digital peacebuilding, there is another important trend to note. Himmelfarb and some of his staff transitioned to start an entirely new initiative, the International Panel on the Information Environment (IPIE). Modelled after the International Panel on Climate, the IPIE aims to address challenges that evolve from digitalization and the increasing use of ICTs and social media by producing and evaluating evidence that can be employed in regulatory and policy processes. This signals a strengthening of the society-centred view on digitalization, attempting to mitigate its adverse effects on conflict and enhancing its peace-supportive potential through support to regulatory processes that aim at changing political institutions and social practices, instead of attempting to simply build 'tools' for conflict prevention and peacebuilding.

The Social Embeddedness of Technology

The brief reconstruction of the emergence of ICT4Peace and PeaceTech presented above is anything but complete, and it can't do justice to the multitude of ideas and voices that have come to define digital peacebuilding over the past two decades. However, it highlights some significant differences in how some of the pioneers of digital peacebuilding hoped to realize the peace potential of digital technologies. The inspiration for these initiatives stemmed from a diversity of places, including the corridors of multilateral institutions, the offices of large technology companies, and the safe spaces of democratic and decolonial movements. All these certainly affected how the planners, designers, and users of digital technologies thought about them. In particular, whether they envisioned peace technologies as proprietary products that could be monetized and sold in processes that would ultimately be led from the 'top-down' or as open-source goods that would emerge, circulate, and evolve among decentralized activist networks from the 'bottom-up'. These different viewpoints didn't lead to entrenched positions. Organizations like USIP also recognized the importance of 'local ownership', for instance. However, the tension between 'top-down' and 'bottom-up' was certainly something that the emergent professional digital peacebuilding community continued to grapple with, and sometimes, this showed publicly in their statements, presentations, or reports, as I discuss further in Chapters 4 and 6.

There is one central takeaway from this cursory stocktaking. When thinking about the role of digital technology in peacebuilding, it is natural to oscillate between a tool-centred and a society-centred perspective. Digital technologies are indeed tools, usually used as a means to an end—they are things to work with to achieve a particular objective. However, peace, and the effort of maintaining or enabling it, are ultimately about society—about people who relate to each other as enemies, friends, citizens, religious or ethnic groups, or holders of different passports. Importantly, all those involved in conflict, and in efforts to end it and build peace, also hold views about how their aims and objectives can be furthered with technology. Yet technological innovation often sparks a deep fascination with new tools and their functionalities, which can lead to a sort of tech fetishism that's also observable in the peacebuilding field. Such fetishism can lead to the assumption that technology can be stripped of the meaning and purpose that we humans attribute to it, as well as of the context conditions that shape its design, use, and impact. The risk with this is precisely that we may pay too much attention to processes of technological change and too little to the contexts that condition how those change processes happen and with what impact.

One common manoeuvre through which digital technologies employed for peacebuilding become stripped of their context is via the rhetorical assertion of technology neutrality. Of course, the claim that technology can be used for all sorts of purposes isn't wrong in empirical terms. There is considerable literature on dual-use technologies that goes beyond the narrow field of peacebuilding (see, for instance, Riebe and Reuter 2019). From this vantage point, it can be convincingly argued that technologies employed by well-intended peacebuilding actors can also have effects that are perceived as aggravating conflict. Indeed, the emerging academic and practice-oriented literature on digital peacebuilding and PeaceTech seems to be rather comfortable with this stance. It tends, first and foremost, to focus on the positive potential of technical tools in the hand of professional peacebuilding actors, while always caveating such accounts with possible unintended consequences and the negative effects of such applications. For instance, Panic and Arthur (2024, 4) acknowledge that employing AI in peacebuilding requires balancing its possible benefits 'with an awareness of its malicious use and unintended consequences'. Moreover, Bell (2024, 44) shows how the plethora of current PeaceTech projects is fuelled by a 'conviction that new forms of digital innovation offer immediate improvement of peacebuilding practices'. However, she also warns against the appeal of 'shiny new things', which justifies her praxeological and normative concerns about how PeaceTech is and should be 'done'.

Yet, merely recognizing that we also have to deal with, and contain, the negative effects of technology may still leave us prone to a position and approach that is ultimately tool-centred: one that is concerned with how positive and negative effects of digital peacebuilding could be balanced during the design and use of tech. However, tools are used in one way or another with a specific intention and purpose. This is, for instance, clearly visible when retracing the early conversations around crowdsourcing, in which new networked forms of data collection and analysis via the internet and mobile devices merged with pre-existing idealization of bottom-up peacebuilding (Tellidis and Kappler 2016; see also Chapter 4). The discussion about tech neutrality thus deflects from a more foundational point. As Avgerou (2008, 135), has argued, ICTs are always situated in a context and are 'socially embedded'. They have been shaped by the environment in which they have been developed and they serve specific preferences. This has been well documented in sociological approaches to technology that emphasize the relationship between the material aspects of technology (the tools or gadgets) and the knowledge, skills, and rational considerations that accompany their use.

For instance, in their work on classification systems, Geoffrey Bowker and Susan Leigh Star demonstrate the power of information infrastructures in ordering the natural world. They show that technologies, for instance computer programs and databases that collect and code information about psychological conditions, work alongside institutional decision-making processes and bureaucratic routines to establish a classification of diseases, such as those documented in the Diagnostic and Statistical Manual (DSM). The manual has powerful effects on the world, for instance, by determining which conditions are recognized by health insurance companies and which treatments will be reimbursed by them. By establishing a typology of mental health conditions that structures the professional discourse of therapists, information infrastructures 'valorize[] some point of view and silence[] another' (Bowker and Star 2008, 5). Importantly, technological innovations such as remote sensors, satellite systems, mobile broadband networks, or machine-supported data analysis provide the backbone of such information infrastructures and propel innovation by enabling the implementation of some approaches—but not others. This may also help us to understand why the outputs of information infrastructures are commonly contested: they are political. We can observe similar phenomena in peacebuilding, for instance, when methodologies for data collection and analysis become the object of controversy between various experts and decision makers, or when they are silently discontinued, as I demonstrate in Chapters 4 and 5. This makes clear that technology not only shapes how we make sense of the world but also how we approach conflict and devise and implement options for conflict resolution.

In the remainder of this chapter, I will show how we can deepen our engagement with the social embeddedness of technology in peacebuilding through Martin Heidegger's discussion of technology as something that is 'ready-to-hand' or 'present-at-hand'. I will employ this differentiation to reflect on how we commonly approach the study of technologies for peacebuilding and will argue that a third approach is warranted, namely one through which we render technology as something that creates a void. More concretely, I suggest thinking of technology as 'missing-from-hand'—a concept I will elaborate to critically interrogate how socially embedded technology creates utility for some approaches and understandings of peacebuilding, while turning out to be insufficient for others.

The Obsession with Technology Utility

The social embeddedness of technology is characterized by a reciprocal relationship between society and technology: we humans can shape the digitalization of peacebuilding, but we are likewise shaped by it. To further explore this relation, I suggest briefly embarking on a more general reflection on how technology affects human existence and our ability to live in peace. To this end, it may be helpful to draw on Martin Heidegger's idea that technology 'enframes' the world. In his reflections on 'modern technologies', written at a time when the impact of industrialization had become manifest across most, if not all, parts of Europe, Heidegger argued that all technologies (even very simple tools, such as a plough or hammer) lift humans out of their immediate relationship with the natural world. Heidegger called this process 'enframing' and attributed to it a productive quality, arguing that it fundamentally shapes how we humans are in the world. As he put it, through this enframing, the world 'reveals itself' and this 'destines being' (Heidegger 1977, 21–6). Looking at early modern technologies, such as hydroelectric plants built into the Rhine River, Heidegger described this revealing as an act of intervening in the natural world that was also productive because it unlocked 'energy' that was concealed in nature (1977, 16.). Of course, in the case of a power plant, this energy generation happens literally through the transmission of hydropower into electricity. However, he suggested that this revealing could also be more broadly understood as productive processes that generate a capacity to act, which, in turn, conditions human existence.

It goes without saying that the same revealing qualities can be attributed to digital technologies, including those focused on information and communication. To interpret his thinking liberally, we could think of technologies as doing work that could be described as both 'mediating' and 'ordering' because they insert themselves between us humans and some natural state, and they thus shape the universe of possible ways of being. I will discuss in the next chapter how this can be concretely studied through the concept of 'apomediation'. Of course, writing before the information age, Heidegger did not put particular emphasis on the discursive or semiotic mediation of our world through ICTs. His point about enframing was arguably much more fundamental, namely that technology inserts itself into our relationship with the world and, therefore, changes how we do what we do and live our lives. And this observation is part of Heidegger's broader concern with understanding human 'being' in a world made up of all sorts of 'stuff' that sometimes manifests itself in human consciousness as 'things' but often shapes our existence in a more latent manner. Key to this endeavour was the differentiation between two ways in which technology can relate to humans, namely as an object that is ready to be used for specific tasks, for example, *utile* (in German, *Zuhanden*), or an object that is present, for example, *ubiquitous* (in German, *Vorhanden*).

As discussed, peacebuilders often treat digital technologies explicitly as tools—as objects that can be used for a specific purpose, such as to counter hate speech and misinformation that fuel armed conflict, to improve the political institutions that govern it, or to enable new methods for participatory decision making employed in dialogues or peace negotiations. This tool perspective on technology can be captured by Heidegger's reading of technologies as having utility, as objects that humans employ intentionally and for a specific purpose. For instance, Heidegger described a hammer as a basic technology that is 'ready-to-hand' as he called it, which translates as the state of *Zuhandenheit* in German. Importantly, when tools have practical utility, they are commonly used without even meriting a thought, just as we employ a hammer, a water faucet, or a light switch. Moreover, this state has an immediate enframing effect on the human who holds and uses it. For instance, tools such as a hammer or a saw socially construct the personae of the handyman who will employ them for a specific action (hitting a nail or cutting a pipe) more likely than for another (weaving a carpet, for instance). The same is true for specialized digital applications developed to conduct online focus group discussions or detect hate speech or misinformation on social media, for example. To put it in more contemporary language, we can't separate technological artefacts from their intended functions, and this intention is telling of what peacebuilding *is*, that is, how it is understood and practised. 'Tools' are thus reflective of, and shape the conduct of those who employ them in support of peacebuilding, and this, in turn, shapes how peacebuilding is understood as a field of practice. As I demonstrate in Chapter 4, for instance, simply deleting hateful social media comments does not enable a deeper engagement with the underlying causes of hate.

As tools that are ready-to-hand, these innovations shape the self-understanding and identity of the peacebuilding sector. For many professionals working in this field today, peacebuilding is no longer (or at least not only) associated with activities such as facilitating community dialogues underneath a tree in a rural village or organizing inter-communal or international summer camps and exchange programmes for youth. It is something that can take place through a participatory social media campaign. Nor is it necessarily something to be carried out by those organizations conventionally associated with peacebuilding, such as the UN Department of Political and Peacebuilding Affairs (DPPA), or non-governmental organizations (NGOs), such as Conciliation Resources, Search for Common Ground, or the HD Centre. Indeed, digital technologies have thus created a new generation and sub-community of peacebuilders—self-described PeaceTech entrepreneurs, digital peacebuilding consultants and trainers, and data scientists. On the other hand, a look into programme and project descriptions of conventional peacebuilding actors reveals that digital technologies today are often central to their activities as well, and that their objectives couldn't be reached without them. For instance, in a 2021 report, the UN launched the digital transformation of peacekeeping, calling for 'data-driven approaches' to support surveillance and reconnaissance, to collect evidence about atrocities against civilians, and to measure the impact and performance of peacekeeping missions (DPO, DOS, and DMSPC 2021). The UN Development Programme (UNDP)'s Digital Strategy for 2022–2025 likewise called for evidence-based decision making and the fostering of 'data usage' (UNDP 2022). This all suggests that digital technology profoundly impacts how peacebuilding is exercised and understood.

Technologies that are ready-to-hand for peacebuilding enable remote work and seamless coordination between colleagues, increase the speed with which information can be obtained, and reduce costs or security risks. However, they also mediate how peacebuilding professionals and parties or stakeholders to a conflict make sense of it and what solutions they suggest, as will be discussed in the next chapter. But the fact that digital technologies are often perceived and used as tools makes them neither separate from society nor neutral. For instance, many who employ digital tools for peacebuilding (or, indeed, in any other fields), will acknowledge that unplanned, unintentional, and spontaneous innovation and adaptation plays an important role. Often, the digital means we employ somehow 'offer' themselves to us and shape how we do things. This is certainly the case with 'off-the-shelf' technologies—from laptops to mobile phones, to video conferencing software and specialized applications using machine learning that turn programmers into peacebuilders. For instance, the recent releases of OpenAI's ChatGPT, and especially the ability to develop customized generative pre-trained transformers (GPTs) in a matter of minutes, has led to a frenzy of innovation in the peace mediation community, as I will argue in Chapter 5.

However, when drawing on Heidegger's idea of enframing, we should avoid simplistic cause-and-effect narratives about how technology conditions human existence or vice versa. To put it simply, his reflections invite us to think of the revealing of the world, and, with it, the shaping of peacebuilding as a field of practice, as something that happens as technology is designed and used, as a co-constitutive process, the results of which cannot be exclusively attributed to either humans or technologies. Heidegger also argued that technologies that are ready-to-hand to some degree also remain mere objects, in the sense that they are waiting to be put to use—they are 'on call for duty' as he put it (1977, 17). As long as they are not put to use, they remain on standby or as part of the 'standing reserve' (Heidegger 1977, 17). This means, by extension, that humans can take an active role in shaping the productive, revealing characteristics of technology because they produce these objectives in the first place and then make them ready for use. On the other hand, Heidegger argued that the revealing of the world isn't 'merely human doing' either, because nature always reveals itself in one way (Heidegger 1977, 18–19). This seems to suggest that he didn't want to reduce the *Zuhandenheit* of technology—its use as tool, and with it, the co-creation of the user—to a result of intentionality or subjectivity (Tzavaras 1989, 369).

Importantly, digital peacebuilding practices come in all shades and colours. Emails, websites, and GIS are used by a wide range of initiatives or activities that can be labelled as 'peacebuilding', ranging from armed peacekeeping by UN or African Union peace operations to liberal economic reform efforts supported by the World Bank, left-wing activism at university campuses, or curation of digital art by initiatives that aim to use online spaces to facilitate an engagement with past injustice and promote reconciliation. This means that while digital technologies are increasingly always ready-to-hand as equipment that can be put to use, their ability to determine the work of peacebuilding seems less directed and static than earlier technological innovations. Indeed, a social media platform or data analytics software is more adaptable and can accommodate many more uses. It can, therefore, also create many different enframings. Many of the tools employed in support of peacebuilding can likewise be utilized for various other tasks in different sectors, such as market research or project management. This means that, while ready-to-hand, digital tools don't compel their users to carry out precisely *one* type of activity in *one* particular way. They can often be used for various activities and have indeed usually been developed for use in other sectors.

The multiplicity of different purposes makes it important for researchers to facilitate a process of reflection through which we can distance ourselves from the immediate utility of digital tools. With Heidegger, we could attempt to obtain an 'objective' view on technology in moments when digital tools in peacebuilding become 'present-at-hand'. In his philosophy, this would be the case once tools (or things) lose their utility and stop functioning. In German, *Vorhandenheit* seems to simply imply that something is *existing* and *available*. German speakers usually don't think about the 'hand' in the term as a limb with which they could grab this object, even though the etymology of the word certainly allows for this inference, that is, something that is in front of your hands. Heidegger aimed to highlight precisely the state of a 'thing' or a 'tool' suddenly seeming distant from us because it has become detached from its immediate utility. Heidegger associated with this present-at-hand state of things the human capacity to (re-)gain an objective view that can enable scientific scrutiny. Likewise, the *Vorhandenheit* of technology manifests in moments of disruptive innovation and change, when practitioners struggle to understand how particular innovations can be leveraged for peacebuilding. Just as a broken hammer invites us to reflect on what it means to be a handyman, an early warning application without internet connectivity may offer deeper reflections on the particular approach to peacebuilding to which it contributes and the purposes it is supposed to serve, and potentially, its limitations. It could invite us to pause and reflect. Indeed, several of the off-the-shelf applications discussed throughout this book are often accepted and implemented prematurely, with the result that they don't match what peacebuilding actors need. Consequently, they are often met with a distanced scepticism and quickly discarded. In such moments of disruption, technologies create surprise and bewilderment and receive considerable attention and scrutiny. They become a topic of strategic and policy conversations that merit the creation of dedicated working groups. In these moments, technology is elevated to a visible object of conversation and inquiry, which provides a moment of pause from its usual state of being ready-to-hand.

In the field of peacebuilding, such moments were noticeable during the introduction of earth observation systems and drones, social media analytics, online conferencing applications, or LLMs, all of which became topical during workshops, conferences, or in practice-oriented reports and handbooks that explicate their utility, impact, and effects on peace and conflict. In those moments, we can learn something from the performing of the 'digital revolution' or the 'digital turn', because digitalization becomes stylized as a watershed moment, and professionals working in the field must adjust to new approaches, methods, and constellations in which digital applications play a seemingly ever-expanding role. It is then that more fundamental conversations—such as those about the neutrality of technology, the globally uneven unfolding of digitalization, and its adverse effects—can begin to take place. In these moments of controversy, not only technology but peacebuilding itself loses its 'taken for granted' status. They thus invite us to reconsider more radically what peacebuilding is and how it should be done. Moments of technological innovation are thus worthwhile studying not primarily because they promise to make peacebuilding 'better' or more 'effective', or because they necessarily lead to change, but because they enable a critical scrutiny of the field as such. This is when technologies can turn into an object of reflection, a sort of 'prism glass' that helps to condense and order various considerations and concerns around the question of what solutions to conflict are needed and how they can be achieved. This was, for instance, clearly visible during the COVID-19 pandemic, when the increased use of video conferencing software created debates about the inclusivity and accessibility of peacebuilding efforts, or triggered questions about what it means to build trust between conflict parties and stakeholders.

However, it is not at all automatic that these conversations really take place. Indeed, there are many examples of blind tech-fetishism or tech-solutionism, which render peacebuilding professionals passive consumers of ready-made, off-the-shelf tech without initiating any critical reflection. Indeed, much of the discourse continues to obsessively ask how digital technologies can have utility for peacebuilding. This is not unique to this professional field. In fact, society and politics broadly conceived prefer to think of technology as something that can be used. This obsession with the utility of technology often curtails critical reflections, not least because practitioners are inclined to demonstrate that their tools 'work' to maintain donor support and advance their careers. Looking at a broken tool seems to have little merit. This is because digital peacebuilding is often conditioned by the projectification and bureaucratization of the sector. The ever-expanding ubiquity of digital infrastructures, applications, and artefacts, even though often leading to disruption and bewilderment, tends to be quickly colonized by a concern with how they can be retained as useful.

Digital Technology and the Voids of Peacebuilding

In light of these reflections, I would like to suggest a perspective that goes beyond engaging with technology mainly in terms of its utility or inutility for peacebuilding. The challenge is to find a way of relating to technology that goes beyond focusing on it as an object that is ready to be employed or as an object that fails to serve its intended purpose. Engaging with these ways of being with technology does enable us to reflect on how digital peacebuilding currently manifests, but it doesn't help us in the search for alternatives. Heidegger's discussion of 'modern technologies' hinted at the socially productive effects of technologies on the world, and it can be read as an invitation to think more systematically about the co-constitution of the technical and the social. However, instead of banking on the hope that some critical distancing from the digital may become feasible in moments of disruption or technical failure, I suggest nurturing a critical engagement with how the digital technologies employed for peacebuilding are always useful for some purpose but not for another, to build on Robert Cox's (1981) critical methodology. Peacebuilding is by definition driven by the objective of building peace. This makes it essential to ask how digital technologies can help in this matter when they are ready-to-hand, and how they 'reframe' what peacebuilding is. This is more straightforward when we get a more detached view of them, because they are present-at-hand. However, I would like to suggest moving towards a reflexive engagement with the digitalization of peacebuilding as a process that, while always driven by somebody's utility motive, regularly produces a sense of void for somebody else and their vision of peace.

The German language knows a third term with the root 'hand', namely *Abhanden.* It expresses the absence of an object (material or immaterial), which is caused by an act of loss. The German '*etwas ist abhanden gekommen*' translates into 'something has become lost'. This state is more than just the absence of something that is needed, rather it entails a sense that something caused it to be lost and missing in the first place—a sense of void. What would it mean to think about digitalization as causing something missing-from-hand in peacebuilding? I suggest introducing this third reading of technology to ask about the shortcomings and inadequacies that become visible in peacebuilding practice as we interrogate how digitalization and the increased employment of digital technology affect the field. A sense of void can emerge because technology is indeed 'missing' in material or functional terms. However, these voids can also point to something that has become missing in the ways we humans conceive and think of and go about building peace as we increasingly draw on digital technologies to build peace. I don't suggest this neologism merely because coining concepts is part of scholarly bread and butter. I do so because it invites us to reflect on how the increasing use of digital technologies in contemporary peacebuilding makes the objective of building peace more difficult.

Importantly, it is not technology alone that creates this sense of void, but the socio-technical relations that emerge as new innovations and tools are employed. What is missing-from-hand, is not merely a better tool, but relations between technology and society that would be more suitable for the work of building peace. Professionals aiming to design and use technology for peacebuilding very often tend to be concerned with everything that their tools cannot do, or indeed fail to do. Social media analytics platforms can scan thousands of comments for harmful speech, but they provide little opportunity to understand and tackle the underlying causes of hate (Chapter 3). Early warning systems are good at collecting large amounts of data in real time, yet these data often fail to trigger effective responses (Chapter 4). AI-enhanced dialogues promise to help establish a representative picture of the needs of hundreds or thousands of stakeholders in real time. Yet, they tend to privilege the collection and generation of data over facilitating encounters that could change how people relate to one another (Chapter 5). We can read these as examples of insufficient or failing technologies, but I would like to suggest that we need to go further. There is often something missing from these digital peacebuilding approaches, but what's missing is not purely technical. There is also something missing in the way we think and go about the work of building peace. This sense of void may have arisen or widened through the increasing use of digital technologies, but it may also have to do with much more fundamental reasons that limit our ability to imagine and implement a pathway to peace.

Asking about how something is missing-from-hand in digital peacebuilding situates this book in the diversity of critical approaches in peace and conflict studies. Critical inquiry, at least the type that has emerged in the footsteps of phenomenological and existentialist thought that Heidegger inspired, the traces of which can be found among many contemporary critical thinkers, takes a primary interest in the absence of things. This is the case in discourse theoretical approaches concerned with the 'gap' that remains between groups of statements (Dingli 2015; Schweiger and Tomiak 2022), in practice theoretical approaches concerned with the disconnects between 'sayings' and 'doings' (see Bigo 2011; Adler and Pouliot 2011), as well as a growing body of research that sheds light on the virtuality, detachment, or superficiality of digital interventions (see e.g. Chandler 2017b; Duffield 2016). Moreover, a concern with lack and deprivation also connects well with theories on peace that aim at understanding peace as more than the mere absence of violence—and consequently, peacebuilding as a transformative activity that requires much more than simply making the weapons go silent. Johan Galtung's understanding of positive peace, just as much as feminist, decolonial, or ecology-oriented approaches to peacebuilding, all define positive conditions for peace, which require a transformative endeavour that changes politics and society (see, for instance, Ide 2017; Blaney and Tickner 2017; George 2020).

That said, I prefer to avoid using theory to prescribe what's missing-from-hand, in line with the concerns of the methodological risks of employing a 'thick' ontology, discussed in the Introduction. Subscribing to a specific definition of peace or paradigm of peacebuilding to reflect on how digital innovation generates voids may unduly reduce the scope of inquiry and thus curtail the range of insights that come to light through empirical inquiry. There is also a risk that we employ outdated theoretical perspectives to new phenomena, which could render this research less relevant. Therefore, this book operates with a parsimonious theoretical understanding of conflict transformation discussed in the Introduction, that is, efforts that support the parties or groups in conflict to seek adversarial, instead of agonistic relations. Using this understanding as a theoretical horizon, the empirical chapters of the book seek to reconstruct and highlight a diversity of voices that expressed specific pathways for how digital technologies could support conflict transformation. At times, these voices include my own. The book thus aims simply to shed light on how concrete uses of digital technologies in peacebuilding make something missing-from-hand, revealing socio-technical constellations that are inadequate from a specific vision of how conflicts could be transformed. At times, the reader will find this approach frustrating because it offers few concrete solutions and steps. Indeed, when looking at the voids of digital peacebuilding, we're confronted with something that has become lost—and what is lost often remains obscure and in the dark. My hope is that staring into this void can inspire new practices of peacebuilding that reflexively engage with the socio-technical in efforts to transform armed conflict.

2
The Socio-Technicality of Digital Peacebuilding

Looking back at the emergence of PeaceTech and ICT4Peace in the previous chapter has made it clear that digitalization and digital technologies have become trending topics over the past few years. We also learned that some influential pioneers of digital peacebuilding have argued for an approach that takes the social embeddedness of technology into account when designing and carrying out digital peacebuilding interventions. The purpose of this chapter is to operationalize this insight for empirical research by asking what it could mean to shed light on the socio-technicality of digital peacebuilding. It's important to note that the suggestions I make in the following should not be understood as prescriptive. They are but one way of grappling with the social embeddedness of technology, albeit one that I found analytically fruitful. In other areas of International Relations (IR), such as Security Studies, insights and sentiments similar to those on digital peacebuilding have led to a focus on the analysis of socio-technical systems and the interactions or interplay between society and technology, broadly conceived (Hoijtink and Leese 2019b; Evans et al. 2021). Naturally, this also triggers scepticism about digital peacebuilding as a distinct phenomenon. Should we make it a separate matter of research? I suggest that we should, given that digital peacebuilding has become a 'thing', not only because of the increasing use of digital technologies by peacebuilding actors, but also because the term is now widely used in peacebuilding policy and practice.

However, this doesn't mean that we should treat digitalization and its effects as 'independent variables' that affect peacebuilding dynamics and outcomes. Our own daily experiences in the digitalized world quickly demonstrate the difficulty of disentangling the effects of its 'digital' components, such as computers and video conferencing platforms, and 'non-digital' components, such as technical objects that have not (yet) been digitalized such as tables and chairs, or non-technical objects that cannot be digitized as they form part of the 'natural' world, such as ideas, beliefs, or emotions. All these, and many more factors, tend to play a role in our daily work, be it research seminars at a university or problem-solving workshops hosted by a mediation support organization. While we can name them, it is the relationship between these components that shapes the dynamics of human interactions and their outcomes, be it research papers or peace agreements. And there are, of course, many more components that often go unnoticed, including global power relations and inequalities and historical path-dependencies that shape socio-technical relations,

Digital Peacebuilding?. Andreas T. Hirblinger, Oxford University Press. © Andreas T. Hirblinger (2026).
DOI: 10.1093/9780198986553.003.0003

and importantly, how we make sense of this entanglement of the social and the technical.

A tool-centred perspective is thus clearly insufficient for shedding light on the shortcomings of digital peacebuilding. This insight, however, brings the need to gain some critical distance from the discursive and performative appeal of the digital revolution, which continues to narrate the promises of individual technological innovations and, with it, distracts from their social embedding. Therefore, I suggest that making sense of the effects of digitalization on peacebuilding requires a way of thinking that simultaneously enables us to turn *towards* and *away* from the digital. This can be achieved by operationalizing the notion of the 'post-digital' to develop a reflexive posture (or attitude) towards the digital in peacebuilding.[1] To this end, I found especially valuable inspiration in discussions of the post-digital in other sub-fields of the humanities and social sciences (for an overview, see Taffel 2016; Jandrić et al. 2018, 2019; Sinclair and Hayes 2019; Peters and Besley 2019). Building on these discussions, I suggest employing the 'post-' prefix *not* to move to a time *after* digitalization, that is, to create a sense of (temporal) distinctness, but to obtain a reflexive attitude vis-à-vis the rhetorically and performatively constructed appeal of digital technologies. The 'post-' before the 'digital' has been used to express a commitment to rewrite and rework the digital by engaging with digitalization in terms of a concern with the impact of digital technologies on the world (Peters and Besley 2019; Sinclair and Hayes 2019). This chapter goes even further by exploring the worldly manifestations of the digital in peacebuilding as a reciprocal relationship between the social and the technical. I argue that, when studying the socio-technicality of conflict transformation, the digital and the non-digital shouldn't be thought of as a binary, but instead, as a dyad in which both elements relationally shape digital peacebuilding.

To this end, the chapter first discusses the benefits of this change of perspective in regards to four interrelated aspects, namely historical path-dependencies that shape the long durée of digital peacebuilding, the global power relations and dependencies that shape digital peacebuilding, the tangible and material world in which digital peacebuilding is embedded, and finally, how to think of digital peacebuilding dynamics and outcomes as the result of decentred and relational agency that is conditioned as much by the technical as by the social.[2] The chapter then asks how a post-digital perspective can be operationalized for the study of digital peacebuilding. To this end, I will undertake three conceptual manoeuvres that prepare us for exploring the socio-technicality of conflict transformation. I will first discuss how we can shed light on the social embeddedness of technology by asking how technologies

[1] Some parts of this chapter include material previously published in Andreas T. Hirblinger. 2023. 'When the Digits Don't Add up: Research Strategies for Post-Digital Peacebuilding.' *Cooperation and Conflict*, 59(3), 425-446. https://doi.org/10.1177/00108367231184727, and Andreas T. Hirblinger, Julie Marie Hansen, Kristian Hoelscher, Åshild Kolås, Kristoffer Lidén, and Bruno Oliveira Martins. 2022. 'Digital Peacebuilding: A Framework for Critical–Reflexive Engagement.' *International Studies Perspectives*, November, ekac015 https://doi.org/10.1093/isp/ekac015.

[2] I have discussed various strategies that may help operationalize a post-digital perspective on peacebuilding elsewhere (Hirblinger 2023), and many speak to the four dimensions discussed in the first part of this chapter.

for peacebuilding and peacebuilding with technology are co-produced. More concretely, I suggest approaching these dynamics of co-production through the lens of 'apomediation', which enables us to ask how peer-to-peer networks of humans and machines produce understandings of conflict and peace, and advance concrete digital peacebuilding approaches. Finally, I present a thinking tool that helps to explore how apomediation dynamics differ globally, the 'digi-political playing field'. Comparing different configurations of the 'playing field' on which apomediated peacebuilding takes place enables us to ask about the combined impact of different manifestations of digitalization, as well as variations in the political context.

The Long Durée of Digital Peacebuilding

Conversations around PeaceTech, ICT4Peace, and digital peacebuilding are usually glossed in promises of innovation. Invocations of the digital turn or digital revolution imply radically altering, or even a 'reboot' of, peacebuilding efforts. However, the use of digital technologies for peacebuilding is neither exactly a new phenomenon nor can it be meaningfully separated from a time before digitalization, as I previously discussed (Hirblinger 2023). The following four sections reuse parts of this earlier work to explain why it is often difficult to draw a clear line between an assumed 'pre-digital' and contemporary 'digitalized' peacebuilding approaches. Handbook definitions of digital technologies tend to differentiate them from analogue electronic technologies by how they generate, store, process, and transmit information, namely in a numerical format. Digital technologies contain a computational component that translates information into a row of zeroes and ones (Janet and Lynn 2008, 221). We usually refer to the aggregates of numerical units as (numerical or digital) data once they represent a particular bit of information or knowledge.[3] While such digital technologies provide new opportunities for making sense of and engaging with the world, it is also true that the distinction between digital and non-digital Information and Communication Technologies (ICTs) often remains somewhat fluid and ill-defined.

The differentiation between digital ICTs and other electronic ICTs (such as old-fashioned FM radios) seems especially artificial when both versions of the same tool fulfil the same function, such as broadcasting television or radio programmes, or enabling the sending and receiving of text messages. Today's 'digital revolution' is led by technologies such as machine learning, remote satellite systems, and virtual reality. Yet, many other technologies, such as calculators, radios, and telephones, which were essential peacebuilding tools for decades, likewise operated through the basic elements of digital technology: discrete and countable units used to store, transmit, or reproduce information (Cramer 2015, 15–16). It may thus be more fitting to think

[3] Of course, there also exist analogue forms of electronic data, such as video tapes, or analogue non-electric data, such as that contained in books. To complicate things, numeric data can also be stored in analogue media, such as in a printed table—but this doesn't allow the same type of data processing as is commonly done by digital technologies.

about the long durée and the gradual, generational change towards increasingly—but never completely—digitized peacebuilding approaches (Schirch 2020). The same caution should arguably be exercised when trying to delineate the 'peacebuilding' part of digital peacebuilding. Contemporary efforts to leverage digital technologies for peacebuilding are part of a long and multifaceted trajectory of technology applications in efforts to deal with conflict, even though these may have been described with different terms. For instance, the establishment of ICT infrastructures has, in fact, long been viewed as an essential component of state-building efforts and as contributing to states' capacities to exercise a monopoly of power and deliver services (Livingston and Walter-Drop 2014). Further, where citizens are not satisfied with state performance, they have long relied on ICTs to organize political opposition, make their grievances heard, advocate for political change, or organize service delivery independently from state institutions (Livingston and Walter-Drop 2014).

This suggests that contemporary digital peacebuilding approaches are not developed and employed on a technical *tabula rasa* but within the historical trajectory of earlier methods of peacebuilding and other forms of international intervention. Indeed, the idea that technology may help end armed conflicts and build peace dates back not only several decades, as I demonstrated in the previous chapter, but, in fact, centuries. For instance, today's attempts to address conflict remotely, such as with the help of satellite imaging services that document atrocities, form part of a long historical trajectory of the international governance of conflict-affected contexts, where the employment of (human) resources on the ground tended to be costly and potentially dangerous. Technology also played an important role in efforts to assert power, be it by imperial or colonial regimes, the nation-state, or international organizations that support the settlement of conflict and contribute to political change. For instance, many colonial regimes have aimed to demarcate and regulate communal land with the help of mapping technologies to reduce or end violent conflict between various ethnic communities, particularly between nomadic and sedentary groups.

The technologies used a century ago may have been much simpler: using pen, paper, and maybe a compass to cartograph areas, for instance. However, they shaped human efforts of preventing and managing conflict just as much back then as do Geographic Information Systems (GIS) today. Moreover, political institutions and governance arrangements that have been shaped by centuries of imperial, colonial, and postcolonial rule continue to yield their impact on digital peacebuilding today. For instance, local governments and institutions of 'traditional' authority, empowered by external interveners as a means of indirect rule, continue to influence the collection and sharing of early warning data in places like South Sudan until the present day, as I will discuss in Chapters 3 and 4. In many ways, digital innovation thus often promises a future free from the many historically evolved and conditioned causes of conflict that is never fully realized. With assumptions about digitalization as a linear progress appearing just as misplaced as those about digitalization as a watershed, reflecting on the temporality of digitalization can help us to gain a deeper understanding of its complicated relationship to contemporary peacebuilding.

The Globality of Digital Peacebuilding

Many readers will agree that there is a global dimension to digitalization that defies neat distinctions between digital and analogue places, or more or less digitalized places. Digital peacebuilding is a global process that involves and connects a variety of actors that are affected by—and affect—digitalization in various ways. Many observers have expressed concerns about the limits of digitalization, especially the global 'digital gap' in infrastructures, access, and literacy between the Global North and South, and between different demographics (Njeru 2009). There is undoubtedly a globally uneven distribution of digital infrastructures and technologies as well as the knowledge and skills to develop and use them, which also affects how digital peacebuilding is exercised. However, when we categorically differentiate between worlds that have turned to the digital vs worlds that have not (yet) done so, we risk falling into the trap of binary thinking that ultimately creates otherness. This is common to modernist thought, which has historically constructed a difference between the modern and the traditional (Goga 2015). Problematically, this binary establishes a disempowering hierarchy of development and legitimizes the assumption that the prospects of digital peacebuilding are ultimately conditioned by homogenous and linear technological progress that those on the lower end of the digital gap simply have to follow. We do not live, fight, and make peace digitally in one world and analogically in the other. Rather, the dynamics of peace and conflict traverse digital and non-digital spaces and times. Thus, when it comes to their effects on contemporary peacebuilding, the presence of the digital somewhere cannot be thought of without the absence of the digital elsewhere.

When digital peacebuilding initiatives aim to involve populations with limited digital access or literacy, in places such as Somalia or Burkina Faso, through the use of 'low-tech' applications (Meier 2021), these efforts are inextricably linked to the 'high-tech' global centres of digitalization, where digital technologies and digital peacebuilding approaches are developed and devised. To circumvent global inequalities in digitalization, for instance by deploying technologies that require little data and little digital literacy, practitioners assess the digital 'readiness' of specific conflict-affected contexts. Transnational data-gathering and data-analysis efforts to fight disinformation or support conflict early warning neatly merge simple text messaging services to crowdsource data from rural populations with computationally demanding machine learning-powered prediction algorithms that require advanced computer science degrees, as I will demonstrate in Chapter 4. This may help avoid direct adverse effects, such as excluding less connected or less literate populations. However, by circumventing rather than addressing the digital gap, such efforts also risk normalizing and reproducing the global inequalities that underpin digital peacebuilding.

An answer to this challenge could be to invest in building the skills and capacities of people affected by conflict and deprivation. Initiatives such as the United Nations Development Programme's (UNDP) PeaceTech Accelerator, discussed in Chapter 3, often aim to support 'local' tech solutions, for instance, by enabling project participants to design mobile applications using open-source application development platforms. However, these projects often foster capacity only in a superficial manner by harnessing and furthering the skills of already motivated and qualified 'tech entrepreneurs' for a short period of time. In places like South Sudan, which is characterized by many complex development challenges, the 'digital gap' cannot be simply 'closed' through a couple of one-day workshops. It will require years, if not decades, of large-scale investment, including boosting primary, secondary, and higher education, providing affordable access to digital infrastructures, and enabling civil society and private-sector-driven technology development. However, the prospects for such policies are dire, given the continued violence and the large-scale economic mismanagement, corruption, and nepotism that have evolved under the government of President Salva Kiir. Digital peacebuilding efforts often unfold on top of the globally uneven distribution of digitalization, which ultimately links to, and is sustained by, much more entrenched global disparities, as well as unjust and often exploitative national policies. Moreover, discrepancies are usually also found between sprawling urban centres, where fast internet connectivity is common, and rural areas where people lack access and skills. These disparities affect how digital peacebuilding is done and with what effects.

Overall, peacebuilding actors are disproportionately concerned with providing digital 'solutions' to places other than those that lead global digital innovation: on the African continent and in countries such as Iraq, Myanmar, or Sri Lanka. In contrast, civil society organizations operating in places like Northern Ireland tend to avoid the 'peacebuilding' label. This reproduces a variant of Eurocentrism that simultaneously declares Western or European technological systems as points of reference while projecting them to the Global South where they can be tested and improved. However, we may want to ask if the notions of a globally homogenous digitalization that underpin this manoeuvre are derived from the liberal rhetoric promoting universal access and connectivity in an effort to open new technology markets, rather than from the actual dynamics of digitalization that play out in a much more heterogeneous fashion, both globally and domestically (Dutton and Reisdorf 2019; Gallagher and Knox 2019). For instance, Chapter 4 explores how the utility of local conflict early warning systems is shaped by context-specific dynamics that differ depending on where they are located. Therefore, as Sri Taffel (2016, 5) put it, we may instead want to speak of, and look for, an 'ongoing series of molecular digital revolutions that continue to enact changes in heterogeneous ways across different geographies'.

The Tangibility of Digital Peacebuilding

Digital peacebuilding is clearly shaped by material, and often tangible, aspects. Surprisingly, however, there is a tendency among practitioners and academics to think of digital peacebuilding as something that mainly takes place in what is understood as a distinct 'space', portrayed as 'digital', 'cyber', 'virtual', and 'online', in contrast to 'offline'. Given the I and C in ICTs, it is natural that attention is first and foremost paid to aspects of information and communication. Moreover, data and algorithms are usually perceived as abstract things that are difficult to grasp, and digital innovations such as social media, the metaverse, and virtual reality often generate a similar attitude. Of course, staff working in UN peacekeeping missions or non-governmental organizations (NGOs) implementing reconciliation initiatives continue to carry tangible devices, such as mobile phones, laptops, or more specialized hardware. They type on keyboards with their bare fingers, mount cameras in trees, launch drones into the air, and set up antennae in crowded marketplaces, and thus bridge the virtual and the real in many ways. More importantly, the lack of data and connectivity becomes tangibly apparent in moments where they create black spots that give rise to physical insecurity or violence that remains undocumented. This is why a post-digital perspective on peacebuilding is crucial: it rejects the conceptual shift implied in the digital revolution towards an exclusive focus on the virtual or the digital (Pepperell and Punt 2000 cit. in Taffel 2016), and it remains critical of a view of the world as reducible to a discrete set of computational problems (Morozov, cit. in Jandrić 2018).

Rather than separate from it, digital technologies thus blend with the lifeworld of those who live in conflict and those who aim to prevent it. They interact with human minds and bodies, creating visceral experiences in increasingly ubiquitous fashions, defining how we go about our everyday professional lives in ways that blur the boundaries between humans and machines (Clayton et al. 2015). Rather than approaching digital peacebuilding as something that happens in separation from the 'real', as the association of the digital with the 'virtual' may suggest, we should thus engage with its very tangible effects. The post-digital asks us to reckon with the fact that 'cyberspace has insidiously insinuated itself into our existence, at every scale and every turn', which makes it impossible to keep up a binary opposition between the 'digital' world and the 'normal', 'real' world (Spiller 2009, 95–6). As Ryberg (2019, 165) put it, the challenge is 'dragging digitalization and the digital—kicking and screaming—down from its discursive celestial, ethereal home and into the mud'.

Asking about the tangible effects of digitalization on peacebuilding seems warranted, not least because peace is, in many ways, much more than an abstract idea that can be stored or transmitted in data. While there are many different approaches and understandings of peace, few would counter the sense that peace is ultimately embodied in the material world. Nonetheless, it remains common to think of digital peacebuilding as something that happens 'online' or 'on social media' and as the task of finding virtual solutions to virtual problems. I discuss this, for instance, in relation to current understandings of hate speech and efforts to fight it, which take place almost exclusively in the virtual domain. As I argue in Chapter 3, we should no longer (or not only) ask about the preconditions of a 'virtual' peace as something that is merely represented or constructed through digital media, because fake news and disinformation are not only 'wrongful representations' of reality. Instead, there is a need to engage with their very real, embodied effects in terms of hatred and violence and the material, organizational, and systemic factors that enable these digital realities to shape both war- and peacemaking. This suggests that it's crucial not to lose sight of the non-virtual aspects of digital peacebuilding: its tangible effects on people's lives, security, and well-being and its embodiment in the experiences of people affected by conflict, as well as in the everyday work of professional staff and volunteers.

The Agency of Digital Peacebuilding

My discussion above makes clear that many of the dimensions uncovered by the post-digital point of view are, in fact, not shaped by technology only but are very much the result of human agency. However, large parts of the academic community, particularly those inspired by Bruno Latour (2005) and other pioneers of Science and Technology Studies (STS), assert that we must also make sense of the agency of things—including that of digital artefacts. While their conversations have unfolded somewhat independently from the digital and often emphasize the mutual constitution or interdependence of human and machine agency, recent artificial intelligence (AI) innovations have triggered the sentiment that technical agents will increasingly replace humans in numerous fields. Unsurprisingly, they have led to speculations that autonomous machines could carry out peacebuilding tasks *instead* of humans. For instance, some observers voiced the prospect that innovations in AI could lead to a future in which robots not only wage war against each other but also make peace (Lamb 2017). If AI can wage war by enabling autonomous weapons systems or social media bots, why could it not also resolve conflicts? While many observers would agree that the achievement of artificial general intelligence that would match the cognitive capacities of the human brain is not a realistic target in the immediate future, others are more seriously contemplating the employment of generative AI that could write entire peace agreements (Rutherford 2021).

We don't have to dive into discussions about the most advanced AI models to note that the emerging digital peacebuilding community tends to be obsessed with the utility, effectiveness, or impact of specific 'tools', 'applications', or 'platforms'. This trend stands in stark contrast to the research and practice of peacebuilding, which has in the past been largely concerned with different approaches, perspectives, paradigms, or agendas that shape peacebuilding—all things that are made and done by humans. (Carey 2020; Richmond and Visoka 2021). Liberal norms, international organizations, state sovereignty, or peace infrastructures, to mention only a few, are implicitly or explicitly considered human-made. Thus, the existing peacebuilding scholarship is largely grounded in an enlightenment notion of human agency, understood as the capacity to make a difference based on free will and intentionality (Passoth and Rowland 2015; Sayes 2014). This stance is also echoed by many practitioners and policymakers, especially among the peace mediation community that continues to think about their work by and large as human-centred, as I argue in Chapter 5.

However, if we were to insist that digital peacebuilding is something that is done exclusively by human agents who use digital tools, we would turn a blind eye to the power of digital technologies. Simply put, if digital technologies simply enact and extend human will, why should we care about them? Could it be the case that the insistence on the predominance, or at least the separateness, of human agency is ultimately a rhetorical strategy that reproduces modernist notions of human dominance over their lifeworld? Could it be that the return to human agency often occurs disproportionally in moments of disenchantment with technology and technological progress, which trigger an Othering of technology—and thus its externalization as distinctly nonhuman? Indeed, where digital peacebuilding initiatives have limited success or undesirable effects, these tend to be attributed to a design failure or the limited functionality, efficiency, or availability of tools, for instance, limited digital infrastructure and access, errors in prediction algorithms, or lacking data. An often-heard claim about online dialogue platforms, for example, is that they do not enable 'meaningful' encounters because the cameras and microphones that transmit our voice and picture do not suffice to create a sufficiently immersive experience that would allow participants to open up emotionally. Yet, we all know what a well-directed movie can do to us.

In this context, a post-digital lens offers us the 'recognition of the increasing sense that any analytic separation of "technology" and "humanness" fails to articulate our contemporary condition', as Knox (2019, 167) has argued. This may mean acknowledging that machines have agency in the sense that they are often indispensable for human action and make a difference to the course of political processes and their outcomes, based on an understanding of agency as a distributed effect that emerges in human–machine networks (for a detailed discussion, see Hirblinger 2022; Hoijtink and Leese 2019a; Rammert 2012). However, while acknowledging that tools may well have agency, a post-digital perspective also challenges us to take a distinctly human-centred look at the socio-technical systems that jointly 'do' peacebuilding. For instance, when investigating peacebuilding dynamics and outcomes, we may want to ask how critical choices about the purposes and underlying functions of technology-supported interventions are made during the design of technologies or trace the decision-making processes that lead to deploying or not deploying a particular tool.

While recognizing that agency in digital peacebuilding, and with it, the impact it may have on conflict-affected populations, is always relationally produced by humans and machines, we may want to strategically emphasize the distinctiveness that underpins human agency, namely a certain degree of intentionality of our actions. With that, we are able to, metaphorically speaking, 'extract' human agency from the labyrinths of the digital to both reassert and rearticulate the relevance of human decision making (Jandrić et al. 2019, 173). Such efforts can help re-humanize digital peacebuilding by focusing on the potential of human agents to observe if particular digital peacebuilding activities leave something missing-from-hand—and then attempt to fill the voids through alternative approaches. From this vantage point, digital peacebuilding becomes perceivable as a result of human intention in the context of structural constraints, for instance, in the form of technology design or a market logic that shapes innovation, and human agency returns into the spotlight of our attention because it is needed to alter them. This does not mean ignoring the many other factors that are not under the immediate control of human agents, including the ones discussed above, such as the globally uneven distribution of digital resources, historical path-dependencies that shape technology use, or unequal material conditions.

However, it helps not to lose sight of the fact that, ultimately, how digital peacebuilding is done and what agency it has in conflict-affected contexts *can* be influenced by us humans.

The Co-Production of Peacebuilding and Digital Technologies

The discussion above makes clear that digital technologies do not build peace on their own—but neither do humans. In response, I suggest thinking of peacebuilding as a process that is shaped by human–machine relations, which jointly influence how peacebuilding is done and with what effects. I have developed this perspective previously with colleagues at the Peace Research Institute Oslo (PRIO) in a co-authored article (see Hirblinger et al. 2022). Reproducing some of these considerations, I first explain why it is important not to start with preconceived notions of technology and peacebuilding but rather to view them in a co-productive relationship. Technology shapes what peacebuilding is and how it is practised, and peacebuilding contributes to technological innovation and change. Rather than thinking of peacebuilding and technology as two separate categories, I thus suggest that it makes sense to explore how they co-constitute each other.

The co-constitutive relationship between society and technology has been explored well in the Science and Technology Studies (STS) literature, for instance, through Sheila Jasanoff's concept of co-production. The concept was initially used to demonstrate that science does not produce objective 'facts' about the world, but that the production of knowledge about the world is closely linked to the production of notions of how the world ought to be. As Jasanoff puts it, co-production is a 'shorthand for the proposition that the ways in which we know and represent the world (both nature and society) are inseparable from the ways in which we choose to live in it' (Jasanoff 2004, 2). The relevance of co-production for the study of technology lies in the insight that technology, as part of scientific exercises, just as much as of policy or practice-oriented endeavours, does not merely produce objective facts about the world. The gathering, analysis, and use of data are intertwined with social and political practices that add meaning to those facts and ascribe consequences to them. As Jasanoff put it elsewhere, 'seeing the world in a particular way (how things are) gets coupled to commitments to particular norms and values (how things ought to be)' (Jasanoff and Simmet 2017, 754). 'Seeing the world' is achieved primarily through epistemic practices, such as efforts to collect data that help to establish facts and evidence. Negotiating how the world ought to be has a normative dimension, which defines values and rules of good conduct. However, it also has an ontological dimension, which establishes how we are and how we relate to the world, in its material and embodied manifestations, but also in more subtle ways, such as through beliefs, assumptions, and narratives about what the world is made up of. In sum, co-production brings attention to the interrelation of efforts to get access to the world in epistemological terms, and to produce it and be in it in ontological and normative terms. As peacebuilding is a process that spans from making sense of conflict to constructing ways of being that allow for non-violent co-existence, co-production is well-placed to direct the study of technology in peacebuilding.

Building on Jasanoff's work, Katja Lindskov Jacobsen and Linda Monsees suggest a narrower reading of co-production for the study of technology that more explicitly focuses on the production *of* technology by society and the production of society *by* technology. On the one hand, this double perspective encourages us to shed light on the processes through which technologies are constituted, for instance, for the purpose of providing safe, trusted, or objective methods to make sense of the world. On the other hand, it brings attention to those processes through which technology mediates our relationship to the world, for instance by providing particular framings of the world and by having effects on social practice (Jacobsen and Monsees 2019, 26–7). Bringing this research perspective to the study of digital peacebuilding, I suggest taking neither technology nor peacebuilding as pre-given objects of research but inquiring into how both are co-produced in efforts to prevent, manage, mitigate, or transform armed conflict. As previously argued in Hirblinger et al. 2022, such an effort can benefit from a critical-reflexive approach that is concerned with the social construction of peacebuilding and technology, and its powerful effects on conflict-affected contexts. Critical reflexivity directs attention to the practices of knowledge production that shape digital peacebuilding, and more specifically, to how claims about both peacebuilding and technology shape digital peacebuilding agendas and outcomes.

To start, peacebuilding is an umbrella term which accommodates multiple meanings and is employed differently by various actors. While we can agree that peacebuilding is commonly associated with external interventions into conflict-affected societies that aim to prevent armed conflict, Michael Barnett and colleagues noted almost fifteen years ago that, as a professional field, it is characterized by a stark diversity of approaches (Barnett et al. 2007). Scholars have often employed the label of 'liberal' peacebuilding to discuss interventions that aim to end conflict through a range of social, political, and economic reform measures with the objective of strengthening liberal institutions (Newman et al. 2009; Paris 2010). Yet, others have pointed to the fact that such 'liberal' interventions tend to be accompanied by authoritarian traits and continued violence that stand in contradiction to their programmatic promises (Kühn 2016), or that liberal interventions are often situated in a trajectory of imperial and colonial forms of rule (Laffey and Nadarajah 2012), which makes them more diverse and more ambiguous than the label 'liberal' peacebuilding suggests. While there now seems to be consensus that peacebuilding tends to be a 'hybrid' process that brings together many different actors and agendas (Mac Ginty and Richmond 2016), research has also pointed to new forms of authoritarian conflict management and authoritarian peacebuilding that may compromise or undermine the establishment of liberal democratic institutions (Barnett et al. 2014; Lewis et al. 2018).

It is also important to note that those actors whose practices are studied under the label of 'peacebuilding' may not necessarily label themselves as 'peacebuilders'.

Of course, the UN Peacebuilding Fund or the UN Department of Political and Peacebuilding Affairs carry the term in their name, and many civil society organizations (CSOs) operating in conflict-affected contexts reference peacebuilding in the description of their mandates or projects. However, there exists a range of additional interventions that are somewhat directed at preventing, mitigating, ending, or transforming conflict, and they merit attention as they are relevant to the overall societal and political effort to build peace. Such interventions include efforts to build conflict early warning systems, prevent polarization and hatred, or foster reconciliation between the constituents of former conflict parties. Because such practices are similar in their approach and objective, they should also be included in the study of peacebuilding, broadly defined. This thinking is also reflected in current international policy. For instance, the 'Sustaining Peace' agenda, promoted by the UN since 2018, suggests thinking of peacebuilding as an approach that should involve all parts of society and should range from the prevention of conflict to the prevention of its reoccurrence (General Assembly Security Council 2018). This means that peacebuilding is not a clearly definable, discrete activity but a field composed of heterogeneous practices that aim to prevent armed conflict.

As I attempt to demonstrate throughout this book, these various understandings, theories, or approaches to peacebuilding shape the use of digital technology in conflict-affected contexts. However, technology also has the potential to alter them. In the chapters that follow, I will explore how technology shapes peacebuilding in many forms: as material objects, infrastructures, and methods, among other, and how it mediates co-production processes in various ways. Epistemologically, technologies provide devices for making sense of the world and establishing representations of 'what is'. For instance, new remote sensors, which record video and audio footage of ceasefire violations, change how monitoring missions document violations. New data sets provide innovative perspectives on the effects of armed conflicts, such as on forms of gender-based violence, which can now be better documented because of the confidentiality and anonymity provided by encrypted mobile phone applications and messaging services. Yet, the design of applications, for instance, when using pre-defined categories of conflict that users can select to report an incident, shapes what kind of data and evidence will be produced, and how those who aim to respond to violence make sense of it. I will explore this further in Chapter 4, which looks at mobile applications used for conflict early warning.

However, these technologies produce more than mere 'evidence' about the world—they also make us relate to, and be in, peacebuilding contexts differently. Data collection and analysis routines make conflict visible in specific ways, but they also reinforce particular peacebuilding theories and paradigms. I will, therefore, explore how technology affects the *ontology* of peacebuilding, thus very much defining what peacebuilding and peacebuilders are made of. For instance, while for decades, or even centuries, local reconciliation efforts involved community gatherings, such as in town halls, churches, or underneath a tree, often embedded in local traditions and institutions, today's efforts to deal with the past increasingly involve decentralized, ICT-enabled processes that may include social media platforms such as Facebook or

Instagram. Where such technologies are used, they change how people *are* in peace processes by producing new views, narratives, and beliefs about what the conflict is made up of. Much has been written about how technology can harden, alter, or challenge existing conflict ontologies, such as through social media narratives about the other. The increased polarization on social media, where conflict constituents and social sub-groups communicate in echo chambers that separate those in conflict by lines of ethnicity, language, class, or religion, is an often-cited example of technology's 'world-making' capacities, because they harden identities. Yet, technology can also destabilize existing ontological frames, deconstruct stereotypes and hateful narratives, humanize the other, and lead to the emergence of joint narratives and views on the conflict. I will discuss this in Chapters 5 and 6.

Finally, if and how such effects of technology unfold also depends on how technologies are viewed and assessed, that is, what possible benefits or harmful effects are attributed to them and based on what normative considerations. This commonly happens through the suggestions or commentary of experts, bureaucrats, or policymakers proposing that particular digital approaches or methodologies are normatively desirable. For instance, the field has seen considerable investment in digital approaches that can broaden participation in peacebuilding and lead to more inclusive processes, a normative commitment shared by many international peacebuilding actors (Hirblinger 2020). Furthermore, normative claims may make certain digital approaches more legitimate, acceptable, and impactful. For instance, if the conflict parties endorse a monitoring body with the epistemic authority to report on ceasefire violations, its findings may gain legitimacy. At the same time, technological innovation commonly changes notions of which peacebuilding approaches are good or desirable, as is visible in the many calls for data- and evidence-based approaches that incorporate statistical methods (Mack 2014), and increasingly, machine learning (Wählisch 2020; Masood Alavi et al. 2022) in the planning, implementation, and evaluation of peacebuilding initiatives.

Apomediated Peacebuilding

How can we more concretely utilize the co-production perspective in the study of digital peacebuilding? And how can it be connected with the four post-digital concerns spelt out at the beginning of this chapter? To answer these questions, I put forward the proposal that digitalization leads to what I call 'apomediated peacebuilding'. The term 'apomediation' was originally employed to make sense of the increasing impact of the internet on personal healthcare decisions, describing how individual users rely on decentralized human–machine networks, namely, 'apomediaries', to obtain relevant information not only about their health conditions but also possible cures and preventative measures. The digitalization of peacebuilding has led to strikingly similar developments because the socio-technical systems that make up digital peacebuilding influence how we make sense of conflict and how we attempt to solve it.

Many digital technologies are essentially media—they are means to enable communication. Yet, these means are 'in-between' those human agents that they connect. In other words, digital technologies themselves mediate. This phenomenon has been described with the term 'apomediation'. It was coined in the field of public health to describe the process of internet users searching for online advice on diseases and medical treatments, for instance. In many areas of life, digital technologies enable users to access relevant information more directly without going through traditional intermediaries, such as doctors, teachers, or advisers who possess specific expertise or privileged access to information. The term was first used by Gunther Eysenbach to describe how the increasing use of web-based technologies changes how patients retrieve information about possible diseases and cures. It aims to capture a new approach to sense-making, through which individuals no longer solely rely on experts (such as health professionals) to obtain trusted information. Apomediaries have been defined as 'tools and peers standing by to guide consumers to trustworthy information or adding credibility to information' (Eysenbach 2007, 166). They form a decentralized network of information exchange in which both humans and machines guide users in obtaining relevant information about a specific topic (Eysenbach 2008, 129).

Apomediation takes place in many spheres of life. Amongst others, the concept has been used in the context of disaster response and humanitarian relief efforts to describe how the rise of social media has led to the replacement of conventional 'information gatekeepers' with information flows regulated by 'network filtering' and 'group moderation' instead (Alexander 2014, 718). When looking at political processes, the challenge is that users often do not systematically search for information, but instead use social media, messaging services, and various other online resources in a much more undirected fashion to engage in politics. However, their factual knowledge, perceptions, and convictions are likewise apomediated by the communications networks in which they are embedded. Some of the pitfalls of this development have recently been prominently deplored in the context of social media impacts on political communication and mobilization, for instance, when producing 'echo chambers' or 'filter bubbles' through recommendation algorithms (Jiang et al. 2019). During armed conflicts, digital technologies shape an individual's understanding of the causes of conflict and possible options to overcome it, and they shape how individuals engage with other actors. Human mediators, for instance, have traditionally occupied a prominent position between conflict parties and stakeholders, enabling communication between them through 'shuttle diplomacy' or by enabling transfers between peacemaking tracks (Palmiano Federer et al. 2019). However, the availability and use of digital technologies integrates them into a more extensive, heterogeneous network through which conflict parties and those affected by it interact.

Apomediation thus alters the role of these human intermediaries who enjoyed a privileged position because they stood 'in-between' consumers and the information they seek. While information continues to be generated and provided by authoritative sources, additional information is made available and brokered by a large variety of humans and machines that expert intermediaries do not have control over. Whereas traditional intermediaries stood 'in-between' the consumer of the information and the source of information, apomediaries 'stand by' and 'steer' users to the information (Eysenbach 2007, 2008). Apomediation describes the manner in which users obtain authoritative information about the world and how this may affect their reasoning processes and behaviour. For instance, research on apomediation has indicated that the use of online media affects reasoning and decision-making processes, such as how stakeholders make credible judgements (Eysenbach 2008, 129). In the end, both intermediaries and apomediaries provide the individual user with authoritative information, that is, information they use to make decisions about the subject matter. However, the former maintain a centralized position, whereas the latter operate in a decentralized manner.

Building on these thoughts, I suggest investigating the interplay of relation-making and meaning-making in apomediaries, that is, the creation of connections between specific users and platforms on the one hand and the creation of specific understandings of conflict and peace on the other. First, as the term suggests, apomediation entails *relation-making* in heterogeneous networks. Online peer-to-peer information exchange in apomediaries differs from traditional communication models that operate through linear, vertical, and horizontal information flows. Connections can be easily established and undone. They are conditioned by a variety of user-centred designs in which user preferences are mediated by visual platform design features, hidden algorithmic content, and implicit information management. Digital infrastructures may create relatively stable connection points, that is, nodes between users (such as through closed messaging groups or mobile applications for data collection), but whether these nodes are activated for information exchange depends on situational demands and path-dependencies (for instance, content recommender systems on social media). Nodes can also be established instantly, such as through accepting invitations, establishing chat groups, joining live discussions, or forwarding content, and can just as quickly be terminated. This points to the relevance of human agency in the rhizomatic shaping of apomediaries.

Moreover, as apomediaries construct and convey understandings of peace or conflict, a second concern should be with *meaning-making*. This entails asking how information that expresses specific understandings of conflict and peace becomes trusted and authoritative. While research on apomediation was originally concerned with the circulation of written information, I suggest broadening our conception to move beyond the semiotics of text and towards multimodal significations, including those contained in pictures, videos, and memes, to capture their emotional and affective dimensions. As will become apparent, meaning-making and relation-making are closely intertwined. Certain relations open the gates to *some* meaning while they close the gates to *some other* meaning. For instance, many of the conflict early warning systems discussed in Chapter 4 rely on the integration of so-called field monitors, yet they condition how field monitors can inform what counts as early warning information through structured data input masks that only allow for the submission of particular content. There are many further examples discussed in this book. However, certain types of meaning also encourages certain relations and not others. As I argue in Chapter 3, highly aggregated social media data about hate speech, for instance, can be used to coordinate conflict prevention policy at the headquarters level, but it is less suitable for inter-personal interventions.

How meaning is generated, provided, and brokered in apomediation networks challenges the conventional role of peacebuilders as 'intermediaries' between conflict parties and stakeholders. We can make sense of this from a discourse–theoretical angle: apomediation undermines those elements of a discursive formation that conventionally have knowledge authority because they can determine the limits of the sayable (Feltwell 2020). However, apomediation does not mean that traditional experts, such as doctors, sales assistants, humanitarian workers, or peacebuilders, are obsolete. For instance, almost all cases in which peace agreements are signed entail physical meetings between the conflict parties, mediated and facilitated by a third party. Yet, such third parties now have to share and negotiate their epistemic authority over a subject matter with other elements in the network that provide, distribute, and broker information. We thus need to better understand how the increasing digitalization of peace processes brings new ways in which conflict parties, conflict stakeholders, and the general population make sense of the conflict—and how this digitally mediated sense-making helps or hinders the resolving of conflict. In fact, the impetus behind the apomediation literature, at least in the field of health, is to find ways in which trusted and reliable information can be supplied by experts within and through the networked environment. Translating this insight into the critical study of knowledge production, and its effects on political dynamics, means asking how information about conflict is generated, brokered, and made credible and actionable in apomediation networks—and through what power struggles such dynamics are characterized. To think of digital peacebuilding in terms of apomediation thus means to situate the research on it broadly within a concern for the knowledge–power nexus common among constructivist International Relations (IR) scholars (Barnett and Duvall 2004, 55). Apomediation exercises a diffuse, productive power because it generates knowledge about peace and conflict that influences peacebuilding approaches and dynamics. At the same time, it is also subject to the power of humans and machines that influence the conditions of knowledge production in apomediation networks.

Against this backdrop, it is vital to carefully revisit how we have conventionally thought about power in international peacebuilding. Many peacebuilding scholars have stressed that peace cannot be built 'from the top', criticizing international organizations for using 'blueprints' that attempt to dictate how societies transition out of conflict (Richmond 2010, 34), for instance. There is also a large body of literature that stresses that peace must be built in an 'endogenous' manner (see, for instance, López López and Ingelaere 2021) or by 'local' actors (Mac Ginty and Richmond 2013). However, these concepts that are often problematic in practice because clear lines between the 'international' and the 'local', or 'endogenous' and 'exogenous' are hard to draw. Nevertheless, it goes without saying that even where such approaches are pursued, peacebuilding dynamics are characterized by an unequal distribution of power and that those who can claim knowledge authority about the conflict (by producing knowledge, disseminating, or using it) also enjoy greater agency in the peacebuilding effort. The idea of apomediation puts the relevance of this international–local binary into question, because digital peacebuilding has the potential to generate authoritative information in decentralized networks that may or may not include traditional intermediaries. Instead, it directs the attention to the relational, networked, and complex communication processes that underpin peacebuilding efforts, yet without romanticizing technology as a vehicle to promote 'bottom-up' or 'grassroots' approaches. While it assumes that traditional hierarchies may be challenged by digitalization, it directs attention to the dynamics within networked information systems and invites study of their dynamics.

In his discussion of apomediation, Eysenbach noted that due to the increasing relevance of peer-to-peer information networks, the power relationship between recipients and intermediaries changes. This may lead to conflicts, as traditional providers of information see their authority challenged (Eysenbach 2007, 163). This is also visible in the peacebuilding field, where peacebuilding actors bemoan the possible negative consequences of social media, for instance, when pointing out that armed conflicts are accompanied by new forms of 'information wars' that play out on social media, in which the conflict parties advance competing narratives (Lanz and Eleiba 2018). They are also aware of the challenges that the web-based information environment poses to their traditional role and undertake efforts to retain their authority by integrating social media into their strategic public communication about the conflict and peace process (UNDPPA and HD Centre 2019). When negotiating an agreement, the formal representatives of conflict parties can no longer be sure that their constituents will seek access to the peace process through them. Digital technologies enable conflict-affected populations to participate more directly in peace processes, in terms of receiving and providing information. For instance, mediation teams now often open additional channels of negotiation or consultation with the help of digital technologies when hosting online meetings or initiating consultations or discussions through smartphone-based messaging services such as Whatsapp (Meier 2021). Such activities do not require physical presence in a country or access to such populations. Therefore, they do not have to be negotiated with conflict parties who may control the territories of the consulted populations.

All this suggests that apomediation changes who and what has agency and authority among those affected by conflict. It can reduce the influence and authority of conventional peacebuilding actors and conflict stakeholders. However, it can also lead to their resurgence. Indeed, as will become apparent in this book, peacebuilding professionals themselves commonly form part of apomediation dynamics—and they try to influence them. They rely on decentralized human–machine networks to obtain information, but also commonly play a role in efforts to determine which information about the conflict is deemed relevant and valid. For instance, as I will demonstrate in Chapter 5, when digital technologies are used by mediators, they shape what mediators know about the agendas and interests of the conflict parties or conflict dynamics. Many political analysts in mediation teams rely on social media platforms such as Facebook and X (formerly known as Twitter) as a source of information that feeds into their analytical reports. Additionally, some increasingly use social media data analytics tools that provide a 'skimmed' view of online discourses through selective text-mining techniques such as sentiment analysis. On the other hand, when mediation support actors employ digital technologies to facilitate online consultations, the simple choice between a text-based or an internet-based messaging system affects who can participate in a digital consultation and who is excluded. This will, in turn, have an effect on the overall composition of the gathered data, as well as on the understanding of conflict and the possible responses that can emerge from them. This demonstrates the role of material objects and the effects of their design on peacebuilding dynamics.

I initially attempted to document and visualize the human–machine relations underpinning digital peacebuilding initiatives through a qualitative network mapping approach. Through structured online surveys and interviews with project staff, I attempted to create temporary snapshots of how information relevant to these initiatives is produced and shared. This involved drawing connections between the projects' digital infrastructures, hardware and software, and their human users. I also experimented with coding these relations alongside various analytical dimensions to understand who or what asserts authority over understandings of conflict and peace, and how. For instance, one could attempt to map powerful nodal points in the network and identify top-down versus bottom-up dynamics by analysing the flow of information and its direction, building on Roche, Propeck-Zimmermann, and Mericskay's (2013) differentiation of information architectures. I asked how humans and machines exercise different kinds of power depending on what they do with the information: do they simply send, host, or receive it, or do they create, process, analyse, or evaluate it? The mapping exercises provided a fine-grained view that provided the basis for a considerable chunk of the research presented in this book.

However, I increasingly perceived that the mapping approach and the software I employed could not accommodate the many manifold observations I made in the course of the research. As I attempted to integrate them into my network drawing approach, the visualizations became increasingly heavy and convoluted. One key challenge had to do with the impossibility of determining the right level of granularity. At times, a whole application mattered, yet at other times, my research unearthed the relevance of individual algorithms. Another concerned the fact that the connections between humans and machines varied to a large degree depending on how users thought about them. The drawings proved unsuitable for accommodating the heterogeneity of subjective views that my human research participants expressed regarding the role of individual technologies and their projects as a whole. More importantly, I found that the visualizations hardened and objectified what I observed in ways that camouflaged the fact that research cannot be disentangled from the researcher's subjectivity and situatedness. In many ways, this attempt at network drawing thus turned into an autoethnographic study of what it can mean to be influenced and constrained by a specific digital tool.

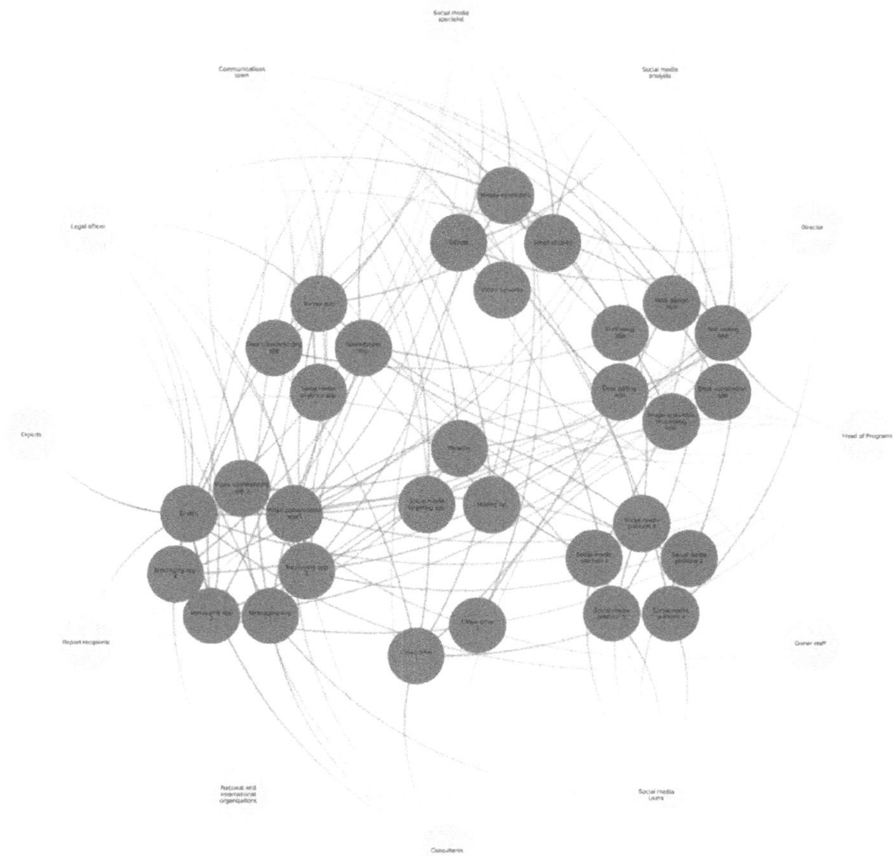

Figure 2.1 Illustrative network drawing of a social media monitoring project, visualizing connections between humans (yellow) and machines (blue)

While the notion of decentred human–machine networks remained central to my research, I ultimately preferred to convey it primarily through the written analysis presented in this book. Nonetheless, the reader may find Figure 2.1 helpful as an illustration of the kind of human–machine relationships that can be at the core of digital peacebuilding projects. The drawing represents a simplified network of one of the social media monitoring projects discussed in Chapter 4, notably without visualizing many of the important factors that condition the apomediation of peacebuilding. The figure categorizes various tools along coarse functional categories and describes them in general terms to avoid disclosing sensitive operational details. It only maps relations between human users and the tools and does not capture human-to-human or machine-to-machine relations. Human agents are represented in the yellow dots and arranged in a circle to help with reading the network in a systematic manner. The tools they used are represented in blue dots and arranged in smaller clusters, differentiated by their functions. Nonetheless, the reader may find the many connections in the figure already somewhat overwhelming and little informative. In particular, the reflections on the co-production of society and technology summarized above can be more adequately conveyed through qualitative written analysis, which can account more thoroughly for the co-constitutive relationship between technology and society that shapes peacebuilding dynamics and outcomes.

The Digi-Political Playing Field

The discussion above suggests that we shouldn't think of apomediaries as pre-given or static but as highly dynamic and context-dependent, shaped by human agency through design decisions, policy, regulations, individual user behaviour, or political interference, to give just a few examples. It is likewise conditioned by tangible and embodied factors, such as the availability of digital infrastructures or the ability of conflict-affected populations to produce or interpret data about the violence surrounding them. Therefore, it is not surprising that the dynamics of apomediation manifest globally in a heterogeneous manner. I suggest starting to make sense of this global variation in the context conditions through the thinking tool (pun intended!) of the 'digi-political playing field'. The empirical chapters of this book will discuss in detail how varying configurations of this digital-political playing field shape the apomediation of peacebuilding by conditioning both meaning-making and relation-making practices.

More specifically, they highlight how apomediation dynamics are influenced by what I conceptualize as the *slanting* of the digi-political playing field, which describes the degree to which governing regimes stifle the possibility of articulating, sharing, and receiving understandings of conflict and peace that depart from their narrative. With 'governing regimes', I refer to the de jure or de facto rulers of the whole or parts of the territory where the armed conflict occurs, and these are usually also the main conflict parties. However, the book's empirical chapters also point to how apomediation is shaped by what I conceptualize as the *adaptability* of the field, that is, the degree to which users can employ digital technologies to cope with and challenge the slanting of the field. We can think about the interplay between these two dimensions as determining the overall slantedness of the field and, thus, whether apomediated peacebuilding is, in metaphorical terms, an uphill battle or not (see Figure 2.2). The slantedness of the field determines an ordinary user's ability to articulate, share, and receive understandings of conflict and peace that diverge from those of governing regimes. The stronger a regime tries to slant the field, the more difficult it is to advance alternative understandings of conflict and peace. The more peacebuilding actors, which often chiefly involve ordinary people, can adapt to the slanting, the more likely that apomediation can advance alternative understandings of conflict and peace and pave the way to approaches that change the political status quo. While I will explore these dynamics through qualitative case studies that draw comparisons between various peacebuilding contexts, we can also draw on quantitative data sets

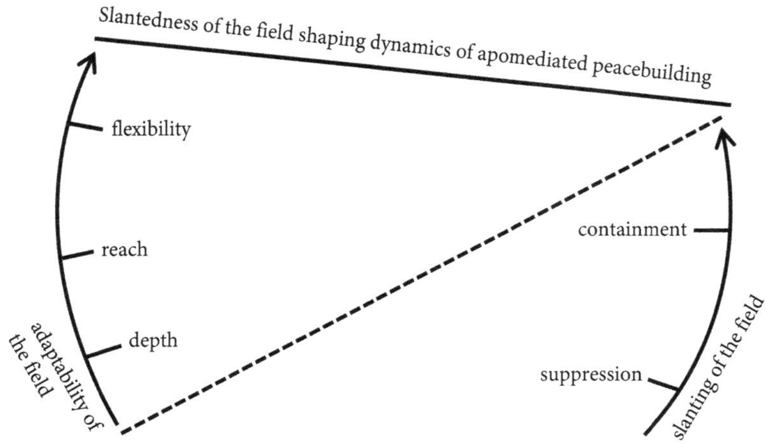

Figure 2.2 Dimensions shaping the slantedness of the digi-political playing field

to get a sense of the variations in the configuration of the digi-political playing field across the three cases on which this book mainly focuses, namely, Northern Ireland, Sri Lanka, and South Sudan. Comparing dynamics between these three cases is particularly insightful because their playing fields differ in the degree of the effort with which governing regimes attempt to slant the field, and the field's ability to adapt to this effort.

Let's look first at how we can make sense of how governing regimes may be slanting the field. The political playing field will be relatively level if governments provide institutional safeguards that enable equal access and usage of ICTs. In contrast, regimes can slant the field simply by failing to provide institutional safeguards or if they or other powerful conflict parties actively undermine apomediation efforts, such as by intimidating peacebuilding actors. We can construct the spectrum from level to slanted by drawing on the existing literature on political regimes and their effect on online political expression. Concerning the digitalization of peace processes, Schirch (2022, 5) has highlighted the need to pay attention to how a widening 'playbook of digital authoritarianism' affects peacebuilding, enabling mass surveillance, internet shutdowns, malware attacks, and information warfare. Therefore, I suggest focusing on the substantive characteristics of the political context that shape apomediation (Lührmann et al. 2018). Some studies of authoritarian regimes have measured such characteristics through indicators representing power relations in the political arena broadly defined, as well as in the media and online realm (Levitsky and Way 2010) specifically. Others have focused on how political regimes tilt the political playing field to secure their survival by reducing the ability of the organized opposition to voice and share dissenting views (Roessler and Howard 2009) or by severing communication links between the opposition and the public (Ekman 2009). Inspired by these works, I suggest exploring how regimes slant the playing field through online and offline tactics that may affect apomediation.

To start with, apomediation will undoubtedly be conditioned by a population's ability to challenge authoritarian regimes *online*. I thus suggest asking how regimes slant the field through the *suppression* of alternative understandings of peace and conflict. We can get a sense of the level of suppression across cases using the V-Dem indicator 'Government Internet Filtering in Practice'. This indicator captures on an ordinal scale how frequently governments censor political information online by blocking access to websites, ranging from never or almost never (value 4), to extremely often (value 0) (Coppedge et al. 2021). During the research for this book, such online filtering commonly occurred in both South Sudan and Sri Lanka, as did partial internet blackouts that caused whole social media platforms and messaging services to go offline for several days at a time. Moreover, just as essential is the link between online and offline capabilities to challenge a regime's narratives of conflict and its approaches to dealing with it. I suggest, therefore, comparing how regimes use *containment*, that is, the effort to hinder that online conversations translate into offline organization and action. We can do so using the V-Dem indicator 'CSO Repression', which measures a government's repression of civil society on an ordinal scale, ranging from no repression (value 4), to weak repression, such as through material sanctions such as fines (value 3), to moderate, which adds legal harassment and restrictions (value 2), to substantial, which adds arrests, imprisonment, disruptions of public gatherings, and violence against protesters, amongst others (value 1). For instance, the very high degree of civil society containment in South Sudan is manifest in the regime's effort to co-opt civil society-led conflict early warning efforts to strengthen the security monopoly of the state instead of enabling conflict-transformative interventions, as I will discuss in Chapter 4. The regime also curtails critical civil society voices on social media. As one interviewee put it, civil society activists are 'frequently complaining that cars without blades are following them, they are watching them in front of their houses, and they have to ask for permissions to organize events'.[4] Taken together, online suppression and offline containment constitute two crucial tools authoritarian regimes employ to tilt the playing field in their favour and stifle online discourse and political mobilization.

Next, I suggest asking how variations in digitalization may influence the apomediation of conflict and peace. I describe this in terms of the *adaptation* of the digi-political field and argue that it shapes how users can respond to authoritarian tactics. Building on existing research on adaptive digital systems and adaptive peacebuilding, I suggest understanding the adaptability of the digi-political playing field as the capacity to respond to authoritarian constraints on apomediation in a self-organizing manner (de Coning et al. 2023; Wright 2016). Similarly, 'digital resilience', or the ability to recover quickly from social media harms, such as misinformation and hate speech, is increasingly viewed as a crucial resource of conflict-affected communities (Ferroggiaro 2021; Reynolds and Parker 2018). To get a better sense of what conditions a field's ability to adapt, it is helpful to briefly look at discussions around the 'digital divide'. The stratified impacts of digitalization are often discussed and assessed

[4] Interview 41, Staff of Diplomatic Delegation, Juba, South Sudan, 24 November 2022.

through this lens of the digital divide, often conceptualized narrowly as an 'access gap'. However, more recent approaches take a more comprehensive view. Ragnedda (2016, 51) suggests differentiating three levels of the digital divide: the first is concerned with access to the internet, the second with inequalities in skills, and the third with the unequal opportunities that arise from those. This tripartite approach can also be employed to digital peacebuilding because differences in digital access and skills conditions will define opportunities to engage in the apomediation of conflict in peace.

Indeed, all three levels of the digital divide can affect the adaptability of the field. The first level determines *if* people can participate in apomediation. In this regard, the overall percentage of internet users may tell us about the relative size of the population that can participate in apomediation efforts. I suggest describing this as the *reach* of the field. Such basic access to internet-based technologies can be measured through the International Telecommunication Union (ITU) indicator 'Percentage of Individuals using the Internet', which serves as a good proxy to compare digital reach across the cases (ITU DataHub 2023).[5] The second level affects *how* people can participate in apomediation, determining if they can cope with challenges such as online repression, censorship, and disinformation. Thus, I suggest comparing the average digital literacy of the population across peacebuilding contexts as an indicator of this *flexibility* of the field. Unfortunately, there is currently a lack of good data to measure this capacity across cases. The ITU is in the process of creating data sets that assess digital literacy in a variety of aspects, including online safety. At the time of research, these data are available for only a few cases, which points to a considerable global discrepancy in the ability of ordinary users to protect themselves online. According to the available data, 49 per cent of the Canadian population knew how to change the privacy settings in their browsers in 2023, in contrast to about only 8.7 per cent in Côte d'Ivoire. As a proxy that allows us a better comparative picture for our three cases, we must currently rely on the World Economic Forum's Global Competitiveness Index, which captures 'digital skills among the active population', such as computer skills, basic coding, and digital reading on an ordinal scale from 1–7, with 1 being not at all and 7 being to a great extent (Schwab and Zahidi 2020, 18).[6] Importantly, this indicator only captures the skills of a smaller sub-group of users who employ digital technologies professionally and was created through an expert survey asking whether the skills were 'sufficient', which means that the answer will likely be subjective and context-dependent. The values also don't capture how the broader population may respond to potential online harms. Nonetheless, the comparison of this capacity across cases still provides an informative proxy indicator about the perceived digital skills of a critical part of the population that needs to be involved in conflict transformation efforts (see Chapter 6 for more details about the role of educated and relatively privileged social media users).

[5] In the case of Sri Lanka and Northern Ireland, the percentage of internet users was sourced from National Statistics Offices, whereas for South Sudan, the data is based on an ITU estimate.

[6] The index lacks data for South Sudan. Therefore, values from Uganda are used as a proxy.

Finally, we should be concerned with a field's ability to generate complex understandings of conflict and use apomediated conflict data in a way that enables suitable responses. I suggest describing this in terms of the population's *depth* of the field. Where the depth is relatively low, apomediation may struggle to advance the nuanced understanding of conflict and peace, which stand in the way of a transformative response, as will become evident in Chapters 4 and 5. In contrast, deep engagement enables digital peacebuilding activities in which rich and complex data and a variety of digital artefacts can be employed to facilitate work on the underlying causes of conflict. For a population's ability to interact with, and contribute to, complex understandings of conflict and the intricate question of how conflict can be addressed, general literacy is a good measure. To this end, we can draw on UNESCO data,[7] which show that the percentage of literate people aged fifteen and above is 99 per cent across Europe and comparably high in Sri Lanka, yet much lower in South Sudan. However, it is important to note that a high score on this index doesn't mean that everybody can equally take part in apomediation. A substantial part of the adult population in Northern Ireland has low reading skills (Shiel and Gilleece 2015), which conversely affects media literacy, critical reading skills, and exposure to a greater variety of information.

Figure 2.3 visualizes the digi-political playing field for the three cases with quantitative data collected in the years leading up to the qualitative research presented in this book, using the sources discussed above. For this, I used values from the year closest to the onset of the qualitative research in the three countries, 2021, in which data were available for all cases. I visualize these data in a parallel coordinates plot displaying the five values that I found helpful to indicate the configuration of the digi-political playing fields across cases. The values for depth (digital literacy) and reach (internet

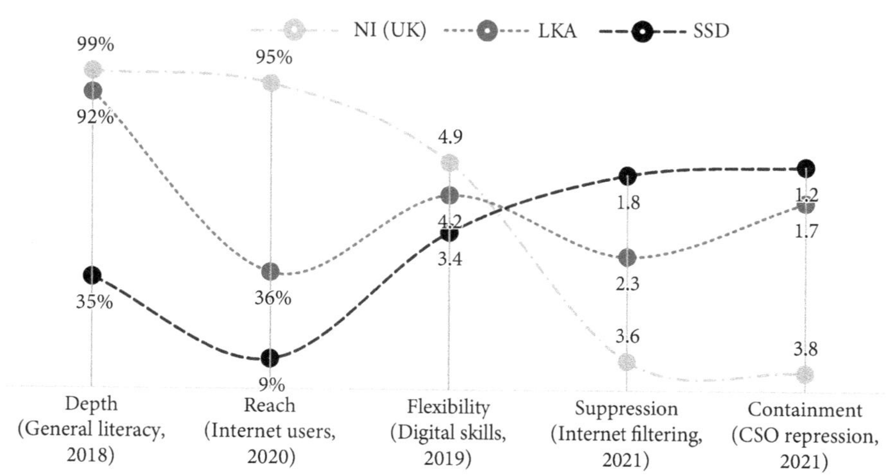

Figure 2.3 Digi-political playing fields of South Sudan (SSD-red), Sri Lanka (LKA-violet), and Northern Ireland (NI (UK)-green)

[7] This data set does not have country-specific data for the United Kingdom. Therefore, I use the aggregate value for Europe and Central Asia as a proxy.

users) are displayed in percentages, whereas the values for digital literacy (flexibility), suppression (internet filtering), and containment (CSO repression) are shown on an inverted numeric scale that corresponds with the ordinal values mentioned above.

Reading these values from left to right, we can get a sense of how the digi-political playing field influences apomediation. The depth of the playing field influences the degree to which ordinary people can advance and engage with a diversity of understandings of conflict, especially whether they can employ online communications to learn about and generate information about their often very complex causes. Next, the reach of the playing field captures the degree to which these various understandings can be shared across the population, while the flexibility of the field tells about the degree to which they can do so in spite of hurdles stemming from a governing regime's efforts to slant the field. These are described in the last two columns on the left. The value of suppression indicates the degree to which online interactions may be curtailed, whereas the value of containment tells of the degree to which governing regimes may take steps to prevent the translation of apomediation dynamics into offline action. It is important to note that the different values have no relative relationship to one another since some of the indicators are based on ordinal scales and some are based on numeric scales, and there is no research that puts these quantitative data in a direct relationship. Exploring how the different dimensions of the playing field and the indicators employed matter will thus remain a task for the qualitative research presented in this book.

However, at this point, we can compare each value across cases to theorize about how the different dimensions of the playing field may condition apomediated peacebuilding. For instance, apomediation in Northern Ireland and Sri Lanka will benefit from almost similar levels of general literacy, which could help with advancing a diversity of complex and nuanced understandings of the conflict and what would need to be done to overcome it. However, this is far less the case in South Sudan. The picture looks somewhat different when it comes to the population's ability to share such understandings widely and to overcome potential hurdles stemming from higher levels of political repression because Sri Lanka's field has less reach and flexibility. In contrast, Sri Lanka and South Sudan converge a bit more when it comes to the governing regimes' likelihood of suppressing the apomediation of conflict and peace online and containing its translation into offline action. Overall, we can compare the relative course of the three graphs, which suggests that apomediation efforts that aim to advance understanding of conflict and peace, and approaches to resolving the conflict that is not in line with the governing regimes' vision, will face much more of an uphill battle in South Sudan, than in Sri Lanka, where some hurdles exist, and in Northern Ireland, where apomediation can take place without any substantive restrictions put in place by the government. Alternatively, we can read it from right to left, exploring, for instance, how, in contexts with a highly slanted playing field, efforts to advance alternative understandings of conflict and solutions to peace depend on the flexibility, reach, and depth of the field. For instance, social media activists in Sri Lanka find it relatively easier to circumvent oppressive digital practices and spread their messages more broadly than do activists in South Sudan.

It's important to note that this way of using quantitative data to narrate how apomediation is shaped by the digital playing field comes with several additional caveats. First and foremost, using quantitative data to compare the dynamics, especially indicators that only help us to approximate what is concretely taking place, may help with establishing an overall comparative picture that shows the relevance of looking at how digital peacebuilding is shaped by digital, as much as by political and social dimensions. However, it is of little help when the objective is to take a closer look at the socio-technical dynamics that shape apomediation. We might also simply miss out on other important details. When taking a closer qualitative look, many other factors that shape apomediation come to light. Most notable among them is the design of the technologies employed, such as recommender algorithms on social media or top-down approaches to information extraction. Equally important are the assumptions, including those about the digi-political playing field itself, and the corresponding visions and ideas that drive digital peacebuilding initiatives and determine which technologies and methods are employed. Furthermore, there is no guarantee that people will actually make use of the opportunities that come with apomediation. Instead of exploring alternative ways of looking at the conflict and exploring how digital technologies could be employed creatively to resolve it, they may prefer to remain stuck to established narratives and beliefs. For instance, Chapter 6 makes clear that apomediated peacebuilding is something that needs to be actively pursued by committed professionals and, most importantly, ordinary people who want to see conflicts transformed. This underlines the importance of taking an in-depth and qualitative look at digital peacebuilding, which the subsequent chapters will offer.

The next section provides a first overview of how the dynamics of slanting and adaptation shape the digi-political playing field and thus shape apomediated peacebuilding through a brief discussion of Northern Ireland, Sri Lanka, and South Sudan, in which many of the digital peacebuilding projects discussed in this book take place.

Comparing Different Fields

In South Sudan, the percentage of the population using the internet almost doubled between 2013, the year when the civil war broke out, and 2017, when international peacemaking efforts had resulted in a shaky peace agreement, namely from 2.2 per cent to 4.1 per cent.[8] Nonetheless, South Sudan continued to be one of the countries with the least developed communications infrastructure globally, with digital access and digital literacy remaining extremely low. This also affected how peacebuilding organizations conducted their work. Many of the staff working for international organizations, embassies, or civil society groups, especially those based in the country's capital, Juba, perceive that most parts of the country are not yet 'ready' for

[8] All statistics are from the International Telecommunications Union (ITU) Indicators Database, retrieved through https://data.worldbank.org/indicator/IT.NET.USER.ZS, Accessed April 26, 2026. The exact indicator is 'Individuals using the Internet % of population'.

digital approaches. They therefore often relied on conventional means of engagement, including legacy media such as the radio, or in-person activities, especially in rural areas. In practice, this meant that the apomediation of conflict—and apomediated peacebuilding—often involved only a small fraction of South Sudanese society and mainly relatively well-educated and well-equipped elites located in urban areas and in the diaspora. However, even this very low level of access can influence dynamics of peace and conflict. Indeed, South Sudan's armed violence, between organized armed groups, but also between parts of the civilian population that engage in cattle raiding and conflicts over land and natural resources, has been catalysed by the selective availability of mobile internet connectivity through which violence can be coordinated and legitimized. Digitally connected local intermediaries, such as community youth leaders who communicate with urban elites and the diaspora via social media and messaging applications, often played a crucial role in the instigation of violence.

Importantly, peacebuilding organizations that employ digital approaches usually also create such digital intermediaries, as they distribute mobile devices, data vouchers, and digital skills among their networks of volunteers that they integrate into their various activities, such as local conflict early warning or the fight against disinformation. Indeed, while poorly connected, South Sudan has seen myriad efforts to establish local conflict early warning networks using mobile connectivity, as will be discussed in Chapter 4. These efforts create patches of connectivity and empower certain parts of the local populations by making them part of remote peacebuilding activities. Overall, at the time of writing, the *reach* of the playing field in South Sudan remained very limited. As a consequence, digital peacebuilding organizations often struggled to establish apomediation networks that could effectively counter the influences of conventional elites, as I will argue in Chapter 3. Furthermore, efforts to 'crowdsource' early warning information were shaped (and at times curtailed) by the continued influence of political institutions and authorities that sustained their power through conventional, 'offline' means, as I will discuss in Chapters 3 and 4.

South Sudan's playing field was also characterized by very high levels of political *suppression* and *containment*. For instance, South Sudan's social media discourses were marked by ill-defined yet commonly known red lines that curtailed what understandings of conflict could be generated and shared. For instance, as I will discuss in Chapter 6, it was risky and indeed often impossible to employ social media to discuss South Sudan's underlying causes of conflict—many of which have to do with the centralization of power in the hands of a small elite. Over the past decade, South Sudan has considerably extended its online surveillance capacities, including through the interception of mobile communication and the monitoring of social media to 'to identify, intimidate and arrest and detain people perceived as government critics', as an Amnesty International Report (2021) put it. Furthermore, in May 2021, the government introduced temporary cybersecurity legislation in the form of the Cybercrimes and Computer Misuse Provisional Order, and a parliamentary process to establish permanent legislation started in 2023 (Radio Tamazuj 2024). These efforts formally granted South Sudan's security services the right to take action against online dissent and were viewed by journalists and civil society as an attempt to institutionalize restrictions on free online speech.[9]

Therefore, it is hardly surprising that most social media users will not criticize government officials or the incumbent regime directly. Individuals who call out power holders risk repressive reactions, and the large number of attacks, arbitrary arrests, and disappearances demonstrate the risks of criticizing the regime. In July and August 2021, there were sparse attempts by a limited number of civil society activists from the People's Coalition for Civic Action (PCCA) to use social media for the organization of peaceful anti-government protests, which led to arrests, intimidation, and internet shutdowns. International and national peacebuilding organizations and large parts of the diplomatic community also tend to avoid open criticism or associating the regime with the causes of conflict, as their presence and operations in the country are conditional on the regime's goodwill. While not facing the same repercussions of political repression, several members of the diplomatic community have been declared *persona non grata* after criticizing members of the government for their limited willingness to implement political reforms. The regime also commonly suppressed critical internet and social media resources, such as those of independent media houses and international human rights and pro-democracy organizations.

The increase in connectivity has been even more dramatic in Sri Lanka, where the percentage of the population with access to the internet rose from 1.5 per cent in 2002, when the formal peace negotiations faltered, to 44.5 per cent in 2021,[10] the year before the country witnessed large-scale popular unrest organized and coordinated on social media. This has undoubtedly meant that online platforms, social media, and mobile devices have been more easily employed in support of peacebuilding and reconciliation efforts since the end of the civil war in 2008, and there was a stronger tendency overall to utilize digital technology for peacebuilding in Sri Lanka than there was in South Sudan. However, it clearly also played an even bigger role in the instigation of conflict. In 2009, when the Sri Lankan government had defeated the LTTE militarily after the peace process had broken down, the number of internet users had grown considerably since the beginning of the peace process. In the years after, reminiscent of popular protest movements during the Arab Spring, the country witnessed a boom in online and social media activism and campaigning, which led to the ousting of President Mahinda Rajapaksa in 2015, who had been accused of authoritarianism and corruption. This paved the way for a short period of political liberalization, as well as progress on post-conflict reconstruction and reconciliation. Access to the internet reached 35 per cent in 2017, and this rise was accompanied by a period of increasing political polarization and ethno-nationalist populism, leading to the return to power of the Rajapaksa regime in 2019.

As in South Sudan, the political context of Sri Lanka curtails what can be said about conflict. In the period immediately before the popular protests of 2022, civil society

[9] Interview 41, Journalist, Juba, South Sudan, 24 November 2022.

[10] ITU Indicator 'Individuals using the Internet (% of population)', accessed through https://data.worldbank.org/indicator/IT.NET.USER.ZS.

representatives would exercise restraint in directly criticizing the government for its divisionist politics or past human rights abuses, largely because of fears of repression. While the levels of *containment* were lower than in South Sudan, many civil society activists and political opponents continued to fear political repression, given the country's history of human rights abuses against civil society activists and political opponents. Dissident voices among the Tamil minority, as well as civil society activists, lawyers, and journalists, continued to be censored through the application of the Prevention of Terrorism Act (PTA), which allows arrests without warrant for unspecified 'unlawful activities, including speech-related offenses'. The PTA also provided the pretext for the Rajapaksa regime to *suppress* alternative voices on social media, including through internet filtering and social media censorship. For instance, politicians and social media activists who wanted to call for a stronger devolution of power and a federal system to end decades-long minority grievances would be stifled under this legislation. As in South Sudan, efforts to introduce 'cybersecurity' legislation to further legalize such efforts were underway during the research for this book. The so-called 'Online Safety Act' was passed in January 2024, but this legislation already had chilling effects on civil society in the years prior.[11] The Act grants broad powers to a so-called 'Online Safety Commission' to determine 'prohibited statements' while remaining vague about what kind of content falls under this definition, which, according to many critics, could be used to suppress dissent (Amnesty International 2024).

However, Sri Lanka's slanted playing field was fiercely challenged through the political upheaval that unfolded in 2022. Faced with worsening living conditions, thousands of Sri Lankans joined country-wide mass protests calling for an end to the Rajapaksa regime in April 2022. The protests became known as the 'Aragalaya'—the Sinhalese word for 'struggle'—and were largely driven by a disgruntled middle class that perceived the impacts of the economic crisis as an existential threat. Seeking cover and anonymity among the hundreds of thousands of co-protesters, this well-educated and networked movement formed a critical mass that could at least temporarily tilt the political playing field, which led to the tumultuous overthrow of the government in July 2022. Social media and other forms of digital communication played a key role in the mobilization and coordination of these protests, and high levels of digital literacy enabled the activists to circumvent the regimes' efforts to suppress dissident voices. I will discuss in Chapter 6 how, in this context, everyday social media interactions also provided an opportunity, albeit temporarily, to work on the decade-long causes of conflict and violence in Sri Lanka.

Northern Ireland is an exemplary case of how online polarization can shape understandings of peace and conflict in contexts with a relatively level and highly adaptable digi-political playing field. Digital and general literacy and digital access in Northern Ireland are among the highest in the world. Since the signing of the Good Friday Agreement for Northern Ireland in 1998, which ended decades of violent

[11] Interview 83, Director, Civil Society Organisation, Colombo, Sri Lanka, 9 February 2023.

conflict between supporters of Unionist and Republican movements, the percentage of internet users has risen from 13.7 per cent to 96.7 per cent in 2019 (World Bank 2024). Today, more individuals exchange information through internet-based technologies than through conventional, analogue means of communication such as the radio, television, or a landline telephone. Furthermore, there are no known hurdles imposed by the governments of the United Kingdom or Northern Ireland on access to the internet, nor is there any noticeable censorship or filtering of online content. However, online interactions about peace-related matters are often very divisive, harmful speech is rampant, and Northern Ireland has seen the emergence of social media influencers who generate their following through the creation and sharing of divisive content.

Likely because of the nature of the armed conflict in Northern Ireland and the strong involvement of the UK government, international organizations played a limited role in peacebuilding efforts, with the exception of the European Union (EU)'s funding for the PeacePlus program (SEUPB 2021). Moreover, while local civil society organizations implemented similar activities to peacebuilding efforts in other contexts, many referred to their projects as 'community' or 'social' work. This may be a case-specific characteristic, mainly due to donor support for peacebuilding decreasing since the Good Friday Agreement. However, other cases with a similar digi-political playing field and more recent histories of tensions or violence, such as the United States or Spain, also lack the type of professional peacebuilding sector that can be found in contexts with less level and adaptive playing fields. This may have to do with the magnitude of violence and an international system that favours intervention in places that are politically less free and infrastructurally less developed. Importantly, the comparable lack of a professional peacebuilding sector doesn't mean that the risk of violent conflict between Northern Ireland's communities has been overcome. In fact, at the time of research, Northern Ireland had a dysfunctional power-sharing arrangement, caused by a conflict over issues regarding the Irish border following Brexit. These considerable tensions at times resulted in localized violence (O'Neill 2021), which has caused many observers to worry about a return of the 'Troubles' (Gladstone and Robins 2021).

Putting Apomediation to Work

The remaining chapters of this book will shed light on the apomediation of peacebuilding in a variety of contexts that differ in the configuration of their playing fields. Some are highly slanted and not very adaptable, such as South Sudan. In comparison, some are relatively level and quite adaptable, such as Northern Ireland, while others, such as Sri Lanka, sit in between these two extremes. The latter case also demonstrates that these two parameters are not fixed but change over time and sometimes in a dynamic manner. As I will argue in Chapter 6, civil society organizations and ordinary citizens in Sir Lanka had the potential to counter the regimes' predominant social media narratives on the conflict, but it required building political momentum to do so at scale. This momentum was temporary and did not bring a lasting transformation of the antagonisms that underpin violence in Sri Lanka.

However, the different theoretical considerations and concepts presented in this chapter are not meant to provide an explanatory theory for why digital peacebuilding fails or succeeds. Moreover, they are clearly not intended to be translated into hypotheses that can be tested, verified, or falsified. Given the heterogeneous and path-dependent nature of armed conflicts, peace processes, and digitalization—and, of course, global politics more broadly—such an attempt would be futile. I also don't suggest thinking about this chapter as providing a conceptual 'toolkit'. Given the scepticism I have expressed about a tool perspective and our ability to isolate those tools from their social embeddings, it would be a little ironic to do so. Instead, I would like to think of this chapter as providing a set of lenses that can be employed to develop a context-dependent view of the empirical processes we study. This view has evolved in dialogue with many other concerns I raise in this book, which have emerged through my own reflections, in conversation with other academics, and most importantly, through learning from the many research participants I have engaged with. Depending on the research task, the different conceptual lenses can be combined flexibly to shed light on various aspects of digital peacebuilding. A good starting point for such endeavours would be to explore how digitalization decentres the production and flow of authoritative understandings of conflict and how this shapes the approaches to dealing with conflict. I have referred to this as the apomediation of peacebuilding. The central question triggered by this conceptual lens is whether the digitalization of the field indeed alters who has authority in peacebuilding or whether it may help to reproduce the power and influence of conventional intermediaries instead. It's important to note that there is no guarantee that shifts in authority over understandings of conflict and peace, and in the agency involved in peacebuilding, makes the ideal of conflict transformation easier to achieve. In fact, the opposite may be the case when digital approaches give more power to new actors, such as social media platforms, or technical artefacts such as machine-learning algorithms. Thus, the end-point of such an exploration should be to ask about the voids such dynamics produce, that is, by focusing the inquiry on what missing-from-hand in digital peacebuilding, as I explained in Chapter 1.

However, many other relevant issues can be explored along the way. These include the four central points I described at the beginning of this chapter. There is the question of agency. Apomediation asks about how understandings of conflict are generated in human–machine peer-to-peer networks. However, it also begs the question of how peacebuilding outcomes emerge due to such dynamics. Which human–machine relations determine which understandings of conflict are used as the basis for the efforts to prevent, end, or transform it? How do, for instance, PeaceTech engineers promote specific software solutions that ultimately determine how local crowdsourcing networks collect certain types of early warning data, while leaving other valuable information aside? There was also the question of the tangibility of digital peacebuilding. The concept of apomediation connects offline and online practices, and it also considers the role of material factors such as infrastructures and devices. However, it enables us to take our concerns beyond the 'virtual' world and towards the embodied experiences of populations living with conflict and violence. For instance, political regimes that aim to contain specific technology uses will shut down infrastructures and starve civil society activists or volunteers of internet data, as I will demonstrate in Chapter 4.

There are also historically ingrained structural aspects to hatred and armed violence that have grown over the long durée. Apomediated peacebuilding can cause these to disappear from the databases and dashboards that determine preventive measures, as I show in Chapters 3 and 4, especially when data become highly structured and aggregated, such as social media or conflict early warning data. However, it may also enable new ways of exchanging views on conflict between antagonistic groups, for instance through social media memes that contain specific types of humour, as I demonstrate in Chapter 6. Importantly, digital approaches cannot only be ignorant about the long durée of conflict, but they can also be shaped by it. This 'past in the present' becomes visible, for instance in the resurgence of modernist obsessions with evidence-based approaches that promise an 'objective' view on conflict, as I argue in Chapter 5, but likewise in the longevity of (post)colonial and authoritarian governance arrangements that insert themselves into seemingly innovative and revolutionary digital peacebuilding approaches. Finally, the apomediation framework may also inspire us to explore how digital technologies could be employed in ways other than the predominant approaches that focus on data and evidence. My discussion of subjunctive technology affordances in the context of dialogue and reconciliation efforts in Chapters 5 and 6 points towards this direction, as do my explorations of everyday social media practices and the role of the 'critical digital mass', which I discuss in Chapter 6.

All the examples highlighted in this chapter, and many others discussed in this book, are shaped by the context-dependent constellations of the digi-political playing field in each case. At the very least, it is my objective to demonstrate that. Importantly, there is no linear cause-and-effect relationship between the different properties of the playing field, such as its reach or the level of political containment. However, working with these concepts gives us the opportunity to sharpen and condense our analysis and take a closer look at some of the most important aspects that shape the apomediation of peacebuilding in a globally heterogeneous manner. At the same time, the qualitative analysis in the following chapters, based on four years of participatory observations, more than 150 interviews, and considerable amounts of desk research, aims to present the intricate details behind the individual constellations of each peacebuilding case. The remainder of this book will take a closer look at these and many other dynamics that shape contemporary digital peacebuilding.

3
Preventing Conflict
Countering Harmful Social Media Speech

Few would disagree with the claim that language and speech matter for both instigating conflicts and terminating them.[1] Consequently, peacebuilding as a professional field has long been occupied with promoting forms of communication that can de-escalate tensions between individuals and groups (Hawkins and Hoffmann 2015; Bodie et al. 2023). In contemporary conflicts, this work has increasingly shifted to online and social media spaces. As became apparent in the previous chapter, one central driver of the apomediation of conflict and peace has been the rise of social media, which enables decentralized communication flows between ordinary citizens. These networks circulate a considerable part of the content through which ordinary citizens obtain information about political and security developments and, by extension, make up their minds about who is friend and who is foe. This chapter will focus on what has become one of the most prominent areas of digital peacebuilding, namely, efforts to prevent some of the most extreme adverse outcomes of apomediation, namely the production, circulation, and consumption of social media content that nourishes antagonisms between individuals and groups. Such content is broadly referred to as 'hate speech' or 'harmful speech'. While many peacebuilding initiatives aim to counter harmful speech on social media, parts of this chapter reproduce my previously published work (see Hirblinger 2024) to make clear that harmful speech cannot—and should not—be treated as a purely digital phenomenon. Peacebuilding initiatives that intend to tackle the antagonisms that manifest, and are reproduced, through harmful speech will necessarily have to engage with the socio-political context in which such speech is produced, spread, and consumed.

Peacebuilding actors that aim to counter harmful speech naturally tend to operate in contexts that are characterized by violent conflict or are at risk of it. Social media platforms provide an excellent tool for conflict parties and extremists who intend to maintain or deepen divisions through harmful online narratives and hate speech that create antagonisms along ethnic or religious lines, while also curbing online freedoms to stifle political dissent and mobilization for change (Jones 2022; Yilmaz et al. 2022; Garbe et al. 2023; Hattotuwa and Wickremesinhe 2022; Bradshaw and Howard 2019). It is not uncommon that incumbent regimes are directly involved in

[1] Some parts of this chapter include material previously published in Andreas T. Hirblinger, 2024. 'Fighting Hate Speech on Social Media: The Challenge of "Deep Work" in Digital Peacebuilding.' In *Social Media and Peacebuilding. How Digital Spaces Shape Conflict and Peace*, edited by Stephan Stetter and Anna Reuss. Rethinking Peace and Conflict Studies.

Digital Peacebuilding?. Andreas T. Hirblinger, Oxford University Press. © Andreas T. Hirblinger (2026).
DOI: 10.1093/9780198986553.003.0004

instigating hatred on social media, which they exploit as a strategy of divide and rule. In places such as South Sudan and Sri Lanka, as well as many other contemporary conflict zones, violence erupts because people have grievances with their government, which makes them more prone to join violent protests or armed insurgencies. Such grievances are often caused by a lack of opportunities to participate in politics or because governments use political oppression and force to govern their populations (Cederman et al. 2013). While both cases saw various armed insurgencies attempting to oust authoritarian incumbent regimes, organized violence between groups identifying with ethnicities or religions is also common, such as inter-communal riots in Sri Lanka or cattle rustling in rural South Sudan. However, ethnic or religious hatred on social media is not a given but rather, the result of what could be described as a strategy of 'digital divide and rule' meant to secure regime survival (see also Chapter 6).

Indeed, as will become evident in this chapter, the spread of harmful speech on social media is often part and parcel of a regime's effort to keep the playing field tilted while maintaining a loyal followership and justifying unfair, exploitative, or discriminatory policies. Cleavages between different groups and threat perceptions are commonly nourished through hate speech and rumours produced and shared by a network of social media influencers loyal to high-level political leaders—or even by the leaders themselves. In the course of South Sudan's civil war, hate speech was spread on social media, but also conventional media such as the radio, and even printed pamphlets left on doorsteps, which repeatedly triggered violence against specific ethnic communities or foreign nationals (UN News 2014; OHCHR 2016). Influential political elites and government officials often play a direct role in the production and spread of such hateful content from the comfort of their offices. In South Sudan, government officials have reportedly shared hateful language on their personal social media pages while simultaneously exercising their duties as diplomats or members of parliament. In Sri Lanka, the so-called 'Social Media Monks' are another strong example of how prominent and influential figures can be at the forefront of spreading harmful speech, in this case, while playing an important spiritual and ceremonial role in Sinhalese society and politics (Orjuela 2020).

Hate speech is widely acknowledged to be based on three essential features (Parekh 2012, 41). It is directed against an individual or a group of individuals based on an arbitrary feature, it stigmatizes this target by ascribing undesirable qualities to it, and renders it a legitimate object of hostility. Hate speech is also often compounded by disinformation, misinformation, or malinformation. These three conceptions of false information all rest on the metaphysical realist assumption that a truthful representation of reality, and with it, a differentiation between correct and incorrect information, is indeed possible (David 2022). Disinformation refers to false information that is intentionally produced and shared to manipulate the reader. Misinformation is false information that may cause harm accidentally, and malinformation usually has a degree of truth to it but is exaggerated or misleading. Today, the link between hate speech and false information is obvious on social media, and this has also been acknowledged in international policy, such as the UN's Plan of Action against hate speech (UN Secretary-General 2020, 7). The conflation of hate speech and false information is commonly summarized under the terms 'harmful speech' or 'dangerous speech', defined as rhetorical techniques that increase the risk of violence and harm to individuals or groups (Dangerous Speech Project 2021; Sellars 2016). This explains why efforts to counter harmful speech can be understood as a peacebuilding effort because they contribute to the prevention of armed conflict (Barnett et al. 2007).

Stereotypical depictions of groups, degrading language, or the partisan representation of events are in many ways an important part of the apomediation of conflict and peace. This is because harmful speech that is produced, circulated, and consumed in online peer-to-peer networks contributes to popular understandings of conflict. For example, social media posts may represent specific political, military, or security events in ways that cast blame on specific individuals or groups and create the narrative that they have contributed to, or even caused, the conflict. Harmful speech thus brings conflicts to the fore but also causes them through insults, degradations, and incitements to violence against an 'Other'. However, it also purports that the same conflicts can be solved through acts of violence against individuals or groups associated with this Other. In this process, harmful speech often distracts from the more abstract, complex, or underlying causes of conflict, such as social injustice, systemic corruption, or historically grown identity cleavages, by putting forward simplified narratives that attribute responsibility to individual actors or groups. Moreover, affect-generating content, such as slurs or abusive terms that often form part of harmful speech, not only provides the emotional undercurrent that strengthens or justifies such claims but also fuels conflict in perceivable, visceral, and sometimes tangible ways. Harmful speech is thus a good example of how apomediation does not merely construct causes of conflict at a discursive level but also in ways that affect the practices and experiences of conflict-affected populations in a tangible manner.

Importantly, we can, vice versa, think of efforts to fight harmful speech online as part of the apomediation of peace that aims to erase, reduce, or transform understandings of conflict that are likely to cause harm, violence, or suffering. Over recent years, donors and implementing organizations have increasingly recognized the fight against online hate speech as a conflict prevention and peacebuilding activity. This points to a shift in how they perceive the work of preventing and ending armed conflicts characterized by a growing social media dimension. For instance, the United Nations Strategy and Plan of Action on Hate Speech describes social media hate speech as a 'menace to democratic values, social stability, and peace', and at the same time, promotes technology use to counter it, including through partnerships with social media companies and civil society (UN Secretary-General 2019, 2–3). So far, we know little about how such new coalitions leverage technology to fight hate speech and how this shapes peacebuilding as a professional field, which is what this chapter will explore.

As suggested earlier, discussions about digitalization are often couched in ahistorical narratives of disruptive and fast-paced technological progress that radically changes conflict dynamics and peacebuilding approaches. This tendency also affects contemporary efforts to counter harmful speech that focus primarily on social media. Harmful speech is understood today, by and large, as an online phenomenon: something that happens in posts, videos, and pictures, something that can be shared and forwarded and go viral, and something that can also be clicked away and deleted. However, by predominantly rendering it a digital artefact, we risk ignoring the fact that harmful speech clearly pre-dates the internet age. Indeed, while not referred to with the same term or definition, harmful speech is probably just as old as human conflict and violence. What is certain is that concerns with hate speech date back well into the twentieth century, when they evolved in the context of state-orchestrated mass atrocities, as I will discuss in the next section. Hate speech harms through both its *immediate* role in inciting violence (Timmermann 2008; Townsend 2014; Al-Zaman 2019; Daniels 2018) and through *intermediate* efforts to undermine the dignity, equal citizenship, and social standing of minorities in society (Waldron 2014, 65, 92–3). Thus, while fighting hate speech is viewed as instrumental to efforts to prevent immediate violence, it is also intricately linked to efforts to tackle and overcome the historical, cultural, and structural causes of violence, including ethnic hatred and racism. Consequently, approaches to countering it can be influenced by a broad array of assumptions and considerations, which connect to various peacebuilding approaches and objectives, ranging from the short-term prevention of violence to long-term conflict transformation that tackles its underlying causes.

I discussed in the previous chapter that there is much to gain from studying how technology and peacebuilding are co-produced. A promising, more concrete research strategy entails asking how certain technological solutions drive how we make sense of the problems they are meant to solve (Hirblinger et al. 2022; Jacobsen and Monsees 2019). As I will argue in this chapter, this also affects efforts to counter harmful speech, where there has been a confluence of diverse, historically evolved problematizations of hate speech with more recent technological solutions advanced to address them. Faced with the increasing impact of digitalization, the academic literature on hate speech has seen a shift from a preoccupation with its structural, 'offline' causes towards solutionist approaches that are more concerned with automatic hate-speech detection and classification of its digital manifestations (Tontodimamma et al. 2021, 176). This has partly been driven by the fact that many tools for social media analysis that were originally developed for the private sector monetization of social media data have become ready-at-hand for the peacebuilding sector. However, instead of simply taking this trend towards digital approaches for granted, I believe it is important to reflexively engage with how digitalization affects the struggle against harmful speech and to ask what becomes missing-from-hand when they are employed for peacebuilding purposes. Therefore, this chapter first sheds light on how the problem of hate speech is discursively constructed, and how, in the context of contemporary conflicts, social media and its affordances have become implicated in such problematizations, as well as in the search for solutions. As will become clear, digital technologies are anything but neutral tools that simply enable or catalyse efforts to counter harmful speech. This is because the solutions that can be digitally enacted speak back to certain problematizations of conflict and less to others. Concretely speaking, drawing on my research with several social media initiatives aimed at countering harmful speech in South Sudan and Sri Lanka, I will show that the digitalization of harmful speech renders the 'deep work' on the underlying causes of conflict more difficult. This is especially true when it interacts with conflict-affected contexts characterized by a slanted digi-political playing field with limited reach and depth.

How Harmful Speech Causes Conflict

The problems identified with hate speech are diverse, as are the assumptions that premise the fight against it. However, many of them arguably arose after the advent of mass communication in the twentieth century, when spreading hate speech on the radio and television provided the ideological groundwork for nationalist and fascist agendas that resulted in large-scale violence and destruction in Europe and many other parts of the world. In the aftermath of the Second World War and the Holocaust, policymakers aimed to tame the instrumentality of hate speech in mobilizing for mass-scale violence and genocide through legislative means (Imbleau 2011). The 1948 UN Genocide Convention prohibited the 'direct and public incitement to genocide' and the 'advocacy of national racial or religious hatred that constitutes incitement to discrimination, hostility or violence'.[2] Yet, despite these efforts, hate speech continues to yield a considerable potential to catalyse mass violence, as has been documented in Rwanda, Bosnia, Iraq, Israel, and Palestine, to mention only a few cases.

While the concerns about the role of hate speech in mobilizing for ethnic violence and genocide have thus largely remained unchanged, social media has replaced the radio as the technology of concern. In Myanmar, for instance, Buddhist ultranationalists used Facebook to normalize hateful views, stoke fear, and facilitate acts of violence against the Rohingya and other Muslim communities, leading to widespread destruction, loss of life, and displacement in 2015 (Fink 2018, 43). Indeed, a fact-finding mission led by the Office of the UN High Commissioner for Human Rights concluded in its 2018 report that 'the role of social media [was] significant' in fuelling the atrocities committed against Rohingya and other minorities in Myanmar,[3] and Amnesty International (2022) found that 'Facebook's algorithmic systems were supercharging the spread of harmful anti-Rohingya content'. Similar problematizations can be found in many other contexts, such as Sri Lanka, where an independent human rights impact assessment on Facebook's role in instigating anti-Muslim violence found that the 'platform contributed to spreading rumours and hate speech, which may have led to "offline" violence' (Article One Advisors 2018, 4).

[2] United Nations Convention on the Prevention and Punishment of the Crime of Genocide, December 9, 1948, https://www.ohchr.org/en/instruments-mechanisms/instruments/convention-prevention-and-punishment-crime-genocide, Accessed March 28, 2023.

[3] Human Rights Council, Report of the Independent International Fact-Finding Mission on Myanmar, UN Doc. A/HRC/39/64, para. 74 (August 24, 2018).

Besides its instrumentality in instigating mass violence, a narrower hate speech problematization can be found under the banner of Preventing Violent Extremism (PVE). Since the commencement of the global 'War on Terror' in 2001, scholars have pointed to the radicalizing effects of propagandistic hate speech on social media platforms by jihadist organizations such as Al-Qaeda and the Islamic State (Zelin 2015; Klausen 2015). More recently, there has been growing concern with far-right violent extremism, encouraged by the widespread normalization of online hate speech that provides a fertile breeding ground for Islamophobia (Kenny 2022). Importantly, far-right or nationalist hate speech is often explained through individual or technical causes rather than factors that are more deeply ingrained in society. For instance, its violent offline manifestations are frequently depicted as the work of 'lone wolves' or psychological processes underlying the radicalization process (Koehler 2014). However, the 'lone wolf' moniker obfuscates the fact that hate speech and violent extremism are 'overwhelmingly correlated with and caused by offline political and economic conditions', as Richards and Wood put it (2020, 112). Similarly, Halverson and Way (2012, 141) contend that violent extremists are more 'shaped by the sociopolitical, economic, and historical contexts in which they operate' than by the mere appeal of hateful speech—an aspect we will come back to repeatedly in this chapter.

Hate speech has also been problematized more broadly concerning its potential to undermine democratic institutions and processes. For instance, advocacy groups and scholars have pointed to the role of social media in spreading hate speech and fuelling tensions in the context of elections, such as in Kenya, Cameroon, and Nigeria. Examining the latter case, Ezeibe (2021, 931) argues that political elites employ hate speech 'as a political campaign strategy to capture state power' by insulting, discrediting, and stigmatizing members of the political opposition and stimulating violence against their followers. Concerns about the detrimental effect of hate speech on democratic processes are also evident in the recent literature on the instrumentality of social media in populism, which problematizes populist hate speech against the backdrop of the crisis of democratic representation (Norris 2019). It suggests that radical right-wing populists successfully leverage social media platforms to build a support base, interact with their followers, and criticize political elites (Kessel and Castelein 2016) because 'social media serve as direct linkage to the people and allow the populists to circumvent the journalistic gatekeepers' (Engesser et al. 2017, 2)—a clear example of apomediation. Many populist discourses have also been found to be mainstreaming far-right extremist ideologies, such as the 'great replacement theory', and thus to nourish racism. For instance, far-right and populist politicians in Western Europe, such as Marie le Pen, propagate populist hate speech that draws on racist ideologies of 'cultural' compatibility that were central to white supremacist groups for decades (Davey and Ebner 2019, 8). It is this concern with the role of hate speech in reproducing discrimination and racism that brings us back to the original conceptualization of hate speech as racist speech, as I will discuss next.

Indeed, when looking beyond the narrow policy and academic discourse concerned with preventing violence or safeguarding democratic processes, we find that the term 'hate speech' featured prominently in US debates about free speech and, relatedly, in efforts by US legal scholars to foreground the role of speech acts in the marginalization of vulnerable groups. While the origins of this debate can be traced to at least the early 1920s, it gained steam in response to racist incidents on US university campuses in the 1980s, when increasing demands to regulate hate speech resulted in a debate about the value of the liberal right to freedom of expression versus the interests and freedoms of vulnerable and subordinated groups (Walker 1994). According to Heinze (2016, 15), it was after these incidents that US-based critical race theorists 'wrote the phrase "hate speech" into the legal lexicon'. Arguing from the viewpoint of critical race theory, this group of scholars stressed that hate speech formed part of a broader cultural discourse that reproduced racism and white supremacy. Hate speech thus became viewed as the flip side of the liberal right to freedom of expression because it could undermine equal citizenship (Murphy 1995; Heinze 2016). Indeed, in their introduction to the 1993 edited volume *Words that Wound*, Lawrence, Matsuda, Delgado, and Crenshaw equated the term 'hate speech' with 'racially abusive speech' and 'subordinating speech' (1993, 12) and argued that regulations against hate speech were necessary to 'protect the rights of those who have been and continue to be denied access to the full benefits of citizenship in the United States', including the Afro-American population, women, and sexual minorities. Importantly, in their view, the struggle against hate speech had to be closely linked to the efforts to 'confront and oppose dominant societal and institutional forces that maintained the structures of racism' (Lawrence et al. 1993, 3). This entailed 'examining the relationships between naming and reality, knowledge and power', with the effect that their 'work presented racism not as isolated instances of conscious bigoted decision-making (sic) or prejudiced practice, but as larger, systemic, structural and cultural, as deeply psychologically and socially engrained' (Lawrence et al. 1993, 5).

This holistic problematization of hate speech stressed the connection between speech acts and historically grown, as well as culturally and institutionally engrained, racism and violence and emphasized the need to go beyond a mere concern with language. Racism then becomes understood as 'more than race hatred and prejudice' because it is 'the structural subordination of a group based on an idea of racial inferiority' (Matsuda 1993, 36). In this view, hate speech is harmful because it is a 'mechanism of subordination, reinforcing a historical vertical relationship' (Matsuda 1993, 36), which made 'fighting words' (i.e. hate speech) meaningful as part of the struggle against racism that had to involve 'all levels' (Matsuda 1993, 38). Thus, the fight against hate speech had to be part of a larger struggle 'for a vision of society where the substance of freedom is freedom from degradation, humiliation, battering, starvation, homelessness, hopelessness, and other forms of violence to the person that deny one's full humanity' (Lawrence et al. 1993, 15). In this reading, hate speech emerges as but one element of the larger social, political, and economic apparatus that reproduces racism and, by extension, violence. Problematically, often, these aspects only form the silent background of the hate speech created, shared, and consumed on social media. They are also the most difficult to tackle through technology-focused response strategies, as I will demonstrate in the next section.

All this suggests that online speech is closely linked to the deeper drivers that usually have political, economic, social, and cultural dimensions. For instance, in

South Sudan, harmful speech circulating on social media plays an important role in the maintenance of inter-communal violence over communal land or cattle raiding between various pastoralist communities. These conflicts seemingly take place at the local level and are perceived to be of an 'ethnic' nature because they involve groups identifying with different 'tribes', 'clans', or 'sections'. However, they are commonly instigated and sustained by national political elites that mobilize their constituents to maintain a political following, punish political opponents, and undertake armed campaigns meant to secure access to land and natural resources (Dawkins and Gaere 2014; Wild et al. 2018). They form part of an extractive and exploitative governance arrangement in which the country's economic elites compete over control of natural resources and political power through the maintenance of feuding clientelist networks (Pinaud 2014). In the mobilization for such violence, the spread of disinformation plays a pivotal role—especially about impending attacks on other communities, since these cause a security dilemma that renders preventative attacks seemingly rational.

For instance, many people with access to mobile phones with internet connectivity join community-based WhatsApp groups. The inclusion criteria for these groups are typically a shared geographic origin, such as from a specific village or county, or the association with a specific section or clan of an ethnic group. This means these groups are also open to community members based in urban areas or abroad, and it is these members who yield an increasing influence over conflict dynamics by spreading hate speech or rumours about impending attacks that often turn out to be false. There are also many Facebook groups and pages where harmful content is shared in ways that create hatred against specific communities.[4] As one social media analyst described it:

> There are a lot of personalities, especially politicians, who use these various social media platforms to promote their agendas . . . So, mostly South Sudanese in the diaspora are using social media to incite people back home. There are scenarios whereby someone goes live in Australia, . . . and people end up fighting in the village—a communal fight in Lake State or in Jonglei because it was triggered by someone abroad. This is because a lot of people who are illiterate believe that these elites share content that is accurate.[5]

All this suggests that harmful content produced by political elites at home or abroad can instantly be picked up by social media users in remote rural locations, which then plays an instrumental role in the mobilization for violence.

The International Response to Online Hate Speech

The role of WhatsApp groups in mobilizing for inter-communal violence in South Sudan demonstrates a more general point. Harmful speech today is predominantly

[4] Interview 56.
[5] Interview 15.

something that manifests itself online and on social media. Nonetheless, leading international organizations employ understandings and definitions of hate speech and disinformation that are more broadly applicable. For instance, the 2019 UN Strategy and Plan of Action on Hate Speech defines it as 'any kind of communication in speech, writing or behaviour, that attacks or uses pejorative or discriminatory language with reference to a person or a group on the basis of who they are, in other words, based on their religion, ethnicity, nationality, race, colour, descent, gender or other identity factor' (UN Secretary-General 2019, 9). This broad definition does not specify how hate speech is communicated. Consequently, it also doesn't emphasize the prevalence of hate speech in online spaces or social media. However, the UN's dedicated website for the Hate Speech Plan of Action reveals a strong focus on the digital manifestations of hate, for instance, when stating that 'the growth of hateful content online has been coupled with the rise of easily shareable disinformation enabled by digital tools' (United Nations 2023b). It also stressed that the unprecedented challenges of the scale and speed of online hate dissemination make 'understanding and monitoring hate speech across diverse online communities and platforms . . . key to shaping new responses' (United Nations 2023b).

Many leading peacebuilding organizations are indeed now focusing their efforts on countering harmful speech online, but at the same time, they often maintain a diversity of objectives that reflect the various problematizations of hate speech discussed in the previous section. This can be seen in international policy, such as the guidance document for the implementation of the UN's Plan of Action (UN Secretary-General 2020, 3), which emphasizes the potential of hate speech for 'undermining peace, stability, sustainable development and the fulfilment of human rights for all'. Moreover, in 2023, the UN Security Council issued a resolution recognizing 'that hate speech . . . can contribute to driving the outbreak, escalation and recurrence of conflict, and undermine initiatives to address root causes of conflict . . . and peacebuilding efforts' (UNSC 2023, 3). International policy thus reflects concerns with both the immediate harms that harmful speech can bring, such as instigating violence against minorities, as well as its connection to the intermediate harms, such as racism and socio-economic injustice.

Based on this broad take on how speech can cause harm, implementing organizations seem to adjust their problem framings freely, depending on the specific targets. For instance, documents by the UNDP, the UN Peacebuilding Fund (UNPF), or international non-governmental organizations (INGOs) such as Search for Common Ground claim that hate speech drives 'violent extremism' (UNDP 2022, 4), fosters 'violence in election and transition periods' (UNU CPR 2024, 6), and underpins 'sexism and misogyny' (UNDP and SfCG 2021, 13). This suggests that, at least on a discursive level, peacebuilding actors often link the online manifestations of harmful speech to socio-economic context factors that have often preceded it and continue to prevail somewhat independently from the internet and social media. Furthermore, the UN Secretary General's (2022, 1) report acknowledges that 'the devastating effect of hatred is sadly nothing new' and references the German Nazi magazine 'Der Stürmer', which served to dehumanize the European Jewish population, or 'Radio Télévision Libre des Mille Collines (RTLM)', which contributed to the incitement to violence against the Tutsis in Rwanda (UN Office on Genocide Prevention 2023, 2). At the same time, hate speech is increasingly acknowledged as something that manifests online and is transmitted through digital data. For instance, the UN Secretary General's report acknowledges that in the digital age, 'hate speech is further enabled and amplified exponentially through digital technology' (UN Secretary-General 2020, 7), and 'social media and other platforms are exploited as platforms for bigotry' (UN Secretary-General 2019, 1). However, digital technologies and social media platforms are mainly described as catalysts of violence rather than as an underlying cause. They amplify the 'extent and magnitude' (UN Secretary-General 2020, 18) of hateful expressions because of their 'ability to spread information at a speed never seen before, reaching potentially huge audiences within a few seconds' (UN Office on Genocide Prevention 2023, 2).

There is also a broad acknowledgement that hate speech is often underpinned by more fundamental, underlying causes of conflict. For instance, the UN Strategy and Plan of Action (UN Secretary-General 2019, 2) notes that tackling hate speech 'requires a coordinated response that tackles the root causes and drivers of hate speech, as well as its impact on victims and societies more broadly'. Similarly, a report by the UN Secretary-General submitted to the United Nations General Assembly (UNGA) in 2023 highlights that 'approaches that try to "solve" disinformation without addressing the conditions conducive to its occurrence and spread will not succeed' (2022, 3). In response, UN bodies such as the Office for Genocide Prevention recommended that civil society organizations conduct 'comprehensive research on the link between the offline root causes of hatred and online manifestations of hate as well as between the online hate and the offline harm' (UN Office on Genocide Prevention 2023, 19). They were also recommended to 'identify good practices in order to design multifaceted and grounded responses to hate speech' (UN Office on Genocide Prevention 2023, 19). Similarly, the UN Secretary-General's Plan of Action calls for a commitment to address the root causes and drivers of hate speech and foster inclusive and just societies, for instance, by raising awareness for 'human rights, non-discrimination and tolerance' (UN Secretary-General 2019, 4). All this suggests that, on the face of it, international policy aimed at supporting the fight against hate speech is supportive of engaging with many of the intermediate harms that hate speech can cause, as well as the deeper causes of conflict, of which it is often only the symptom. However, as will become clear in the remainder of this section, once hate speech is treated predominantly as an online phenomenon that can be datafied, this affects how peacebuilding actors choose to respond to it. Overall, this ultimately leads to a response that appears to be rather 'virtual', in the sense that it is often only weakly connected to those parts of the reality of conflict-affected contexts that are embodied, tangible, or long-lasting.

The Datafication of Hate

It is evident that many organizations concerned with the fight against hate speech today have increasingly invested in digital approaches based on the reasoning that digital problems also require digital solutions. For instance, according to the UN Office on Genocide Prevention (2023, 2), 'just as social media provides the means for disseminating hate speech, it can also provide the tools for tackling it'. Moreover, the UN Secretary-General's (2019, 3) Strategy and Plan of Action stresses the importance of digital approaches when arguing that the 'support [of] a new generation of digital citizens, empowered to recognize, reject and stand up to hate speech' should be a guiding principle. The plan also calls on all UN entities to strengthen their capacities to monitor and analyse hate speech and 'keep up with technological innovation and encourage more research on the relationship between the misuse of the Internet and social media for spreading hate speech and the factors that drive individuals towards violence' (UN Secretary-General 2019, 4). Much suggests that this focus on digital approaches has become predominant. For instance, most interventions on countering hate speech showcased by the UN Peacebuilding Fund (UN PBF) in 2023 were concerned with the digital aspects of hate, such as enabling 'access to dynamic data and analytical tools for national stakeholders to identify and systematically track online hate speech' (2023, 3). As such, when digital tools become 'ready-to-hand' for organizations that monitor and respond to hate speech, they not only start to define the character of digital peacebuilding but also have the potential to shape the peacebuilding sector altogether, as illustrated by the United Nations University Centre for Policy Research's (2024, 55) suggestion that:

> Many of the most innovative and impactful approaches [to online counter-hate-speech]—including partnering with social media companies, use of online tracking tools, and engagement with youth influencers—could also be used in other peacebuilding programming more generally.

In the course of this policy shift, new coalitions of actors have formed, aiming to contribute to peacebuilding by reducing the prevalence of hateful online content. The coalitions include civil society organizations and their volunteer networks dedicated to the fight against social media hate speech and disinformation, as well as the content moderation teams of major social media platforms. They also include a wider group of clients, including international organizations, diplomatic missions, and foundations that support such efforts by creating demand for data on harmful speech, either through direct funding or the financing of specific analytical products or counter-speech campaigns, as will become clear further below. While the review of international policy makes clear that hate speech has indeed become a central concern for international conflict prevention and peacebuilding, many organizations that form part of these coalitions will not necessarily self-identify as peacebuilding actors. For instance, initiatives such as DefyHateNow South Sudan or Hashtag Generation in Sri Lanka are often run by young digital natives who are first and foremost occupied with promoting a healthy or civil social media space.

While these organizations may reference 'peace' in their mandate or mission statement and may even see their role as contributing to peacebuilding efforts, they often do not operate based on clearly formulated theories of change or understandings of peace like more conventional peacebuilding organizations do. Instead, they aim to build peace by detecting and countering harmful speech online. For instance, DefyHateNow South Sudan's leadership described the organization's vision as 'bringing lasting peace among the communities' through the mitigation of hate speech, misinformation, and incitement of violence[6]. The countering of harmful speech on social media would contribute to this goal because of a 'connection between offline and online communication'.[7] While thus taking quite a narrow approach to building peace, the assumed causal link between social media spaces and offline behaviour enables such initiatives to tap into funding sources and other donor support that would otherwise be spent on conventional peacebuilding activities. This includes funding from bilateral development organizations or the UN PBF (EU Commission 2023; UNU CPR 2024). For instance, between 2017 and 2022, the UN PBF invested 58.2 million USD in a total of twenty-four projects aimed at countering hate speech (UNU CPR 2024, 49). The respective reporting on such projects also makes clear why digital approaches to hate speech countering receive increasing attention and funding from international donor communities: digital hate speech detection and deletion is easily quantifiable, something which resonates well with the requirement for peacebuilding projects to be more 'evidence-based' and to demonstrate the reach and impact of their activities (Bell 2024). An example of this is provided below:

> From 2020 to 2021, in Côte d'Ivoire, more than 2673 pieces of false information on social media networks were reported by newly trained young bloggers through the use of technological tools which contributed to an overall reduction of inflammatory discourse on social media, according to the final evaluation. (UNU CPR 2024, 50)

Furthermore, social media platforms such as Facebook directly contribute to the funding of local initiatives such as DefyHateNow to compensate for their time spent on social media monitoring and reporting to them, as well as for digital literacy and public awareness campaigns.[8] All this contributes to digital peacebuilding work that is first and foremost concerned with sanitizing social media discourses from harmful speech. Importantly, rendering harmful speech a social media phenomenon and channelling the response through coalitions that involve companies that host the platforms on which apomediation takes place privileges some problematizations of harmful speech over others. Its manifestation as a digital artefact that populates the online space and spreads through individual users' sharing and forwarding practices speaks well to specific technological solutions that are suited to reducing its risk of causing immediate harm. However, as I will show, the coalitions of NGOs, platform-based content monitoring teams, and international organizations are less well-placed

[6] Interview 29, Chairman of the Board, DefyHateNow South Sudan, via videotelephony software, 8 February 2022. Interview 31, Executive Director, DefyHateNow South Sudan, via videotelephony software, 2 March 2022.

[7] Interview 31.

[8] Interview 15, staff member, DefyHateNow South Sudan, via videotelephony software, 20 September 2021.

to counter those aspects of hate speech that speak to its intermediate, socio-political, historically grown, and often structural underpinnings. This shapes how the work of building peace on social media is perceived—and how it is carried out.

As I have argued elsewhere (Hirblinger 2024), the work of fighting hate speech online is underpinned by several stances that tend to isolate harmful speech from its socio-political context by rendering it a digital object. This rendering makes datafied hate speech suitable for efforts to reduce immediate harms, but ultimately precludes efforts aimed at addressing the more intermediate harms, such as social inequality or long-standing cultural prejudices or racism. Among these stances is the tendency to view hate speech as something that can be reduced to text data. While social media is multimodal—combining different types of media, including text, pictures, visualizations, audio, and video—contemporary efforts to fight harmful speech prioritize engagements with written text. I have referred to this distinctive stance as *lexicalism* because it is rooted in the assumption that hate speech is constituted by words, can be analysed in the form of words, and can be countered with words. This does not mean that social media monitors ignore other forms of content. However, text data can be processed and analysed much more easily and efficiently. Therefore, many organizations use voice-to-text and Natural Language Processing (NLP) engines to translate relevant data into a unified text format. Once all data are available in text format, searching and detecting harmful content is fairly straightforward. The Hashtag Generation team, for instance, employs a list of keywords containing hate speech terms to monitor the Sri Lankan social media space. The more recent innovations in generative artificial intelligence (AI), especially Large Language Models (LLMs), have further contributed to the popularity of text data. DefyHateNow, Hashtag Generation, and many other organizations have also been involved in efforts to create hate speech lexica in local languages that could be used to automate the detection of hate speech through machine learning approaches. Automated detection is used directly by social media platforms such as Meta, which employ these lexica to analyse and filter content. Furthermore, peacebuilding organizations increasingly use text classification approaches for social media monitoring (Hawke et al. forthcoming; Meta 2020). Many organizations have also joined efforts to employ LLMs to further automate hate speech detection.[9] All this furthers a lexical approach to hate speech, thus reducing it to mere words.

This focus on the lexical elements of hate speech and its reduction to text data enables a systematic and efficient identification of relevant topics across many different sources. However, it necessarily leaves other dimensions of hate invisible, such as the power relations between the speakers and the communities they identify with, long-established cultural attitudes underpinning individual posts, or the structural and socio-economic factors that may serve as catalysts for hate speech. In the everyday workflow of detecting hate speech in text, none of these can easily be made visible. This is partly because harmful speech on social media is seldom elaborate and well-argued. Instead, posts on platforms such as X or Facebook tend to be short

[9] Interview 7, Staff Member, Hashtag Generation Sri Lanka, via videotelephony software, 27 May 2021.

and don't spell out the implicit assumptions that premise or justify the hateful content. It is, of course, generally possible to make the drivers of hate, such as falsely held beliefs or long-standing grievances, to mention only a couple, visible in text. Academic research, nuanced news articles, or engaging novels demonstrate this capacity of text to account for the tangible and embodied experiences, as well as the socio-political context in which hate speech emerges. However, such detailed narrations are most often absent from social media. Thus, when it comes to tackling the underlying causes of conflict, suitable digital methods often remain missing-from-hand, and this has much to do with rendering text into data, as well as the partnerships between civil society organizations and social media platforms, as I will argue in the following section.

The Virtuality of Preventing Conflict on Social Media

The tendency to strip harmful speech from its context is partly the result of the partnerships established between social media platforms and monitoring organizations. These partnerships are an excellent example of how the meanings that emerge in apomediated peacebuilding are conditioned by the relations that apomediaries enable, simply because social media platforms provide or influence the analytical methods that peacebuilding actors employ to detect harmful speech. For instance, organizations such as Hashtag Generation or DefyHateNow commonly use social media mining and analytics applications to search public social media content, and these tools are often directly provided by the platforms. This was the case with Meta's CrowdTangle (2021), which was used during the time of research for this book, but discontinued in August 2024. Its successor, Meta Content Library, offers comparable search features. There are also independent alternatives, such as the social media listening tool SentiOne (2024), yet they nonetheless rely on the platforms' collaboration in the form of access to a web interface or an Application Programming Interface (API) through which social media data can be retrieved. This means that monitoring organizations are dependent on social media platforms for accessing the harmful speech data they want to analyse, and they can often only access it through the platforms' applications that have specific design features. It is, therefore, hardly surprising that their approach to countering harmful speech also largely prioritizes the platforms' objectives, namely reducing the risk of immediate harm from harmful speech.

Applications such as CrowdTangle help produce an aggregate view of relevant public social media content, enabling the assessment of social media conversations from a bird's-eye perspective that would be impossible to achieve through manual content screening. They thus drastically reduce the human resources necessary to monitor social media.[10] However, the specific analytical features of these applications were developed primarily for market research, that is, for monitoring social media content

[10] Interview 15.

that is of particular relevance for promoting a brand or a product. They are one of the many services through which social media companies can monetize social media data, that is, by providing insights into aggregate trends in online conversations related to the item of concern. By employing the same tools, this market research logic is somewhat transplanted onto initiatives that aim to fight online hate speech—and this leads to an obsession with aggregate trends in narratives that could signal a change in public attention and interest in certain topics. Harmful speech on social media usually occurs in large quantities. It thus must be monitored, analysed, and responded to at scale. Tools like CrowdTangle make precisely that possible, by displaying trending topics or by allowing monitors to trace the popularity of specific search terms over time. Aggregating text data containing hate speech enables an efficient engagement with the macro dynamics of hate speech, for instance, by mapping topics and observing trends over time. In contrast, individual social media experiences and their backgrounds matter relatively little when trying to sell a product. It's therefore not surprising that this information isn't forthcoming when monitoring harmful speech with such tools. Admittedly, this also has to do with the platforms' privacy policies and efforts to prevent the use of private social media data for political manipulation after the Cambridge Analytica scandal caused a massive public outcry in 2018 (Cadwalladr and Graham-Harrison 2018). Consequently, tools like CrowdTangle do not screen content posted on private pages or closed groups and monitors cannot trace any comments or reactions attached to public posts back to individual users. This means that harmful content becomes disconnected from the social media user who authored it and, thus, the context from which it emerged. Instead, these applications enable an aggregate analysis.

Using these applications, social media analysts will select, download, and store relevant social media data, which is a necessary step for further analysis. In addition, many teams also search for hate speech employing personal social media accounts, and they often engage networks of volunteers to collect relevant posts. For instance, DefyHateNow works with many journalists who report harmful speech to the core team via text messages or emails. When human monitors browse and read social media content just like other social media users do, they can potentially obtain a situated view of individual conversations, observing posts and replies in context, making sense of who writes and who responds. They could even directly engage with other social media users. However, such actual encounters do not commonly take place. They would be time-intensive and they are often non-commensurate with how the monitors perceive their work, namely as one of passively observing the conversation. Instead, monitors tend to engage with the content at a distance, employing several analytical metrics to assess the immediate threat that the post may cause, that is, its potential to cause violence, such as its virality, believability, and ability to cause physical harm, and to then act on content that meets such criteria.[11]

Instead of engaging with the deeper, underlying dimensions of harmful speech, many monitoring initiatives are content with ensuring that it will be deleted. They

[11] Interview 29.

simply report posts with harmful speech to the content management teams of the relevant social media platforms, to which they often maintain a direct communication channel. DefyHateNow and Hashtag Generation, for instance, are part of Facebook's 'trusted partner network'. The platform's content moderation teams will review the content they flag on the platform with priority, take it down if it violates the platform's community standards, and provide feedback. The simple logic behind this approach is that immediate harm from hateful speech can be avoided if the content can't circulate. However, flagging and deleting potentially harmful content also means that the engagement with harmful speech remains at a superficial level. The content removal protocols underpinning these reporting and deletion efforts tend to be tailored to the platform's community standards that guide content moderation. This means that the outcomes of the organization's monitoring and reporting efforts will be conditioned by the platform's views of what constitutes harmful speech and why it is harmful. This, in turn, means that the community standards implemented by a content moderation team, not the considerations or policies of local organizations, primarily determine what preventative measures are carried out. Removing hateful content will curb the most visible expressions of harmful speech while somewhat severing it from its other underlying dimensions, such as long-established patterns of prejudice or discrimination, deeply ingrained socio-economic inequality, or histories of trauma, to mention only a few. As the harmful content is reported and deleted, it vanishes from the platform surface, and with it, any digitally visible trace of the deeper dimensions of harmful speech. Deleting content may prevent or mitigate risks associated with the immediate harms that social media hate speech may cause, but it only treats the direct manifestations symptomatic of hate.

Moreover, the partnerships aimed at removing harmful content seem to be only partially efficient at times. Several social media monitors I spoke to reported that the platform's content management teams often respond very slowly to their requests or don't follow the local team's assessments.[12] One reason for this may be that content moderation teams tend to be chronically understaffed, especially in countries that produce little revenue from advertising, and this is undoubtedly the case for poor countries at war (Elswah 2024). Social media companies neglect marginal markets and allocate insufficient resources to them, which is evident in their limited physical presence in most countries of the Global South, particularly on the African continent, where Facebook only opened its first office in 2015 in South Africa, followed by a Nigerian office in 2021 (De Gregorio and Stremlau 2023). Such limited capacities lead to limited responsiveness. For example, when Facebook posts fuelled Azerbaijan's war with Armenia over the disputed territory of Nagorno-Karabakh in 2020, one whistleblower reported that 'Azerbaijan fell into a gap: neither the Eastern European nor the Middle Eastern policy teams claimed responsibility for it, and no operations staff—either full-time or contract—spoke Azeri' (Wong and Harding 2021).

Among the other options for responding to hate speech are digital literacy campaigns and training aimed at increasing the personal ability of users to detect and

[12] Interview 56, Staff Member, DefyHateNow South Sudan, Juba, 30 November 2022.

counter harmful speech while they click, scroll and swipe through their social media channels. Such efforts are also encouraged by international policy. For instance, the UN Special Rapporteur on the promotion and protection of the right to freedom of opinion and expression during armed conflicts (2022, 26) stressed that it is essential to 'build social resilience against information manipulation by empowering rights holders and civil society'. Therefore, 'more attention should be given in fragile situations to media information and digital literacy' as well as to 'community-based fact-checking, and education programmes'. To this end, many organizations dedicated to fighting harmful speech have organized webinars and online training sessions, and produced educational content to increase the 'capacity' of ordinary citizens to abstain from consuming and (re)-producing it.[13] They also publicly debunk harmful speech containing disinformation to promote broad-based awareness of the prevalence and dangers of harmful speech and work with social media influencers to increase their reach. The idea is to spread 'resources' among volunteer social media activists, which they can then employ 'on their own' to reduce the risk of violence, as one organizer of such capacity-building training sessions put it.[14]

Such efforts aim to equip some nodal points of the apomediation network with the necessary awareness and skill to interrupt the circulation of understandings of conflict that can lead to further violence—in other words, harmful speech. By increasing the digital literacy of ordinary internet users, these initiatives thus contribute to enhancing the adaptability of some parts of the digi-political playing field. Nonetheless, the kinds of engagement with harmful content that I observed in the course of my research often only had a mitigating effect by helping individual users to critically engage with content, stop them from forwarding it, and report it to the content management teams of the platform. Some initiatives also aim to equip their volunteers with the skills necessary to post 'counter speech'. Critically, such strategies confine the fight against harmful speech to the online realm. Moreover, as will become clear in the next section, such efforts struggle particularly in contexts where the depth and reach of the digital playing field are limited, such as in South Sudan, or when harmful speech is deeply rooted in historically grown animosities. One can thus equate such capacity-building efforts to spreading 'social media hygiene', as some observers have called it (Rainie 2023). However, in the same way that hand washing during the COVID-19 pandemic did not undo the original transmission of the virus to humans, for which many scientists blame underlying causes such as environmental destruction or the lack of sufficient preparedness or capacity in many health systems, digital hygiene measures will not eliminate the underlying causes of the pandemic that is social media hate speech, but merely temporarily inhibit its spread.

Initiatives aimed at countering harmful speech online often shy away from directly confronting the politics that cause such conflict and violence.[15] Surprisingly, at first, fact-checks published by such initiatives in South Sudan and Sri Lanka primarily

[13] Interview 31.
[14] Interview 31.
[15] See also Chapter 6 for a more detailed discussion of the politics of which harmful speech forms an essential part, which I describe as a form of 'digital divide and rule'.

focus on mundane topics or events that usually do not have much political significance. During the COVID-19 pandemic, for instance, many organizations focused on debunking misinformation regarding the spread of the Coronavirus. Such efforts are laudable from a public health perspective, responded to an immediate need, and were, in many cases, backed up by donor funding.[16] However, since the pandemic, organizations such as DefyHateNow in South Sudan have primarily published fact-checks to correct gossip that is of minor political importance for conflict and violence in South Sudan or apolitical altogether, such as about local authorities banning women from wearing miniskirts, the managing director of a state-owned oil company owning a Norwegian Hotel chain, or WhatsApp being discontinued. This demonstrates a wider trend, namely, that while many fact-checking organizations receive funding earmarked for peacebuilding activities, their public countering efforts often focus on posts with social and cultural content instead of political or security issues. One could argue that there is a theory of change behind this approach. While engaging with content that won't cause much public controversy, the fact checks are intended as learning examples that can sensitize social media users about the prevalence of false information and provide a resource to foster digital literacy and critical thinking skills. Such efforts can make the playing field on which apomediation takes place more adaptable. Using this indirect approach, fact-checkers will be less likely to be subjected to political repression than they would be when calling out powerful political leaders. However, this also means that the harmful speech that is the most impactful will often remain unchecked, including speech from members of the incumbent regime.

Even if one were to attempt it, debunking political speech is often very challenging. This is because it tends to be formulated in a way that renders countermeasures focused on facts and evidence ineffective. One reason why analysts may end up not fact-checking political leaders is that they miss the 'data', that is, information and evidence to debunk the claim. For instance, an analyst operating in South Sudan mentioned that speech of high-level government officials often does not even contain much information that could be fact-checked. As he put it, 'most of the claims that tend to fuel conflict are not backed up by data', and in consequence, it is 'very hard for a fact-checking organization to be able to undertake research to debunk that'.[17] This was also echoed by analysts in Sri Lanka, who lamented that there are often 'a lot of grey areas' in politicians' statements.[18] When engaging in the political struggle necessary to disarm harmful speech and its underlying drivers, efforts to fight political hate speech on social media may end up being hampered by their evidence-based approach. Where evidence and data are missing, and political speech conveys beliefs and opinions rather than facts, fact-checking initiatives won't be able to engage with harmful speech. If they attempted to do so, they would arguably lose the veil of neutrality that comes with the evidence-based approach. In authoritarian settings,

[16] Interview 29.
[17] Interview 56.
[18] Interview 27, Staff Member, Hashtag Generation, Colombo, 20 January 2022.

remaining neutral and data-driven appears to be a strategy that allows these organizations to remain on relatively good terms with the incumbent regime. The fight against harmful speech would, of course, be possible without having a sufficient evidence base by instead appealing to morality or theory (in the context of the struggle against racialized oppression, for instance). Therefore, the evidence-based approach to harmful speech risks justifying complacency—and thus ignoring one of the most important underlying causes of many violent conflicts, namely authoritarianism.

This is not to understate what's at stake. Engaging with the deeper politics that fuel harmful speech clearly comes with serious risks. In cases where political leaders engage in harmful speech, publicly debunking false information can quickly have negative consequences for those who do it. Civil society organizations operating in such environments must often tread carefully to avoid facing reprisals, which means they will avoid directly confronting powerful actors such as government officials or political leaders. This also pertains to denouncing hate speech or debunking the disinformation they share. When asked why they wouldn't fact-check high-level government officials, in the same way in which Donald Trump's posts received scrutiny in the American media during his first tenure, for instance, one analyst replied that:

> I think it would get too political. We would be just shut down; we would no longer be operating. . . . When you're trying to go too political and too deep, it backfires . . . that's why people don't go too much into political issues.[19]

Calling out high-level government officials may still be acceptable at times when the focus is on technocratic topics, such as education policy.[20] However, in practice, many monitors shy away from engaging with more politically sensitive posts that would question the political support for incumbent regimes in a more profound manner. The efforts of authoritarian governments to intimidate journalists and media organizations and obstruct their work serve as a worrying example of what could happen to fact-checkers if they target politically sensitive issues. Moreover, as one analyst noted, local informants of the fact-checking teams often worry about the potential negative repercussions of sharing information that could upset influential leaders and thus decide not to share relevant information with the fact-checking team. However, given the evidence-based approach, harmful speech cannot be debunked without sufficient data and no sources to cite.[21] Sometimes, this means ignoring some of the most delicate and potentially detrimental social media events.

The activists I spoke to also often aim to avoid being associated with one specific community or party to a conflict, regardless of whether the harmful speech is related to a local case of cattle raiding or a national insurgency. For instance, monitors in Sri Lanka tend not to focus their work on content from politicians or political actors because that comes with the risk of being perceived as biased. One monitor described the dilemma in the following words:

[19] Interview 18, Staff Member, DefyHateNow South Sudan, Juba, 8 December 2021.
[20] Interview 29.
[21] Interview 39, Staff Member, DefyHateNow South Sudan, Juba, 23 November 2021.

> We are trying to be apolitical about it. Not apolitical, but not partisan. Because, to be very honest, when you look at the disinformation narratives, the target is mostly the NPP [National People's Power] or the JVP [Janatha Vimukthi Peramuna] [and they are] mostly coming from the ruling party. So, if we really pick each narrative, we are going against the ruling party.[22]

The NPP and JVP are leftist opposition alliances that have been vocal challengers of the Rajapaksa's Sinhala-Buddhist ideology. Unsurprisingly, they have turned into two of the main victims of online hate campaigns. Under the current regime, as the analyst argued, there is a threat against civil society organizations that are perceived as partisan, which is well documented through examples of surveillance and censorship, amongst others. Repeatedly calling out the ruling party for the harmful speech its members share online could ultimately mean that 'you might not be able to work'.[23] Fact-checking initiatives based in settings where civil society and media freedoms are better protected, such as in Northern Ireland, may not perceive a similar risk. Yet interestingly, those initiatives are often accused of political bias and become the target of hatred by the political parties or individuals whose information they debunk.[24] A common strategy for avoiding allegations of bias is to make sure that all efforts to counter harmful speech are based on solid evidence and to be very 'factual' and thus 'non-partisan'. However, as I will demonstrate in the next section, establishing the apomediation networks necessary to obtain sufficient information is, unsurprisingly, difficult in contexts characterized by conflict and violence.

Patchy Connectivity and Resilient Intermediaries

Fact-checking information is technically demanding and requires skills, equipment, and connectivity. Having entered into the evidence-driven paradigm, initiatives attempting to fight harmful speech will require sufficient quality data, which they regularly struggle to obtain. There are several reasons for this. In the many peace-building contexts where socio-economic deprivation contributes to armed conflict, violence, and insecurity are, of course, also likely to happen in pockets with patchy internet access (Goldsmith 2020) or during wilful internet blackouts instigated by repressive states (Gohdes 2020). This means that to obtain sufficient information to debunk false information or call out hate, monitors would require data precisely from the places where obtaining such data is the most challenging. The limited reach of the digital field, partially a consequence of extractive authoritarianism, thus makes fighting even the immediate harms of hate speech difficult at times. As one South Sudanese monitor described it: 'The challenge for us in Juba is that you can't get the exact information. Someone will be in Unity [State] and write something on their

[22] Interview 27.
[23] Interview 27.
[24] Interview 107, Managing Director, FactCheckNI, via videotelephony software, 27 June 2023.

network on social media. For us to verify exactly what is happening is very hard.'[25] A few months before the interview, there were clashes between the government, allied militias, and insurgent groups in Southern Unity State that left more than 170 civilians dead (UNMISS 2022). Unity State has been a flash point for violence over access to land and natural resources, and this violence is linked to the competition of elites at the state and national levels (De Simone 2014). Reconstructing precisely what happened required several verification missions led by UNMISS peacekeeping forces, with a detailed picture of the violence only emerging several months after it had taken place (UNMISS and OHCHR 2022). Verifying or falsifying social media content to reduce the risk that rumours and false information could lead to further violence is almost impossible in such situations.

Fact-checking initiatives cannot undertake verification efforts in the same way that peace operations or ceasefire monitoring missions do. Instead, they tend to rely on a network of volunteer informants that enables a crowd-based approach. The challenge of this approach is that merely obtaining *some* information is not enough. Information must be obtained from various trusted sources and must be triangulated before a specific factual claim can be verified or falsified. However, as the case of the violence in Southern Unity State illustrates, relevant events often occur in remote, rural locations, from which obtaining and triangulating quality information is difficult. At the same time, the amount of disinformation is often so high that ordinary internet users have difficulty telling the truth from fabricated information.[26] Monitoring initiatives thus often aim to collaborate with professionals or experts who may have a better view of the situation, such as journalists, NGO workers, or religious leaders. Given their position in society, these individuals often maintain more connections than others within the apomediation network, and they often have the skills to assess the trustworthiness of various claims. However, even these informants are often difficult to tap into in places like South Sudan, where civil society organizations tend to be spread extremely thinly in rural areas. This often makes the triangulation of information, namely, obtaining the same information from at least three trusted sources, a highly challenging task.

In such challenging contexts, South Sudanese monitors often fall back on members of local governments, who are among the few actors that tend to have a good overview of developments in remote and difficult to reach locations. Likewise, development and peacebuilding organizations tend to seek information about, and access to, rural populations via state and local authorities. Fact-checkers tend to follow a similar path. For instance, a social media analyst for DefyHateNow described how he would reach out to the state governor or county commissioners to obtain reliable information if there were rumours about armed clashes in remote states like Jonglei. In the apomediation networks that spread into less well-connected areas to establish which information is true and which is false, local government officials retain

[25] Interview 39.
[26] Interview 56.

an important role as providers of trusted information.[27] Traditional authorities in particular, often referred to as 'chiefs', tend to be viewed by the autocratic regime of Salva Kiir as the 'eyes and ears' of the government because they provide valuable information to officials located at higher levels of government (Sudan Tribune 2013). However, these local authorities usually operate under the helm of government officials and political elites at higher levels. Government officials at the state level thus function as gatekeepers to information because monitors will contact these people 'on the top' to reach out to 'local people', as the analyst put it (Sudan Tribune 2013).

It is difficult to say with certainty to what degree the information these government officials provide is accurate. However, in many local community conflicts, they are not perceived as neutral actors, and at times, they are actively involved in inciting violence, including through content circulated on social media.[28] They also tend to play a role in the gatekeeping of wealth and the distribution of resources, especially developmental assets (Leonardi 2013; Hirblinger 2015), and they are thus often intricately involved in local struggles over power. Unsurprisingly, the monitors I interviewed stated that they had experienced limited willingness from these officials to help with fact-checking claims about ongoing inter-communal conflicts. In their words, 'we find that sometimes the government are not willing to share information about what is happening deep in the villages'.[29] In their view, the reason for this is that these potential informants aim to protect the interests of the community they identify with, and thus refuse to serve as a neutral source of information.

This approach of relying on government officials may be less problematic when the false information contains claims related to less political topics, such as public health, as was the case during the COVID-19 pandemic. However, it is bound to run into issues when the topic concerns armed conflict and insecurity. The central challenge when relying on conventional authorities, such as local government authorities, is that they do not play a neutral role in South Sudan's inter-communal conflicts. As appointed officials, they form part of a tightly knit chain of command, with loyalty and patronage that extends from transnational and national elites to the state and local levels. As such, they are integral to the clientelist networks that (re-)produce societal cleavages and mobilize local communities in their competition over South Sudan's resources, all while doing little to reduce the country's stark socio-economic injustice. Ironically, these actors don't have to do much to undermine civil society's efforts to fight harmful speech. Often, it is sufficient to simply not respond to requests for information. Because fact-checking relies on evidence, harmful speech cannot be countered where such evidence is missing. And when activists react to hate speech without sufficient evidence, they quickly risk being portrayed as biased. Unsurpisingly, they commonly decide not to publish a response.[30]

Taken altogether, the reasons why debunking disinformation in such cases is so tricky often relate to the underlying causes of conflict, namely an authoritarian style

[27] Interview 39.
[28] Interview 45, Staff member, DefyHateNow South Sudan, Juba, 25 November 2022.
[29] Interview 39.
[30] Interview 39.

of government, socio-economic inequality, and marginalization. Efforts by the political elite and diaspora groups to spread hate speech that triggers community tensions and violence at the local level ultimately also help with maintaining clientelist systems of resource exploitation that stand in the way of a more equal and redistributive economic development. Moreover, elite entrepreneurs benefit from the relative scarcity of digital infrastructures and patchy digital access, making triangulation and verification of information difficult. The spreading of harmful speech in patchy social media networks helps to control rural populations, foster ethnic resentments, and, when beneficial, mobilize for inter-communal violence that ultimately helps to maintain a rentier system in which political elites can exploit the country's resources. The low levels of development, especially in rural areas, that come with this form of flagrant authoritarianism reproduce the conditions under which harmful speech can flourish, namely limited digital access and literacy.

The Deep Past of Harmful Speech

There is a further reason why countering harmful social media speech through monitoring, deleting, and debunking hate and false information is an ineffective peacebuilding strategy. Behind the harmful social media content often lie years, if not decades or centuries, of conflict and this 'deep past' continues to yield considerable influence on contemporary conflicts. Moreover, attached to this past are often competing historical narratives that cannot simply be disarmed or reconciled through the sharing of better information or evidence—partly because coming by this evidence is often difficult for social media activists and partly because facts and evidence are by no means a silver bullet against hatred and violence. One of the most prominent topics that fuel community tensions in Sri Lanka is the various religious groups' heritage and ownership of historical temple sites, which emerge because Sinhala-Buddhist representatives lay claim to sites historically used by various communities. Claiming such sites as Buddhist temples contributes to efforts of writing Sri Lanka's past as one in which Buddhism played the predominant role, casting aside the influence of a diversity of Tamil, Hindu, and Muslim cultures and their relevance for contemporary Sri Lanka.

One of these sites is referred to by the Sinhala population as *Kurundi Viharaya* and by Tamils as *Kurunthoor Malai*. It is a temple site in the Northern Province of Sri Lanka, which was long inaccessible to Sinhala communities due to war and violence. The site is at the heart of the Mullaitivu District, which was for several decades at the centre of the insurgency of the Liberation Tigers of Tamil Eelam (LTTE) and widely considered the Tamil homeland. Local Tamil communities continued to worship in the ruins during this time. From an archaeological point of view, the site is a good example of Sri Lanka's multi-faith heritage, which is often difficult to entangle due to a past where group identities were less well-defined and much more fluid. In the years since the formal end of the war, the site has become accessible again and Buddhist-Sinhala groups have started to demand that the Buddhist heritage of these sites be preserved. Sinhala nationalist politicians have created considerable media attention for their narrative while silencing histories that stressed the locus of diversity and co-existence that these sites constituted in pre-war Sri Lanka. Controversy around the site was sparked on social media, following the creation of a Presidential Task Force for Heritage Management in 2020. The task force was led by a Buddhist monk and lacked representation from other groups. A couple of years later, the task force began excavating and restoring works at the site, documenting the archaeological evidence of century-old Buddhist culture to undermine narratives of the area as the Tamil homeland and claim several hundreds of acres of land around the temple. This caused Tamil leaders to bemoan the project as an effort to 'Sinhalize' the region (The Economist 2023) and the critical online outlet *Groundviews* described it as an effort to weaponize archaeology to rewrite history (Gunasekara 2023). Indeed, as the issue received increasing public attention, social media monitors observed a sharp uptick in harmful speech related to the temple site.[31]

This case demonstrates clearly that contestations over the cultural heritage of these sites also have implications for contemporary Sri Lankan politics, and they relate to some of the most polarizing issues between the country's Sinhala majority and the Tamil minority, namely the devolution of power to sub-national government, inter alia, regarding the administration of land rights. Tamil grievances have historically formed around what the community perceives as the illegitimate appropriation of their indigenous lands by the Sinhala majority, and these grievances persist in the face of many large-scale agricultural and development projects that have been initiated since the formal end of the war in 2008, often with the involvement of the country's military (Goodhand 2010; Åkebo and Bastian 2020). For several decades, Tamil claims to self-government have been tied to the thirteenth amendment of Sri Lanka's constitution, which was passed by the parliament in 1987 as a result of an agreement between India and Sri Lanka but was never fully implemented. The amendment included a broad range of measures to devolve power, including through the creation of Province Councils, which would, amongst others, have power over land questions (Verité Research 2013). One of the sections that remains to be implemented concerns the devolution of power. For some segments of the Sinhala-Nationalist political elite, this would be the same as accepting the Tamils' claim to a homeland in North-Eastern Sri Lanka (Seoighe 2016).

Given this context, when social media content frames historical sites as Buddhist temples, they contribute to a narrative that undermines the Tamil community's right to land, self-government, and their demands to be viewed as indigenous to North-Eastern Sri Lanka. They also strengthen the portrayal of Sri Lanka as a Sinhala-Buddhist state instead of a multicultural entity in which different groups have historically co-existed and intermingled, and where, in fact, telling one group from the other was often not straightforward. Ultimately, social media controversies around these sites can make violence more likely because they question the legitimate interests and rights of minority groups. An example of this is the YouTube page of a

[31] Interview 122, Staff, Hashtag Generation, via videotelephony software, 21 June 2024.

nationalist TV channel and known media ally of the Rajapakse regime that discussed a fallout between (now former) President Ranil Wickremesinghe and Director General of Archaeology Professor Anura Manatunga. The conflict was about the extent to which the department could claim land around the temple site. The following user's comment encapsulates well how most of the discussion centres on Tamils as destroyers of Sinhala heritage, clearly undermining their rights to land:

> Mullativu is the least populated district. Why do they need land there? They aim to obscure the historical presence of Sinhala Buddhists in the north by acquiring land and burying archaeological evidence. Ranil [Wickaramsinghe] and the Tamils cannot assert the north as a Tamil homeland when Buddhist shrines from ancient times are discovered there. (Ada Derana 2023)

Such issues also lead to heated social media debates that are rife with abuse and have, at times, spiralled into offline confrontations and actual acts of violence. For instance, in the case of the Kurundi site, in August 2023, a ceremony planned at the temple by local Hindu farmers was disrupted by Sinhala protesters who had been mobilized from the Southern parts of the country, and the police had to intervene to prevent violent clashes. In similar cases, such mobilizations have led to riots and violence against minority groups.[32] The challenge with countering harmful speech related to such historical sites is that the ambiguities and complexities of Sri Lanka's cultural heritage aren't easily resolved, and establishing what evidence and information exists is certainly not more straightforward on social media. In fact, the opposite is often the case. Social media conversations tend to be characterized by brevity, not only in terms of the posted content but also in terms of the time that users will spend reading and responding to it. This makes nuanced elaborations of online conversations very difficult (Jaidka et al. 2019). To fact-check claims about cultural heritage sites, social media monitors would also have to have access to expert knowledge. Yet, teams usually do not have the necessary knowledge or sufficient access to experts to engage with the claims, nor can they invest the requisite time.

Ambiguity and Myth in Harmful Speech

Even if initiatives aimed at countering social media harmful speech were able to rely on experts, it is by no means clear that this would make a difference. In deeply polarized contexts such as Sri Lanka, expert opinions are not readily accepted as objective authoritative voices. Commissions that could gain impartial credibility because they act scientifically and independently do not exist. Even when the members of such committees are carefully selected to represent different constituencies and interest groups, allegations of bias tend to spread quickly on social media. For instance, observers who do not identify with the Buddhist-Sinhala majority often perceive the

[32] Interview 122.

commissions and departments dealing with Sri Lanka's historical sites as catering to the majority's interests. However, experts critical of the government's cultural agenda are likewise quickly alleged to be biased—and their scientific findings are discounted as mere 'opinions'. In fact, the challenge appears to be not so much about the hard facts—hardly anyone would contest that sites like Kurundi exhibit traces of an ancient Buddhist culture, such as inscriptions—but more about how these facts are interpreted and politically exploited. A prominent scholar who regularly engages in these controversies put it like this:

> 'Many issues cannot be addressed through fact-checking exercises because most controversies were related to subtle interpretations. [They were] not about facts'.

For instance, when asked how to characterize Kurundi, you could 'come up with hundreds of interpretations ... If you were given the task of coming up with an objective account, by well-qualified historians, it would be very doubtful if you get one interpretation.'[33]

Thus, when experts aim to present facts and evidence on social media, comments and replies to their posts tend to be promptly filled with allegations of bias, alternative interpretations, and hateful language. The case of another historical site in Sri Lanka's hilly central region, Kuragala, proves instructive in this regard. For centuries, this site has been associated with Buddhist meditation practices, Hindu worship, and Islamic Sufi traditions. In contemporary Sri Lanka, however, Kuragala is contested by Sinhala-Buddhist and Muslim communities, with the Sinhalese accusing the Muslim pilgrims of vandalizing and appropriating a Buddhist site as theirs (McKinley and Xavier 2023). In 2023, a lion statue, a symbol of Sinhala supremacy, was placed on top of Kuragala as a part of a project entitled 'Yali Pibedena Kuragala' (Reviving Kuragala), which prompted Muslim communities to accuse the Sinhalese of erasing their heritage (Tamil Guardian 2022). In response, a former Secretary to the Ministry of National Integration, Reconciliation and Official Languages, laid out evidence to document the Muslim minority narrative on the site on X and called for an inclusive dialogue. One illustrative comment he received was as follows:

> All of the evidence you have posted is just fabricated information [sic]. We are not boddy [sic] idiots to believe those fake documents and evidence. Just keep those on your chair and sit down on it.... You can do anything but nobody can erase the history and hide the truth by destroying evidence. Definitely the truth will come out by the time [sic]! (Moulana 2022)

Efforts to promote an evidence-based conversation on social media are further complicated by the fact that much of the Sinhala-Nationalist narrative rests on the *Mahavamsa*, the Pali language chronicle written in the style of an epic poem (Scheible 2016). This makes scientific facts neither a prerequisite for a justifiable political

[33] Interview 121, Senior Lecturer, Colombia University, via videotelephony software, 19 June 2024.

worldview nor a sufficient means for challenging it. Instead, what is perceived as true and real becomes determined by an accepted system of beliefs that spans from the past to the present and into the future and renders certain visions of society more desirable. Expert opinions and archaeological evidence collected around sites such as the Kurundi temple may support less straightforward historical pictures, such as one where Tamil descendants who originated from the Indian sub-continent converted to Buddhism. They might also highlight the difficulty in ascertaining the social circumstances of the religious practices that unfolded at temple sites such as Kurundi, pointing to the impact of European colonial archaeologists who contributed to the reification of group identities through a racializing gaze and thus disavowing the hasty association of such ancient Buddhist cultures with contemporary Sinhala-Buddhism.[34] However, while the existing evidence could produce such a nuanced picture, the harm on social media is done through the ideological charging of facts and their appropriation by Sinhala-Buddhist ideologues: because there is evidence of a Buddhist past, the supremacy of Sinhala-Buddhist nationalism and the violation of the interests and rights of other groups becomes morally defensible.

Controversies surrounding cultural sites such as Kuragala and Kurundi dismantle the fact- and evidence-based approach to countering harmful speech that is so commonly applied because it cannot tackle the long trajectory of historically formed beliefs and convictions that manifest on social media today. Notably, the same dynamics characterize many other conflicts in which ethnic and religious identity and questions of indigeneity and belonging play a role, because social media conversations tend to intermesh not only the factual and the fictional but also what is true and what's *believed* to be true. Indeed, users who identify strongly with a religious community or with another collective identity will often have little interest in querying the evidence base of their convictions. The deep past thus provides a form of mystifying, convenient ambiguity on which group beliefs, convictions, and identity can be erected.

Even so, social media initiatives may be able to reduce the risks of immediate violence stemming from such controversies—and they commonly attempt to do so. For instance, in August 2023, Hashtag Generation's social media monitors fact-checked social media rumours that a group of Hindu worshippers had forcefully gained entry to the Kirundi temple site. Defending the claim that they didn't, their fact check quickly caused a heated discussion, in which prominent Sinhala-Buddhist social media activists accused the monitors of bias, of putting 'Sri Lanka's national heritage at risk', and for 'being paid by the extremists' (Hashtag Generation 2023). Other social media users indirectly called out Hashtag's fire-fighting approach, which does not directly engage with the deeper context on which these social media contestations are premised, by pointing out that 'It's interesting that they forget archaeology while fact-checking at an archaeological site' (Hashtag Generation 2023). Such claims, of course, require that one implicitly accepts the Sinhala-Buddhist narrative about the

[34] Interview 121.

temple site in the first place—and they further entrench it. In polarized social media landscapes such as that of Sri Lanka, even fact-checking falsifiable information about contemporary events opens up opportunities to reproduce ideologically charged historical narratives that do not require much empirical truth to thrive. Ultimately, the back and forth about facts and evidence, mixed with morally charged commentary and hate speech, risks shoring up the support for extremist positions instead of reducing it. This is why politicians aiming to secure a seat in government or parliament commonly engage in such controversies to increase their following.[35]

An alternative way of responding to such historically charged, harmful content may be to step out of the struggle over truth and evidence altogether. This is the approach that some Sri Lankan scholars and activists have chosen to follow. As a known university lecturer and activist put it, 'Whenever I say something [on social media], a lot of people criticize me, so one of my tactics is to be as controversial as possible'. Since his audience would already know his position and views, there was no need to provide his own interpretation and back it up with facts. Instead, he chooses to be 'as controversial as possible', take a 'deconstructivist' approach and 'question . . . the problematic nature of the existing interpretations'.[36] Using provocations of humour to playfully engage with different interpretations of reality, instead of attempting to assert a single definition of what is true and real, as is currently the case with the evidence-based responses to harmful speech, may prove to be a more powerful weapon against deeply ingrained hateful ideologies such as those spread by Sinhala-Nationalist social media influencers. I will discuss this further in Chapter 6.

Conclusion

This chapter has shed light on two civil society initiatives in South Sudan and Sri Lanka that aim to support conflict prevention by monitoring, detecting, and countering harmful speech on social media. Such initiatives form part of the emerging spectrum of digital peacebuilding activities, and they also increasingly tap into—and thereby begin to replace—conventional peacebuilding activities. The mix of hate and false information that constitutes harmful social media speech is an excellent example to demonstrate the power of apomediation. The hate speech that was used to mobilize for and justify the genocidal campaigns in Nazi Germany in the 1930s and 1940s, and in Rwanda in the 1990s, was directed in a centralized, top-down manner, involving legacy media such as newspapers and the radio, enabling one-to-many communication flows. In contrast, while harmful speech on social media may result from organized campaigns, it commonly flows between many decentralized users. The specific understandings of conflict that emerge through social media hateful speech are conditioned by the interplay of meaning-making and relation-making practices. For

[35] Interview 122.
[36] Interview 121.

instance, hateful depictions of Sri Lanka's religious minorities have gained considerable traction with the emergence of the so-called 'Social Media Monks', who reach a large part of the country's Sinhala Nationalists. In South Sudan, urban elites and diaspora social media users play an active role in (re-)producing hatred between populations identifying with different ethnic groups, and they foster perceptions of insecurity and mobilize for violence. While the intergroup animosities that characterize the conflict in those two countries pre-date the internet age, there is little doubt that today, they are kept alive and fester in decentralized social media networks.

Using apomediation as a heuristic lens and asking about the interplay of relation-making and meaning in apomediaries is arguably even more insightful when we shed light on what understandings of peace—and approaches to peacebuilding—emerge in the initiatives that aim to counter harmful social media speech. As has become clear, these initiatives commonly involve coalitions of civil society organizations, social media platforms, and conventional peacebuilding actors and funders. I have shown that the technological process that is employed to monitor, detect, and counter harmful speech to some extent leads to the datafication of hate and, with it, to a response that commonly ignores the deep-seated, historically grown, and often structural causes of conflict and violence. At the same time, the particular data- and evidence-driven approach that is employed to counter harmful speech turns out to be ill-suited for contexts characterized by a digi-political playing field that is slanted and has limited reach. Debunking the falsehoods of authoritarian political leaders can be as self-defeating as efforts to triangulate rumours or misinformation when internet connectivity and digital skills are low. Finally, the chapter has also pointed to the fact that a response to harmful speech that is primarily focused on promoting facts and evidence may have little impact when conflicts are sustained by animosities that are decades—if not centuries—old, and which are grounded in myths and cultural beliefs, coupled with the desire of antagonists to view the world in one way and not another.

Contexts such as South Sudan and Sri Lanka demonstrate that when understandings of conflict are apomediated online, epistemic uncertainty can be exploited to further parochial group interests and sustain political domination. This inability to determine that something is the case can be more easily maintained in contexts with limited digital access and low levels of digital literacy, where quality information is difficult to obtain. Harmful speech transports understandings of conflict that put the blame on the Other in the antagonistic relationship. While these depictions are often paired with false information, apomediated understandings of conflict are usually sustained by ambiguity or vagueness rather than flat-out lies and disinformation. Therefore, social media initiatives that aim to counter such harmful speech with a data- and evidence-based approach tend to fall short of shifting the discourse on conflict, which means that the ideological justifications for violence remain in place. Related to this is the tendency of social media efforts to counter harmful speech to focus rather superficially on text data, while leaving the deep-seated, historically grown, systemic, and structural causes of conflict untouched. However, these very causes sustain harmful speech on social media. Thus, they fall short of contributing to a transformative approach aimed at lasting change in relations between the groups affected by harmful speech. What has become missing-from-hand in efforts to prevent conflict by countering harmful social media speech is precisely this: the capacity to work through the veil of bigotry, hate, and ambiguity to tackle the deeper, underlying causes of conflict.

4
Mitigating Conflict
Localized Early Warning and Response

The social media data cumbersomely collected by the many human monitors in Sri Lanka and other conflict-affected contexts are not only used to support the content management of social media platforms or to counter hate speech. Every manual coding of an instance of harmful speech potentially contributes to a vast global effort to collect digital signals of impending conflict and to prepare mitigating action. This process usually begins when analysts store their observations in a quantifiable data format, for instance, with the help of spreadsheet software. At that moment, every single text snippet of hate is transferred—and transformed—from a unique digital artefact that users can engage with in a living social media environment to just one piece of information in a vast data set employed to assess the risk of conflict outbreaks. The Hashtag Generation team, for instance, saves all collected hate speech instances in an online spreadsheet alongside additional analytical information, such as the source of the post, what communities or demographic populations are targeted, the geographical location of the content, and the level of engagement.[1] Behind this effort lies the idea that large amounts of social media data could help with an early response to imminent or incipient violence, be it localized inter-communal violence such as that in Kandy, Sri Lanka, in 2018, or the genocidal violence that was unleashed on the Rohingya in Myanmar in 2016–2017. This reliance on social media data to inform mitigative action is a relatively recent phenomenon.

To inform conflict mitigation efforts in Sri Lanka, the Hashtag Generation team developed several data analysis and visualization tools to share hate speech data with a broader group of interested stakeholders, including other civil society organizations (CSOs), international organizations, and diplomatic missions. The team hosts a password-protected website displaying the collected harmful speech data in a dashboard, which quantifies and aggregates all hateful content detected in a specific period, categorizes it along analytical dimensions, and visualizes it in pie charts, scatter plots, and bar diagrams. This provides a seemingly objective, descriptive, and top-level representation of online hate speech. However, the clean and neat representation of hate speech data in the online dashboard does not support any form of qualitative engagement with the actual social media content, the author, or the target group of the posts. Instead, the emphasis is on the trends that can be recounted through charts and numbers. The dashboard thus produces what a founding member

[1] Interview 116, staff, Hashtag Generation, via videotelephony software, 20 January 2023.

Digital Peacebuilding?. Andreas T. Hirblinger, Oxford University Press. © Andreas T. Hirblinger (2026).
DOI: 10.1093/9780198986553.003.0005

of Hashtag Generation described as a 'helicopter view of things' that speaks to 'almost an obsession with quantifiable metrics', which means that the organization misses out on 'lots of nuance of things that are happening locally'.[2] This is not unique to this specific project. As will become clear in the course of this chapter, many of the various conflict early warning efforts that have leveraged digital technologies in the past decades have a clear preference for this quantified and aggregable data, and they commonly accept that they are missing out on contextual nuance and indeed, at times, risk losing touch with context altogether.

The assumptions underpinning such data-driven approaches to conflict mitigation, and the challenges that emerge when they are implemented, are part of a much longer trajectory of computer-assisted early warning and response efforts that rely on large amounts of observable information, such as from news outlets. Such initiatives tend to be driven by a tech-solutionist excitement about how computational methods processing large amounts of unstructured data could help predict conflict outbreaks. Yet, many early warning initiatives come to realize that this view 'from above' needs to be complemented by a 'view from below' through the collection of structured data by localized networks of human monitors. In this chapter, I will discuss what happens when digital methods are used to datafy information about conflict, that is, to render it machine-readable and thus suitable for aggregated analysis. More specifically, I will explore how this process of datafication tends to produce a skimmed understanding of conflict, leading to a reduction in situated, context-specific knowledge that makes a comprehensive and lasting peacebuilding response difficult. It also increases the risk that basic data about unfolding conflict and violence are more quickly and effectively used by actors and state institutions whose actions are unlikely to contribute to a sustainable termination of conflict.

Lost in Datafication? Bigrams, Trigrams, and the Digital Skimming of Conflict

The large amounts of information made available through dashboards and monitoring reports are today commonly employed to assess and reduce the immediate risks of violent conflict by informing diplomatic interventions, political reform, and policy processes, as well as the design of programmes and projects. For instance, several organizations that monitor social media data in Sri Lanka share their data with the United Nations Development Programme (UNDP), where the data are used to inform the Programme's so-called 'Crisis Risk Dashboard' (CRD or 'Dashboard'). Following several incidents of inter-communal violence, the Dashboard was initiated as a UN internal conflict early warning system (SparkBlue 2021, 16: 49). Its data and regular reports are shared internally at the UNDP but also with other agencies in the UN country team, with staff in the UNDPPA, the Office of the High Commissioner for Human Rights, as well as donor countries (SparkBlue 2021). With its data being

[2] Interview 116.

shared well beyond national actors, the Dashboard is a prime example of a global effort to leverage social media data for crisis prediction and conflict early warning, among other data types, such as armed conflict event data.

As can be discerned from showcases of the tool, such as at the NYU Center on International Cooperation 2021 Conflict Early Warning/Early Action Workshop (NYU Center on International Cooperation 2021),[3] the CRD produces a highly abstracted and detached view. It was developed to monitor and analyse contextual risks by aggregating, visualizing, and overlaying a wide range of information sources, of which social media data were but one. Other types of information processed by the Dashboard stem from risk indices with socio-economic indicators derived from international organizations or national statistic offices, events databases about natural and man-made hazards, and internal information obtained from UNDP field offices. Behind the Dashboard was the idea that this combination of sources would meet the need for 'evidence-based' decision making based on 'clear and compelling data stories and effectively communicating key trends and risks' (Scognamillo and Gutschke 2021, 2). However, many initiatives that have aimed to develop early warning systems have found that these two objectives—improving contextual awareness of ongoing developments in specific locations through localized data, *and* enabling a standardized, evidence-based approach that can help with prediction—often stand in tension with one another. Indeed, according to one of the staff supporting the development of the Dashboard, the intention was to leverage data analysis to better understand 'what's happening on the ground' (NYU Center on International Cooperation 2021). The Dashboard was meant to help establish a harmonized, sustained, and methodologically rigorous UNDP-wide approach to conflict prevention. However, presenting the Sri Lanka CRD at a webinar, the UNDP's Peace and Development Advisor stressed the need for the CRD to integrate local data sources for it to yield meaningful results, but that 'locally generated data sets require funds, time and human capacity, and that's a little bit of a challenge' (SparkBlue 2021, 6: 56). What exactly did he mean by that?

Let us take a closer look. To constantly monitor Sri Lanka's social media environment, the UNDP country team worked with four CSOs to manually monitor social media narratives. Their approach was arguably similar to that of Hashtag Generation, which supplied them with the necessary methodological training (UN PBF 2023). Without the manual labour provided by the analysts working for these local organizations, the social media aspect of the CRD would not be functional (SparkBlue 2021). While efforts to automate the monitoring process through lexicon-based and machine learning approaches are underway, the UNDP's lead technologist reported serious challenges encountered with this approach due to the variety of spoken

[3] A video recording of the presentation is available at https://www.youtube.com/watch?v=GBgU1ge L6Ds, Accessed April 24, 2024. As with many of these applications, further changes and updates to the CRD's methodology and software interface are likely. While the version of the CRD I describe here is likely to be outdated at the time of publication of this book, my concern is less with the functionality of this specific technological application, and more with what it tells us about the patterns that characterize the apomediation of conflict and peace.

dialects in Sri Lanka, at least as of 2021 (NYU Center on International Cooperation 2021). While drawing on the cumbersomely gathered, coded, and aggregated data, the UNDP country team visualized social media trends using quantitative and comparative methods. This included the visualization of hate speech trends by plotting the overall number of hate speech incidents over time and attributing them as much as possible to a particular Province, as well as a target group, such as Muslims or Christians (SparkBlue 2021, 26: 00). Moreover, the developers integrated standard off-the-shelf text-mining methodology. These included 'TF-IDF', which stands for 'Term Frequency-Inverse Document Frequency', a commonly employed quantitative method that helps identify the terms that occur most often across all social media posts (Sammut and Webb 2010). Building on TF-IDF, the Dashboard displays bigrams and trigrams of the most frequent consecutive words, that is, combinations of two or three major[4] words that can be found most often across all social media posts for a given time frame. This approach is a popular text-mining method to generate insights into topics because it is feasible and relatively quick to implement. Yet, it often yields very limited analytical insights.

When employed in the Sri Lankan context, this method yielded results such as 'military campaign' or 'LTTE recruit soldier' (NYU Center on International Cooperation 2021), which were meant to inform the reader about impending risks. However, without any context knowledge, how would one be able to draw conclusions from such word co-occurrences? Would we know whether the respective social media posts discussed past, present, or future military campaigns by the LTTE, for instance? And whether the social media users were actively engaged in recruiting soldiers, if they were opposed to it, or if they discussed the pros and cons of a draft system? Whether the CRD, employing such methods, could really 'monitor a country's digital pulse', as the UNDP claimed in a public webinar in 2021, is debatable. The use of the CRD's data and regular reports was limited at the time of research. In the words of the UNDP's Peace and Development Adviser, the 'consumption of the reports is plainly limited' because 'UN people don't read', but also because the country team was not very good at 'selling' their knowledge products, which reduced the impact at least for country-level programming (SparkBlue 2021).

The reader, and many of the experts involved in the development of the CRD, could object that the methods I have discussed here have been under constant development and have, since the research was undertaken for this book, been replaced by more adequate and nuanced approaches. However, my interest is not in criticizing individual computational methods or judging the effect of individual projects or initiatives. Most of the digital technologies discussed in this book are under constant development, which makes it difficult to make any lasting assessment. However, more than individual tools in constant need of improvement, it is precisely the narrative of continuous development, prototyping, and testing that creates a pretext for the employment of *beta versions* and enables their (often negative) impacts in

[4] Using data cleaning and standard stop lists, minor words such as connectors or propositions are excluded from the count.

conflict-affected contexts. Put differently, the promise of technical improvement justifies the employment of premature solutions. Thus, my aim in this chapter is to describe larger patterns and tendencies that emerge when digital technologies and methods are employed to support conflict mitigation through conflict early warning and response systems. To this end, I will look at the long durée of early warning and response (see also Chapter 2, 'The Long Durée of Digital Peacebuilding'), to reconstruct recurring patterns in the design and application of these systems that pose recurring challenges to anyone seeking to employ them in efforts to mitigate conflict in a transformative manner.

A common part of the tech-solutionist drive towards evidence-based approaches has been the integration of monitors at the local level who can help 'ground' global early warning and thus reduce the risk that these systems fail to read the context. However, as will become clear throughout the chapter, field-based monitors are often reduced to mere data providers—and this itself partly explains the lack of an effective response. However, even when local monitors are envisioned as being more than just data providers, their role in shaping conflict early warning tends to be limited. This is because the apomediation networks through which conflict is conceptualized, monitored, and detected contribute to the datafication of conflict knowledge, which leads to a skimmed understanding of conflict. The digital methods that are readily available, such as the Tf-idf method described above, but just as many other data mining and quantification techniques, often do not lead to a more accurate or comprehensive understanding of conflict, insecurity, or violence, but to a strongly reduced one.

This is, at least partly, because of the predominant ways in which digital methods tend to be employed. The datafication of information and knowledge about conflict leads to what I suggest describing as the 'digital skimming' of the conflict context, which leads to the removal of relevant situated knowledge and the ability to employ it intuitively and without delay. Skimmed data contain only the most basic pieces of information about the conflict, which helps with the quantification and aggregation of information, and consequently, its use in predictive models. Like skimmed milk, digitally skimmed conflicts may be more easily digestible, and they don't weigh as heavily on the data analyst's mind (or belly) as conflicts in their raw form, with all their complexity. However, they also risk lacking many of the important ingredients required to understand and make sense of them—and to take meaningful action. Data skimming means that those closest to the conflict are often relegated to mere information providers; they frequently lack access to the aggregated data stored in the system's backend to take action, and often they don't have the capacity or means to respond.

As I will show in the course of this chapter, this digital skimming of conflict helps explain the functional disintegration of early warning and early response. The literature on this topic has often lamented the 'early warning–early response gap' (George and Holl 1997; Wulf and Debiel 2010) and indeed has discussed how digital methods could help reduce it (see Duursma and Karlsrud in: Hirblinger et al. 2024). However, I argue in the following that the digitalization of information about conflict, and relatedly, the trend towards aggregated data, has made the link between understanding conflict and acting on conflict much more tenuous. I will demonstrate further below that this may not only lead to a delayed or lacking response, but also to the risk that the skimmed early warning information is employed by actors that prefer top-down, hard security responses to conflicts, instead of sustainable de-escalation. The 'gap' that emerges through the datafication of mitigative approaches is thus also telling of a cleavage between the different ways in which digital technology enframes peacebuilding: while data collection for early warning may be organized through crowdsourcing approaches that contribute to localized and participatory peacebuilding, the produced data may providing mitigative responses that reproduce the power of repressive regimes that respond to threats of violence primarily through the use of force.

Tech-Solutionism Between Scientific Positivism and Decolonial Thought

The tech-solutionist fascination with advancements in computing power is by no means only a phenomenon of the contemporary artificial intelligence (AI) boom. It is as old as the first conflict early warning systems—and indeed, much suggests that technological advancements encouraged their pioneering. One of the earliest attempts (if not the earliest attempt) to leverage computing power for conflict early warning dates to the 1980s and involved a group of researchers around Kumar Rupeshinghe. In an academic essay published in 1989, Rupeshinghe claimed that 'the rapid development of new information systems has given impetus to the discussion in early warning capacity'. He continued:

> The world has indeed become a global village, with rapid communication now theoretically possible between different parts of the globe. The possibility exists to store, collect and transfer large amounts of information. Further advances in pocket radio, satellite, and emerging concepts of computer networking have expanded the possibility for early warning capacity.

His writing powerfully demonstrates how, in the late 1980s, the levels of internet connectivity and computational capacities that we have today could barely be imagined. Thus, these first adumbrations of a digitally connected world capture how a community of scientists began exploring how information from distant, conflict-affected places could be immediately consumed by analysts based in Washington, London, or Oslo. The insight that digitalization enables remote communication, allowing information about imminent or incipient conflicts to travel to distant places where it can be analysed and acted upon, underpins the architecture of early warning systems to this day. Rupesinghe was born in Sri Lanka and thus had direct exposure to the conflict there. Yet, he obtained much of his higher education in London, where he also completed his PhD at the London School of Economics. His exploration of the potential of the emerging global connectivity may thus also have been the result of his familiarity with conflict-affected contexts, as well as the approaches and considerations of researchers and professionals aiming to prevent conflict and build peace from afar.

Importantly, the tech-solutionist belief in the possibility of early warning was driven not only by innovations in Information and Communication Technologies (ICTs), statistical methods, and AI, but also by scientific positivism. Before becoming Secretary-General of International Alert in 1992, he worked for ten years at the Peace Research Institute Oslo (PRIO), where he contributed pivotally to a first set of statistical prediction methods. In one of the first academic articles ever published on conflict early warning in 1987, Rupesinghe described an 'increased humanitarian consciousness to develop information systems which could not only establish rapid and early forms of reporting but also develop ways of preventing the repeated occurrences of such disasters' (Rupesinghe 1987, 217). Referencing the work of his contemporary Akira Onishi, Rupeshinghe (1987, 219) described the prospect of a 'highly sophisticated information system' that had become possible because of the 'astounding developments of computers in the 1980s' and particularly 'progress made by global modelling'. Rupesinghe also drew on insights from the 'Correlates of War Project', conducted by David Singer at the University of Michigan (Singer and Small 1972), to nourish the idea that 'by analyzing past wars it would be possible to predict future ones' based on similarities between past and future conflicts (Rupesinghe 1987, 220). This thinking assumes that conflicts unfold because of a certain combination of factors, these factors occur regularly before conflicts erupt, and they can be modelled using statistical approaches. This conviction was pursued especially by several research projects launched at Northern American universities since the 1960s, such as the University of Maryland (Minorities at Risk (since 1993) and the Global Event Data Survey (GEDS)), the University of Southern California (World Events Interaction Survey (WEIS)), and Harvard University (Protocol for the Assessment of Nonviolent Direct Action (PANDA)). Clearly, many applied early warning projects, such as the ones in Northern Ireland and South Sudan discussed further below, have a less scientific ambition. However, they are nonetheless driven by the assumption that conflict can be known and predicted, based on a set of objective indicators that can help to monitor conflict-prone environments.

What followed was a struggle to leverage digital innovation to build tools that could be employed in practice, and as these tools became increasingly ready-to-hand, they shaped how the work of responding and mitigating conflict became understood. Many of these scientific research projects launched in the second half of the twentieth century were driven and justified by a tech-solutionist belief that innovations in computing and statistical methods could indeed solve the problem of conflict prediction. Funded by large budgets from the United States federal government, these initiatives aimed at building models of political behaviour and testing theories about the occurrence of violent and nonviolent events (Engel 2018a, 4), for which they relied on 'event-data analysis' which involved the monitoring of global news wires and the identification and coding of relevant 'events' (Krummenacher and Schmeidl

2001, 16). However, because of the high human and financial investment needed for coding the data, state funding started dwindling in the 1980s (Austin 2004, 6). At that time, other research institutes began experimenting with rule-based AI, machine learning, and Natural Language Processing (NLP) approaches that could further automate the search for observable correlations that would help to predict conflict onset (Krummenacher 2006). For instance, as early as 1984, the Austrian Research Institute for Artificial Intelligence (OFAI) started using machine learning to compute decision trees 'to find the conflict management strategy with the greatest chance of success in a new crisis situation' (Trappl 2006, 8). As Rupesinghe's writings illustrate, these trends in scientific research also inspired the development of applied conflict early warning systems. Variants of the event-data analysis methodology in particular made it into the design of systems such as the Swiss-based FAST (*Früh-Analyse von Spannungen und Tatsachenermittlung*) system or the Continental Early Warning System (CEWS) of the African Union (AU) and promised to enable the accurate and timely provision of early warning signals, as will become clear further below. However, now more than four decades after the first set of such methods were experientially integrated into early warning systems, scientists tend to be more sceptical about their ability to predict the outbreak of large-scale conflicts (Cederman and Weidmann 2017). More importantly for our concern with the apomediation of conflict and peace, they were somewhat at odds with the vision of exploiting the global expansion of ICT infrastructures to develop locally-led and decentralized early warning approaches through which a conflict could be made sense of and resolved by those who are affected by it. While promising networked approaches that would bypass established decision-making hierarchies, this ideal of apomediation can quickly be curtailed by the socio-technical context in which these systems operate, especially by digi-political playing fields that are slanted and have limited capacity to adapt to a regime's effort to control the production and use of early warning information. I will explore this tension in greater detail in the following.

While at International Alert, Rupesinghe also co-founded and chaired the Forum for Early Warning and Early Action (FEWER), where he oversaw many concrete implementation efforts worldwide. FEWER was one among several global initiatives that were created in the wake of the genocide in Rwanda (OECD 2009, 29), which, alongside the ethnic violence in the Balkans around the same time, is often cited as a 'key trigger' for investments into conflict early warning (Babaud and Mirimanova 2011, 7; Nyheim 2015, 10). The increased impetus for conflict prevention and, by extension, conflict early warning was already visible in United Nations Secretary-General (UNSG) Boutros-Boutros-Ghali's '(1992, 204) An Agenda for Peace' which described early warning based on information-gathering and fact-finding as a prerequisite for 'preventative diplomacy'. However, an influential report by the Carnegie Commission, entitled 'Preventing Deadly Conflict' (1997, xiv), claimed that 'policymakers of good will and human decency were confounded and often paralyzed by the rush of murderous events in the 1990s'. While this failure triggered numerous international efforts to develop early warning systems that could leverage state-of-the-art technologies, international policymakers began to stress relatively quickly that early warning needed to take the peculiarities of conflict-affected contexts into account. For instance, the 'Prevention of Armed Conflict' report (2001, 7) issued by UNSG Kofi Annan stressed that 'a deep and careful understanding of local circumstances and tradition' was 'of great importance' for reliable early warning information.

However, some pioneers of conflict early warning considered the objective of gaining an understanding of 'local' circumstances insufficient, and thought that it mustn't be separated from the arguably more important task of enabling those affected by conflict and violence to seek an active role in its prevention. In a scientific note published in 1997, the network members described FEWER as committed to 'ensure a balanced distribution of members to avoid a "northern-centric" structure and approach to reporting on areas of conflict' (Fewer Secretariat 1997, 25). This approach clearly mirrors the considerations Rupesinghe had published ten years earlier. As he argued, a key challenge that became apparent already in these early efforts was to ensure that not only the collection and analysis of early warning data, but also the response to it, would not be dominated by actors at the organizations' headquarters, most of which were (and still are) located in the Global North. This was a matter of avoiding the establishment of a North-dominated global conflict surveillance apparatus that would further disempower and control vulnerable populations located in the Global South. And it was, ultimately, a strategic consideration to make sure that the early warning information would indeed be used, including by those affected by conflicts and disasters: 'Information is useful if acted upon', including by 'the *victims of the impending disasters*' (Rupesinghe 1987, 218, emphasis in original).

This insight can be read as expressing a transformative ambition, namely, of a socio-technical arrangement that gives agency to those immediately affected by conflict and violence. In fact, Rupesinghe advocated exploring the 'potential for building an early warning capacity for those who are vulnerable: in particular, those who may become the victims of state violence and terror' (Rupesinghe 1989, 185). This goal was, however, hampered by the global inequalities in the availability of technology. More precisely, he warned of the 'tendency of those in the North who are (over-)used to computers' to 'talk about establishing a huge monolithic database able to store, monitor and provide possible indicators in a timely fashion' (Rupesinghe 1989, 185). However, such a model, in which 'the rich North can predict and develop the warning capacities for the victims' by 'analyiz(ing) the information from the raw material derived from the South', was not 'viable' (Rupesinghe 1989, 186), amongst others, because databases may have to rely on 'official documents and published public information' while 'the most interesting and relevant information will often be found within networks which use informal means of communication'. He also mentioned language and translation issues, and the challenge that most 'information brokers' will be based in urban areas and have limited access to the countryside, whereas relevant information can often only be found in 'rural communities where key events can be monitored directly' (Rupesinghe 1989, 186). Not least, actors based in the 'Third World' had, in recent years, become increasingly sensitive to the 'traditional one-way information flow' and 'resentful of being made merely sources of raw material'. He concluded:

> Increasingly, the demand from the South is that the early warning capacity must emanate from the South, and in particular from non-governmental networks and institutions. It is the victims themselves, or the nearest to the victims, who should master their own information systems. (Rupesinghe 1989, 186)

Therefore, Rupesinghe advocated for an approach to early warning that would operate through 'highly decentralized databases', use 'satellite and electronic mail communication' to collect and share information, and, most importantly, have 'competence in handling information and exchange within the South itself' (Rupesinghe 1989, 187). This would require the establishment of a 'democratic' network of information exchange between different non-governmental organizations (NGOs) that would also coordinate their responses. Thus, for Rupesinghe, despite the persisting global inequalities in the availability of technology, the spread of ICTs to the Global South provided an opportunity to enable early warning efforts that would break with the colonial continuities of extractive information systems because it would enable new forms of networked communication among equal peers, some of which would be based in former colonized countries. In this vision, international organizations and actors based in the Global North would certainly retain a role in early warning, but only as some among many.

Much suggests that there were many attempts to realize this vision in the decades that followed, including through the establishment of regional and country-based early warning systems, especially on the African continent. Here, the formation of the African Union Peace and Security Architecture (APSA) in 2002 and eight Regional Economic Communities (RECs), as well as Regional Mechanisms that are mandated to act jointly on matters of peace and security (African Union Commission 2024), led to the establishment of several early warning systems. Most notably, the AU's Continental Early Warning System (CEWS), founded in 2002 and operating out of its headquarters in Addis Ababa, was meant to enable the prevention of subnational conflicts by providing information to the chairperson of the African Union Commission, who would in turn advise the AU's Peace and Security Council on impending conflicts and relevant preventative action (Engel 2018b, 12). In comparison, regional early warning systems such as the Conflict Early Warning and Response Mechanism (CEWARN), initiated in 2002 by the Intergovernmental Authority for Development (IGAD) and covering the Horn and parts of Eastern Africa, focused on more specific subsets of conflict, such as cross-border pastoralist conflicts. These systems became linked with national-level mechanisms that were supposed to retrieve data obtained from their network of monitors, such as the Conflict Early Warning and Early Response Unit (CEWERU) of the South Sudan Peace and Reconciliation Commission (SSPRC), which I will discuss further below.

While these mechanisms received considerable financial and technical support from North American and European governmental donors, regional initiatives such as CEWARN took full ownership of the early warning and response operations. One could argue that this demonstrates that the spread of digital communication infrastructures indeed enabled a decentralization of early warning capacities—at least on the international level. Yet, as I will argue below, the spread of ICTs did not prove to be the silver bullet against Rupesinghe's dystopian vision of early warning. It did not in and for itself provide the necessary means to enable practices of conflict prevention that would be led by—and primarily benefit—the most vulnerable.

The View from Above: Quantitative Data Analysis Between Ignorance and Irrelevance

Through the FEWER network, Rupesinghe also became involved with several other conflict early warning initiatives. Among these was the FAST early warning system, initiated by the Swiss Peace Foundation (now swisspeace) in 1998, with the ambition of providing conflict early warning for all world regions, albeit excluding Europe and North America. Nonetheless, FAST was acknowledged by leading experts as being 'one of the leading pioneers of conflict monitoring and early warning analysis' (Meier 2008) and 'recognized by most practitioners as the embodiment of good early warning practice' (Nyheim 2015, 32). Much of the thinking and methodological considerations that shaped the first early warning initiatives, such as FAST, were also passed on to later initiatives, including CEWARN and CEWS. This was partly because of the relatively small number of technical experts who supported the development of early warning systems on the African continent, often employed as consultants by recommendation of partners and donors. While decentralized, the various early warning initiatives, to a great extent, strangely realized Rupesinghe's fears of a 'monolithic' approach to early warning because they often employed similar methods sourced from the same small epistemic community. Importantly, many in this community of experts shared a feverish excitement about the prospects of modelling and predicting conflicts with the help of statistical and AI-supported methods.[5] Consequently, the evolving early warning systems relied heavily on publicly available data such as news sources. This was because the computational models employed required a lot of data, which was difficult to generate by tapping into and documenting informal knowledge of the kind shared during coffee break chats.

Due to this tech-solutionist obsession with modelling and prediction, many of the systems ran into exactly the issue that Rupesinghe had raised in the late 1980s, namely that the most relevant information was hardly ever made available in public sources, or at least not in a timely manner. Admittedly, the three systems discussed here were all conceived before most newspapers offered their reporting online, or social platforms such as Facebook or X (formerly Twitter) became important venues to gauge levels of popular dissatisfaction with governments or levels of social polarization. However, the development of CEWARN and CEWS took many years during which their technical capacities were incrementally expanded, and it thus overlapped with the overall trend towards 'Big Data' analytics that emerged with the spread of digital

[5] Interview 132, Günter Bächler, former Director of swisspeace, via videotelephony software, conducted jointly with Fabian Hofman, 10 May 2024.

technologies and especially online and social media. The perceived potential of Big Data for conflict prevention was well expressed by the authors of a 2013 report published by the International Peace Institute (IPI), who argued that 'Big Data could help reveal key insights into the drivers, triggers and early signs of large-scale violence to support and improve conflict-prevention initiatives', including data from sensors and the internet (IPI 2013, 4). Amongst others, the 'big data opportunities' (Engel 2018a, 4) that emerged through a combination of new information sources such as the World Wide Web and innovations in automated text extraction technologies also led to new investments into event data analysis from the 1990s onwards. These second-generation models relied on NLP technologies to facilitate the automated coding of event data, as was the case for the Harvard-based 'Virtual Research Associates' (VRA), which was founded in 1996 as a successor to PANDA (VRA 2018), for instance.

The early warning systems that have been set up since the turn of the millennium, such as the AU's CEWS and the IGAD's CEWARN, evolved in the context of this new-found optimism, and many were shaped by technical consultancy companies such as VRA. At the outset, CEWS focused on analysing news data from agency services and online newspapers. The systematic monitoring of social media data remained a challenge until at least the late 2010s (Engel 2018a, 122). CEWS' Africa Media Monitor application provided regular updates in ten-minute intervals about the most important current topics, provided a steady stream of news that could be brought to the attention of decision-makers and mapped out geographically, and helped with structuring and processing the information for qualitative and quantitative analysis, including through the use of automated text-mining techniques to extract relevant events. Earlier attempts to employ similar methods to automate the analysis of news data conducted by the FAST team had yielded mixed results. FAST had experimented with an event-mining methodology developed by VRA that scanned the Reuters news ticker—one of the few news sources that could at the time be easily retrieved in digital format. It then automatically coded events that depicted conflict or cooperative behaviour by relevant actors and used the coded data to generate a 'tension barometer' composed of various data graphs. These served to visualize different aspects of a given political system, such as the number of contentious interactions during a specified period, or the overall political stability of the polity, called Conflict Carrying Capacity (CCC), measured in percentages (Krummenacher and Schmeidl 2001, 17–18).

This system promised a seemingly objective and neutral approach to conflict early warning, as two of the experts involved in its development argued. '[A]utomatic coding improves upon the speed of digesting information and allows impartial assessment, as the machine does not make interpretations beyond indicators it has been programmed to track' (Krummenacher and Schmeidl 2001, 20). As one of them put it during an interview, the idea at the time was very much to develop an objective early warning system that could eliminate the 'political' hurdles to conflict prevention (Baechler et al. 1998, 13–14).[6] The 'view from above' thus promised to eliminate any

[6] Interview 132.

epistemic uncertainty that could be employed by political decision-makers to justify a lacking or slow response. It promised to implement a centralization of information flows that would enable the production of rigorous and objective early warning information and its immediate provision to policy- and decision-makers. This, in turn, would provide the grounds for decisive and effective responses that could master (and curtail) emerging risks and crises—a vision that has become epitomized by the idea of a 'control room', 'situation room', or 'crisis room' at the centre of early warning systems. However, eliminating this interpretative wiggle room also required a rigorous approach, which necessarily turned out to be methodologically monolithic—something Southern experts such as Rupesinghe had warned about.

However, in the course of its application, members of FAST also identified major issues with the computer-assisted event-analysis approach. To mention only a few, these included a lack of news coverage in the relevant regions; a bias towards reporting events of national relevance, which meant that important sub-national developments often remained unnoticed; the lack of context sensitivity by foreign journalists, which led to an underreporting of relevant events; an exclusive focus on English language news; and fluctuations in news reporting that made it difficult to track changes in conflict party behaviour (Krummenacher and Schmeidl 2001, 18; Krummenacher et al. 2002, 11). Some of these challenges will have disappeared due to further technological progress, such as stronger NLP and text-mining capacities, especially due to the advancements in Large Language Models (LLMs), and the diversification and digitalization of the media landscape. However, other challenges remain, and new ones have been added. Among the former are the uneven and fluctuating media coverage of conflict-affected contexts and the fact that news is not a suitable source to detect signals of an impending conflict, simply because there is a tendency to report on past events and rarely on events that are yet to happen. While a chain of escalatory events precedes some outbreaks of large-scale violence, it is hardly the case that they follow a generalizable, law-like pattern that would allow for an automated prediction of conflict occurrences (Panic and Arthur 2024). All the while, the instigators of coup d'états, insurgencies, and terrorist attacks that may trigger large-scale armed violence continue to enjoy the benefits of preparing these acts in secret and capitalizing on the element of surprise. This makes it rather unlikely that reliable early warning signals could be picked up in published news or social media. Furthermore, as discussed in the previous chapter, a new challenge is the 'noise' caused by the enormous amount of mis- and disinformation that characterizes social media. David Nyheim (2015, 7), who was the director of FEWER from 1997 to 2003, referred to the ability to establish a 'ground truth', that is, 'facts' that can be confirmed by somebody present in the 'field', as one of the central challenges of early warning systems. The explosion of available news data sources and the interlacing of factual information with mis- and disinformation has arguably made the need for such ground truths even more relevant today.

While based on sophisticated statistical models that required a vast amount of data, these tools thus often provided insights about conflict risks that were already broadly recognized. As one of the interviewees, who was both part of the inception of FAST as well as a recipient of its Updates and Risk Profiles at a later stage, put it, you simply could not present quantitative insights generated from Reuters news data to development cooperation staff on the ground, who usually had a much more fine-grained view of what was going on.[7] This is also mirrored in critical evaluations of CEWS' usefulness in preventing violent conflict, which argue that 'a good analysis of conflict ultimately boils down to simple personal judgement and that the bells and whistles (graphs, local information networks, etc.) of some early warning systems add little value' (Nyheim 2008, 19).

The View from Below: Qualitative Data and the Disciplining of Field Voice

A widely acknowledged risk of early warning approaches that enable a 'view from above' is that they can be out of touch with the specific contexts in which conflicts erupt. The growth in publicly available data and the expansion of computational capacities led to an ever-expanding suite of software packages for remote data collection and quantitative analysis. However, the community of experts gathering in forums such as FEWER had, as early as 2002, reached the consensus that a 'good' early warning system had to be 'close to the ground' (OECD 2009, 34). This insight reappeared repeatedly as the limitations of large data computational methods discussed above became evident, especially the lack of sufficient information. For example, FAST members noted that 'only 25% of events captured by local information networks can also be found on the internet, while around 75% are missing' (GPANet 2010, 29). In a comparative vein, a CEWARN analyst argued that 'all conflict is local, events all occur in certain areas, certain places' and, therefore, 'the more disaggregated the data collection methodology... the better you are able to capture contextual variations and predict or analyze more precisely than you can with aggregated data' (Goldsmith 2020, 85). While the first experiments with early warning in the mid-1990s had largely taken place among bureaucrats and experts based in international organizations and foreign offices, the following generations of early warning systems, to which FAST, CEWARN, and CEWS belonged, all aimed to implement this insight by establishing or incorporating local information networks, composed of monitors located in the conflict-prone areas (OECD 2009, 34–5).

However, the qualitative data collected from these field monitors also required *some* sort of systematic treatment and structuring so that they could be compared to the various kinds of quantitative data discussed above and ultimately employed for predictive analysis. While many of those involved in developing early warning systems viewed such local sources as valuable, they nonetheless aimed to align them with the tech-solutionist drive towards predictive models that would make political decision making obsolete. This meant that monitors were expected not simply to share their views, observations, or interpretations, but to do so based on a structured

[7] Interview 132.

data collection methodology that corresponded with a particular model of conflict and conflict escalation. To this end, several early warning systems once more worked with technical consultancies such as VRA that promised to triangulate the 'view from above' (an international news agency) with the 'view from below' (field-based international organization/NGO staff) through the use of a common data input format and processing approach (Bond et al. 2003, 736). VRA's triangulation technique was first utilized in FAST and later adopted by CEWARN and CEWS (Krummenacher 2006; Cilliers 2005; Goldsmith 2020, 77).

The methodologies through which inputs from the field were tailored to fit the early warning systems can be described as a form of 'data disciplining' because they forced the monitors to deliver their information in a particular pre-structured format. Information not fitting this particular format would be rejected by the system and returned to the monitor as irrelevant or insufficient. In such cases, the ready-to-handedness of the methodology enforced a specific analytical routine and way of looking at the conflict, thus disrupting a more natural and intuitive flow of information. Since the field monitors were, by virtue of their job description, tasked to submit reports to the system, this enforced a uniform data structure. While the country-level information obtained via monitors was of a qualitative nature, its utility and, indeed, usability for early warning required that it could be quantified and aggregated to allow a statistical analysis of trends. For instance, the FAST methodology, which also inspired the CEWARN methodology a few years later, involved assigning a certain numerical value for a defined set of conflict indicators, and these data were aggregated on a weekly or monthly basis to provide insights about trends in the conflict potential of the country in question (Krummenacher et al. 2002, 4). In comparison, CEWARN field monitors were meant to report relevant incidents directly to the so-called 'Technical Committee on Early Warning' at the CEWARN secretariat, using an SMS channel input (Goldsmith 2020, 92). From there, it was added to a pool of data used to feed a dedicated analytical process referred to as the CEWARN Reporter, which was employed to generate regular Situation Reports (so-called SitReps), as well as Incident Reports (IncReps) (Goldsmith 2020, 77). CEWARN field monitors would register relevant information in a reporting template according to a common set of coding rules and structured along a list of more than fifty indicators. These indicators were then grouped into several categories that in sum were supposed to represent a 'near-total description of pastoralist events': 'communal relations, civil society activities, economic activities, governance and media, natural disaster and resource use, safety and security, and social services' (Goldsmith 2020, 78). This approach emphasized 'comparable, data-driven analysis' (Cilliers 2005, 14). This required that all qualitative data obtained from the field be quantified along a scale from 1–10, which provided the basis for the aggregation and visualization of values for each indicator (Goldsmith 2020, 78–81).

This meant that only those events for which all the pre-set characteristics were known could be submitted to the headquarters in Berne. Otherwise, crucial information necessary for statistical analysis would have been missing, including who was involved, what happened, when, and where (Goldsmith 2020, 78–81). Information that did not meet these criteria could not be included and was thereby ignored. CEWARN thus apomediated not only the early warning information delivered from the field to the headquarters, but with it, the very understanding of conflict. This pre-structuring of the field-level data became a powerful tool for the early warning initiatives to regulate (and limit) their scope of operation. In the case of FAST, a 'general decline of living conditions', for instance, seemed of little interest to the system designers (Krummenacher et al. 2002, 7). In the case of CEWARN, the focus of analysis is chiefly on monitoring pastoralist conflicts, which means that armed conflict or insurgencies of a political nature are left out of sight. For instance, a technical officer in the South Sudan CEWERU (discussed in greater detail below) argued that CEWARN would remain 'neutral' regarding national-level political conflicts and frictions between the Dinka and Nuer ethnic groups because it focused on what was happening at the 'local level' (Goldsmith 2020, 171). As will become clear, the goal of maintaining such neutrality was, at times, difficult to achieve in practice.

It follows that while country-level monitoring was used to enable a 'localized' collection of relevant information, it is by no means certain that the local monitors would be able to share information that they deemed relevant based on their context-specific observations. In addition, local actors tend to have little say in shaping the objectives or scope of systems such as CEWARN or CEWS, or the response that their reporting might trigger. The discipline necessary for the structuring of data at times required putting in place specific organizational arrangements and procedures, including oversight mechanisms implemented by country-level coordinators or staff based at the headquarters who, by virtue of their additional methodological expertise and access to additional information sources, could assess and triangulate local data inputs. In other words, in the apomediation networks that form through centralized early warning systems, the role of field monitors is often limited to the provision of information along a clearly structured template, with no involvement in the design of the system. While some of these systems are established and run by organizations located in the Global South, monitors are nonetheless oddly reduced to those 'sources of raw material' that Kumar Rupesinghe warned about in his critical reflections published almost four decades earlier. Due to the systems' data collection routines, the information provided by field monitors turns into ingredients for pre-conceived analytical models, and these models are based on, and reproduce, very specific understandings of conflict that, in turn, enable certain preventative responses and not others. While relying on field monitors and country-based staff to obtain relevant information, these systems have in many ways centralized not only a possible early response but the production of knowledge about conflict as well. In doing so, they have bound analytical capacities in ways that make alternative interpretations of conflict more difficult and, with them, alternative approaches to conflict resolution.

The AU CEWS system likewise employed a network of field monitors to gather qualitative information about events and incidents, and the collected data are largely used to inform early warning briefings presented to the AU Peace and Security Council. Regular reports compiled by the AU's field missions and country-based NGOs would contribute to one of the 'baskets' from which the analysts in the CEWS situation room in Addis Ababa would obtain the relevant information for their reports. However, at the AU, setting up and maintaining these direct country-level information sources often turned out to be challenging. Ulf Engel, who worked as a consultant supporting the CEWS situation room, diplomatically framed this issue as a conditional possibility when writing that 'CEWS is facing technical limitations when and where the CEWS baskets miss out on local information, for instance, because there is no AU field mission on the ground or because local mechanisms are not reporting to CEWS', while also remarking that 'generally speaking, the regional coverage by field monitors is often fairly limited in scope and volume' (Engel 2018a, 122). This challenge was not unique to CEWS. Several years earlier, FAST had struggled with obtaining enough quality data that could be used to code relevant events. This seemed to be particularly the case in authoritarian contexts, where data collection by field monitors, such as local journalists, proved dangerous. For this reason, FAST discontinued its operation in Eritrea at the time in order not to expose its local monitors to severe security risks.[8] However, in other cases, such as the Atrocity Prevention Project in South Sudan discussed further below, this 'silence' could be read as a form of resistance by local actors against the disciplinary power exercised by the system's underlying understandings of conflict, and the data collection templates and input masks through which it is enforced. Regardless of the reasons, without a sufficient amount of coded data, the statistical analysis necessary to identify trends and run prediction exercises could not be carried out (OECD 2009, 40; Nyheim 2008, 22) and this severely affected the functioning of many early warning systems.

Decentred Crowds: Beyond the Politics of Early Warning?

Despite the considerable investments that went into developing large centralized early warning systems, many reached only partial operationality and had limited impact on decision-making processes and preventative action, and some were discontinued. FAST was shut down in April 2008, four years after FEWER had been terminated. While regional early warning systems have proven more resilient, their added value for conflict prevention remains contested. For instance, several observers have noted that CEWS' outputs have often been underutilized by the decision-makers at the AU Peace and Security Council. Putting it in rather diplomatic terms, Engel concluded that 'the level of utilization of CEWS within the AU Commission remains somewhat of an enigma' (Engel 2018a, 123). His observations mirror the widely acknowledged concern that conflict early warning data is often used very little, not early enough, or not at all. This issue became known as the early warning–early response gap (George and Holl 1997), and it continues to haunt many early warning initiatives to this day. In contrast, early warning pioneers such as Rupesinghe were less concerned with the non-use of information but with its top-down use—or abuse—by powerful actors. However, as will become clear in the remainder of this

[8] Interview 132.

chapter, both risks are related to the functional differentiation between predicting and acting on conflict that underpins many datafied conflict prevention efforts.

However, among a sizeable community of scholars and practitioners, innovations in digital technology have been viewed as having the opposite effect. In particular, the proponents of 'crowdsourcing', which emerged first among a community of 'crisis mappers' in the early 2000s, emphasize how networked digital infrastructures can enable spontaneous and organic forms of collaboration that no longer rely on hierarchical organization. In subsidiary systems, individual actors at the lowest levels may theoretically attempt a preventative response and escalate upwards once it looks unlikely to have the desired effect. In practice, however, alerts about grave or potentially impactful events or developments tend to be escalated upwards. In contrast, crowdsourced approaches would attempt a networked response that operates without chains of authority. This is meant to resolve the problems stemming from slow or irresponsive high(er) level decision making, while triggering collective responses and a 'surge capacity' (Whittaker et al. 2015) strong enough to effectively address the identified risk. Crowdsourcing is also meant to enable data collection without the need to establish costly field presences (OAS and UNDP 2016, 15), enable 'real-time information' (Global Disaster Preparedness Center 2017), produce faster responses by 'outpacing traditional channels of information' (European Union Institute for Security Studies 2014, 130), and help put into practice a 'collective political will' (GPANet 2010, 59). Most importantly, as Patrick Meier, one of the veterans of crowdsourcing approaches, put it early on, combining the 'crowdsourcing of information *and* response' will enable a 'shift away from traditional hierarchical approaches (that characterize the majority of current conflict early warning/response initiatives) and foster a more distributed approach to conflict prevention' (Meier 2008). The suggested benefit thus goes beyond merely avoiding the information brokers at the centre of conventional early warning networks, such as CEWS or CEWARN, who could stifle the response to an emergent crisis or conflict, for instance, by obstructing the flow or utilization of information. Crowdsourcing also promises to broaden the scope of observations and analysis, reducing response time and enabling broad-based and decentred responses to impending crises and conflicts.

Crowdsourcing approaches thus promise to cut out the high-level political decision-making processes that are often said to hinder an effective early response to conflict, and they thus seemingly present a meaningful alternative to centralized and hierarchical early warning systems. Importantly, narratives about the wisdom of tech-enabled crowds and agile self-organizing peer-to-peer user networks also suggest that crowdsourcing approaches could somehow avoid the problems associated with unresponsive political institutions. They thus seem to provide an escape from what is commonly viewed as one of the central dilemmas of early warning systems, namely that they ultimately advance technical solutions for deeply political problems. Nyheim, for instance, argued that 'violent conflicts ultimately are solved through political will—technical solutions should not replace political ones' (Nyheim 2015, 7). Moreover, the literature on early warning has time and again pointed precisely to the fact that the early warning–early response gaps are at least partly due to a lack of political will by political leaders (see, for instance, Baechler et al. 1998, 1; Nyheim 2008, 47; Gnanguenon 2021, 171; Accord 2023).

It is unsurprising, therefore, that the appeal of digitally enhanced crowdsourcing continues to inform some of the more contemporary efforts to reform or reinvent early warning efforts. Many systems today encourage their field staff to contribute to more immediate and direct preventative responses that do not pass through the centre of the network, by communicating their observations 'horizontally', that is, with the relevant officials and authorities at the national and sub-national level. According to Nyheim, these systems 'often merge the functions of information collector and responder, and work on the principle of subsidiarity', meaning that responses are launched first at the lowest level and then escalated if necessary (Nyheim 2015, 17). In consequence, the autonomy and scope of action of local responders usually remain well below what the proponents of 'localized' early warning and conflict prevention may want to see. As will become clear further below, the introduction of a hierarchical, subsidiary approach comes with the need to centralize data collection and analysis, and to coordinate the response. This increases the likelihood that better-resourced and more powerful actors in the apomediation network will have a comparative advantage in shaping the response dynamics.

A good example is the reinvigoration of the South Sudanese branch of CEWARN. In the years that followed South Sudan's independence in 2011, various efforts were undertaken by the newly formed government, supported by various international donors and implementing agencies, to establish conflict early warning networks in the country. This also included the roll-out of IGAD's CEWARN system since two of its monitoring 'clusters' reached into the newly established country. In 2013, South Sudan's CEWERU was constituted and chaired by the South Sudan Peace and Reconciliation Commission (SSPRC). This initially entailed the identification and training of suitable monitors, so-called Early Warning Officers (EWOs), as well as Community Peace Response Mechanisms (CPRMS), sometimes simply referred to as 'peace committees', in each county. The initiative also supported the establishment and training of technical teams at the state level, which led to the creation of state-level CEWERUs. To enable the effective collection of early warning information, it also distributed communication equipment, including long-range radios, satellite phones, and mobile phones.[9] Furthermore, due to the limited coverage of these systems in large parts of rural South Sudan, early warning messages were also submitted 'on paper'.

The CEWARN system in South Sudan reached some operability, despite technical equipment regularly disappearing or becoming dysfunctional, and the collection of early warning information from the country's many remote areas proving cumbersome, as described by former staff. While some information was transmitted by local field monitors via the established communication infrastructure, processing and aggregating this information at the state level was fraught with hurdles. Early warning reports were often still produced on paper and physical copies sent to the SSPRC in Juba. While this commonly caused delays and breakdowns in the reporting

[9] Email exchange with CRS staff, 12 December 2022.

routines, many state governments also opposed sharing sensitive details with their national-level counterparts. The SSPRC, which was meant to lead the country's early warning effort, was created through a presidential decree and reported to the Office of the President. Since Kiir's government had steadily undermined the country's federal system and opposed further demands to devolve power to the state and local governments, many government officials at the sub-national levels had little motivation to volunteer early warning information to them. CRS discontinued its support to CEWERU soon after the outbreak of the civil war.

After a fragile power-sharing arrangement reduced large-scale armed clashes from 2018 onwards, the idea of establishing a country-wide early warning system regained traction. However, almost ten years after the first efforts had been launched, donors and implementing partners no longer envisioned such a system as operating through the cumbersome and expensive network of radios and satellite phones but expanding the system with the help of a mobile phone application that would effectively enable a crowdsourcing approach to conflict early warning. In 2020, this idea came to fruition as UNDP's South Sudan Office launched an 'Innovation Challenge' to build a prototype for such a conflict early warning mobile application, supported by the UNDP's Accelerator Lab, which particularly targeted youth. More than fifty candidates participated in the first round of the competition, in the course of which three teams were selected in a competitive process to receive support for the development of a prototype, in close consultation with the SSPRC. The winner of the competition was invited to launch the so-called 'AlertMe' application in 2021 and it was subsequently rolled out in various states, where it was meant to replace other transmission routes, such as radios or handwritten notes. The UNDP's support activities were part of its 'Peace and Community Cohesion' project, which inter alia aimed to enable 'infrastructures for peace', including through strengthening South Sudan's CEWERU. 'Infrastructures' were thought of as involving all levels—national, state, and local levels, all the way to the community level.[10] The design process was depicted as an effort to showcase the potential of the South Sudanese to design complex digital applications, such as an early warning application.

The process was heralded as a success story on how to integrate local populations in the design and implementation of tech-driven peacebuilding approaches and as an example of how young people can come up with 'local solutions for local problems', according to high-ranking government officials (UNDP South Sudan 2021). This thinking also resonated with some of those involved with the implementation of the project, who wanted the application to enable a stronger linkage between early warning and early response at the local level. As one interviewee put it: 'One of the challenges in the country is getting early information to the early responders and the mobile application was one of the ways to respond to that'.[11] However, such assumptions about crowdsourced early warning are often based on a somewhat artificial differentiation between the technical and the political. The theories of change that

[10] Interview 57, two staff members UNDP South Sudan, Juba, 1 December 2022.
[11] Interview 57.

underpin crowdsourcing are based on a highly abstract and idealized tech-solutionist assumption about the self-organizing capacities of digital networks that oddly ignores how an interplay of social, political, and *technical* constraints conditions the flow of information among different nodes in these networks, independently of who initiates these flows, and with what intention. An alternative view, which I aim to present in the remainder of this chapter, acknowledges that the technical and the political can never be meaningfully differentiated because the design and employment of technical systems are fundamentally political. This becomes evident once we stop reducing politics and the political to the mere instance of decision making based on information (for which political will is in fact required) and instead consider how the production, circulation, and use of early warning information is part and parcel of the very politics that shape dynamics of peace and conflict. The datafication of conflict is itself a political act. It circumscribes a response that is not just purely emergent, decentred, and intuitive, but that is to some degree delegated and organized in socio-technical hierarchies that will encourage one type of response more than another. I will explore this insight in greater detail in the remainder of this chapter.

Missing the Gap: Intuitive Responses in Rudimentary Digital Networks

A look at other local and community-organized early warning initiatives may help envision a different approach, namely one in which the technical and political are considered inseparable from the outset and the lines between early warning and early response are blurred. To this end, several civil society initiatives launched in Northern Ireland during the 1990s 'Troubles' can provide instructive learning examples. The sectarian conflict between Unionist and Republican groups, that led to several violent episodes in the decades before the signing of the 1998 Good Friday Agreement, was regularly reignited during the so-called 'Marching Season'. Especially in urban areas and at the 'interface' areas where Catholic and Protestant neighbourhoods met, youth identifying with the two communities would regularly engage in violent clashes (Jarman and O'Halloran 2001). As mobile phone networks spread and handheld phones became more affordable and popular, several local CSOs initiated community-based early warning and response networks that used simple mobile phones to exchange information about youth mobilizations and imminent violence via text messages and calls. I will argue in this section that this rudimentary form of digitalization enabled a quick and nimble response to unfolding events, mainly enhancing pre-existing social networks without altering them, and overcoming the early warning–early response gap.

At the forefront of these efforts was the Community Development Centre (CDC) in Belfast, a local NGO committed to challenging sectarianism and promoting dialogue and engagement in an area suffering from high unemployment and relative deprivation (The Irish Times 1999). In Derry and Dunloy, similar efforts were initiated by the Peace and Reconciliation Group (PRG) and the initiative Women

Together. One important difference between these initiatives and the various field monitoring and crowdsourcing networks discussed previously is that they employed digital technologies to enable communication across pre-existing social networks composed of volunteers and community activists, instead of building new early warning and response mechanisms. Following a particularly violent episode during the 1996 marching season, members of the CDC considered how civil society could play a more effective role in predicting and de-escalating inter-communal violence should similar tensions arise in the following years. Of particular concern was keeping the 'lines of communication' open between representatives of the two communities. To this end, the CDC established a network of conflict monitors composed of existing activists already known to the staff of the Centre and equipped them with mobile phones to enable more effective communication between them. Compared to the other early warning initiatives, the expenses for technological equipment were minimal. A UK government-sponsored initiative, Making Belfast Work, initially purchased fourteen mobile phones that were distributed to the volunteers, and the network was subsequently expanded to thirty monitors with phones, covering twenty-five interface areas in Belfast.

The monitors were selected due to their position and status within their community, often based on a pre-existing engagement with conflict prevention, de-escalation, and crisis management, and some of them were working with other relevant societal institutions, such as local councils or the police. This meant that these information networks were inseparable from the politics that caused the flare-ups in communitarian violence. Indeed, as two observers of the initiative put it:

> Monitors [were] local community activists, members of community groups and often people with well-known political affiliations. They are not in any sense independent. Rather, their ability and capacity to have an influence on the local situation comes from their position and status within that community. (Bryan and Jarman 1999)

The networks followed a comparatively simple protocol to exchange information and coordinate preventative action, notably avoiding a centralization and aggregation of information through the kinds of data processing hierarchy that characterized systems such as FAST or CEWARN. The monitors simply used their phones' preconfigured calling and texting functions to keep other members of the network updated about any relevant developments they observed, to reach out to local authorities and the police, and to exchange information about ongoing developments and dissipate rumours. Given that these efforts took place before the advent of more sophisticated smartphones, they lacked operating systems and applications that would have enabled stronger information moderation, such as the AlertMe application discussed further below. Compared to the much more recent early warning efforts in South Sudan also discussed below, the limited technological capabilities at the time simply didn't allow for a stronger apomediation of early warning information at the data transmission interface, which also meant that the digital technologies employed could not play a more disruptive or interventionist role.[12] The simplicity of early mobile phones thus precluded any labour-intensive analytical methodology or decision-making routine that could have reduced the chance of an immediate response.

However, not all was down to the particular affordances of mobile phones. In fact, the monitoring efforts were just as much conditioned by how the community volunteers conceived of their work. In many ways, within that network, early warning and response were never functionally separated in the first place. A report of the CDC's initiative described a somewhat fluid set of practices involving answering calls and verifying information to improve communication within and between communities, as well as with local authorities. However, monitors also patrolled the streets at night, engaged with crowds, encouraged them to disperse, or were indeed 'dealing with minor violence'. One former monitor described how, in successful de-escalation cases, the network would function peer-to-peer, with each monitor assessing and calming the situation on their side. If this did not work, the situation report was escalated upwards using laminated cards with the phone numbers of all relevant community leaders printed on them. These individuals would then talk to each other on the phone and sort out the matter. Once, this happened when both community leaders of an interface area were on holiday in neighbouring resorts in Spain, from where they could call off rioting in Belfast once they were notified via telephone.[13] The newly established lines of communication enabled by the monitoring network also helped to re-establish inter-community trust and, at times, led to the initiation of in-person meetings between the community representatives, which allowed for 'direct dialogue to be resumed between some areas for the first time in several months', according to some observers (Bryan and Jarman 1999).

The focus on improving the communication between the communities and potentially enabling dialogue was related to a specific problematization of the conflict that underpinned the early warning efforts in Northern Ireland, which differed considerably from that of other initiatives. Many of the centralized systems discussed above were driven by a desire to predict conflict in order to prevent it, and the technical solutions had to be fit for this particular purpose. They were thus driven more by a tech-solutionist interest in the ability of ICTs to establish global networks of information exchange that could be used to identify and monitor recurring data patterns that would help to predict conflict outbreaks. In contrast, the CDC and other initiatives did not share this techno-scientific fascination with employing ICTs for evidence-based modelling. Instead, they were based on a much more mundane tech-solutionist consideration, namely that mobile phones could enable communication within and between the different communities when it was most needed. This corresponded

[12] Centralized text messaging-based early warning systems were introduced only a few years later through applications such as Frontline SMS, which was developed in 2005.

[13] Interview 99, Catherine Turner, Professor of International Law and Deputy Director of the Durham Global Security Institute, via videotelephony software, conducted by Fabian Hofman, 27 July 2023.

with a problematization of conflict that framed the violence between the various communities as the result of, or at least as exacerbated by, a breakdown in communication (Bryan and Jarman 1999, 4). One observer described these dynamics in greater detail:

> [B]reakdowns in communication helped perpetuate fear, suspicion, mistrust and violence. People in many areas felt increasingly isolated and were uncertain about what was going on in other parts of the city. As rumours spread readily of attacks in other areas, of threats of attack and of people being burnt out of their homes, more people came onto the streets to prepare for the worst. This in turn fuelled rumour and fear on the other side and so helped the cycle to continue. (Jarman 1999)

Indeed, anecdotal evidence from Belfast and Derry suggests that dispelling rumours, calming fears, and responding to minor incidents of violence helped reduce tensions and kept things from escalating. However, it also makes clear that the monitors had minimal impact on conflict dynamics once the tensions between the communities had indeed escalated. Assessing the effectiveness of early warning initiatives tends to be incredibly difficult, not least due to the impossibility of measuring something that has not occurred and, relatedly, the common challenge of establishing a counterfactual. However, my concern here is primarily with the socio-technical arrangement that characterized these community-based early warning efforts in Belfast and other parts of Northern Ireland. In this case, digital technologies enhanced an existing social network of activists around the CDC and other community-based initiatives. More importantly, it strengthened these social networks and facilitated conflict-mitigating efforts that went beyond the mere reporting and analysis of early warning data.

The Northern Irish case can thus be said to have 'missed the gap', because it didn't create a functional differentiation between early warning and response, and thus avoided the skimming of contextual information that happens when more sophisticated methodologies are employed that require structured data. Once the mobile phones were distributed among the monitors, they were simply ready-to-hand—mere tools to be used in the work of communicative de-escalation—in ways that were in line with the volunteers' established interests and practices. In contrast to the more advanced systems discussed below, they enabled a response that reflected the pre-existing inclinations of the civil society activists engaged in the monitoring work, and they didn't cause a stop, pause, or diversion of information that comes with the functional differentiation between data gathering, analysis, and response that underpins contemporary data-driven approaches. However, it's also important not to romanticize technological simplicity. Another vital factor for the relative success of the CDC's initiative is that the digi-political playing field in Northern Ireland was level, since there were no actors who aimed to shape the initiative for authoritarian ends. This will become clear when we look more closely at several initiatives in South Sudan.

Making the Gap: Data Analytics and Devised Responses

The flow of information between the conflict monitors involved in the efforts to prevent inter-communal violence in Belfast happened much more horizontally than in systems where analytical capacities are centralized. The objective in this comparatively flat network was less to produce authoritative analysis that could help measure conflict risks and more to leverage the spontaneous exchange of information between various nodes in the network for an immediate response. The primary aims were to enable communication, and joint action, cross the two divided communities. Similar thinking underpinned the so-called 'Localized Atrocity Prevention Project', initiated by the South Sudanese Branch of Search for Common Ground (SfCG) in 2022. According to a project document, the project would be 'led by locally based embedded observers' and was meant to 'strengthen local atrocity prevention efforts through community-led conflict monitoring, reporting and timely intervention'.[14] One of its objectives was to 'empower key community stakeholders to identify and address atrocity risks'.[15] A country-based staff member put the ambition somewhat more modestly when explaining that the project aimed at enabling local community members to 'take the lead in identifying issues . . . and identifying solutions or at least a way forward' to address violent conflicts in their area.[16] Compared to the initiative in Northern Ireland discussed above, the Localized Atrocity Prevention Project differed in one respect: digital technologies, including mobile phones and a mobile data collection methodology, were introduced to initiate a particular socio-technical arrangement meant to improve conflict prevention. The technologies weren't employed simply to enhance an already existing peacebuilding approach but to create an entirely new one. Moreover, the use of digital tools to collect and analyse data led to a gap between early warning and response, because reports, data, and visualization turned into mediating objects that established a difference between understanding conflict and acting on it.

The project operated in three locations, namely the national capital, Juba, as well as two state capitals, Bor and Bentiu, to reduce tensions between population groups identifying with different ethnic communities, and especially between local communities and people living in the area of former Protection of Civilians (PoC) sites. These sites had been established by UNMISS in the early phase of the war to protect certain ethnic groups from violent attacks, but were redesignated as Internally Displaced Persons (IDP) camps in 2020 due to a change in the UNMISS mandate, which meant that they lost their special status and protection by UNMISS. While the military factions involved in the South Sudanese civil war did not identify with different ethnic communities, ethnically framed violence had been common in Juba and many other parts of the country. In this context, the project aimed to prevent violence between

[14] Search for Common Ground, 'Localized Atrocity Prevention: Strengthening Local Atrocity Prevention Efforts in South Sudan, Conflict Monitor Training Guide', April 2022, p. 4.

[15] Search for Common Ground, 'Localized Atrocity Prevention'.

[16] Interview 33, Staff, Search for Common Ground South Sudan, via videotelephony software, 22 September 2022.

populations identifying with different ethnicities, which could unfold in hot spots such as water points, food distribution centres, or contested grazing areas. However, it also ended up reporting, and at times responding to, other forms of criminal violence, including homicides, robberies, and sexual violence. Pursuing this broader mandate arguably allowed for some degree of protection, given that a project focused solely on monitoring ethnically framed violence might have been perceived as politically sensitive by the South Sudanese authorities.

The SfCG country team began by establishing a community-based conflict early warning network by identifying monitors in the three locations, inviting them to several training sessions in peacebuilding, conflict prevention, and early warning, and equipping them with smartphones and monthly credit for mobile data. Many monitors were selected because of their pre-existing engagements in CSOs, including for peacebuilding and conflict prevention. Yet, they also received dedicated training to build or improve their conflict analysis, dialogue, and mediation skills.[17] To facilitate communication and enable the direct exchange of information between the monitors, the project also established a WhatsApp group for each location. Comparable to the efforts by the CDC in Northern Ireland, this arrangement was also meant to coordinate direct preventative action. However, the SfCG's office in Juba also acted as a central analysis hub, to which the volunteers were asked to submit early warning reports whenever they observed an incident in their location that could indicate a heightened risk of violence. To this end, the team employed the widely used survey software, Kobo Toolbox, which enabled the collection of qualitative and quantitative data based on a questionnaire, which was designed by the country team with the support of monitoring and evaluation (M&E) experts based at SfCG's headquarters. Links to the questionnaire were shared with the conflict monitors via the WhatsApp group, which the conflict monitors could save on their phones and employ when necessary.

While the project's focus was first and foremost to enable community-based early warning and response, the data collected via the Kobo Toolbox were centrally collected and analysed by SfCG staff members in Juba. Moreover, the data were shared with staff in locations outside South Sudan, who had access to the collected data via Kobo's cloud service, including the organization's programme coordinator and colleagues specializing in M&E. Like the other systems discussed above, this meant that local-level information could quickly be extracted and employed to inform programme design and policy at the organization's HQ.[18] On the face of it, this arrangement also enabled a degree of 'quality control', as a staff member at the HQ would work 'closely' with the country-based project leader to craft the reports. However, the involvement of international staff in the design of the questionnaire and the drafting of analytical reports meant that the understandings of conflict that emerged in the monitoring data were not only based on the perspectives and observations of

[17] Participant, focus group discussion with conflict monitors, September 2022.
[18] Interview 30, candidate of the UNDP Accelerator Early Warning Competition in South Sudan, via videotelephony software, 22 February 2025.

the local conflict monitors, since the apomediation process also involved technical staff at SfCG's country office and HQ.

While the Atrocity Prevention Project thus treated its local volunteers as more than mere providers of 'raw material', it nonetheless enacted a functional division between early warning and early response through a centralized unit that defined the standards of relevant information and served as a central analysis hub. The team based in the SfCG Juba office used the collected data to help coordinate a joint response based on regular 'conflict snapshots'. One of these short reports obtained during the research consisted of an analytical narrative, a quantitative breakdown of conflicts reported by the monitors during the specific time period, and various other analytical exercises, including a discussion of 'peace enablers' and 'peace spoilers' alongside a table that lists and quantifies different actors falling in these categories, as well as a list of recommendations based on quantified observations, such as about what percentage of the reported cases could be mitigated.[19] The structure and content of the report clearly demonstrate a desire to employ community-based monitoring to enable a form of evidence-based conflict prevention and peacebuilding approach based on established facts, such as what the 'leading causes' of division were in each of the three localities.

As was the case with the other systems discussed above, this type of structured and quantitative reporting required regular data collection to make trends visible. At the time of my fieldwork, the monitors were encouraged to submit reports at least every month to maintain the flow of information and allow for quality control and further improvements to the collection mechanism.[20] Importantly, this data collection routine was received with mixed feelings by some monitors, who insisted on reporting only if there were indeed reportable events, which meant that reports could be submitted more or less often.[21] The monitors were also asked to 'move around and observe things', but many felt it was more reasonable to ask their friends and acquaintances to share any relevant information and then report it. In a comparable manner to the other systems discussed above, the questionnaire could only be submitted after all questions had been answered. Besides a quantitative element to collect data on the number of incidents, people involved, casualties, and interventions made, the reporting form included several open-ended questions, in order not to pre-structure the monitor's answer, such as: 'Who is involved and what are the underlying motivations?' or 'Why is the conflict happening?' While less prescriptive than the input masks of FAST or CEWARN, this approach nonetheless overwhelmed many conflict monitors, who complained that the form was too complicated and that they struggled to answer all the required questions.[22] The monitors found it difficult, for instance, to determine the 'interests' of

[19] Search for Common Ground South Sudan, Conflict Snapshot, 31 August 2022.

[20] Participant, focus group discussion with conflict monitors, via messaging software, 19 September 2022.

[21] Participant, focus group discussion with conflict monitors, via messaging software, 19 September 2022.

[22] Participant, focus group discussion with conflict monitors, via messaging software, 19 September 2022.

conflict actors—something that was listed in the project's training guide and is common in many Western conflict analysis frameworks, such as the Harvard negotiation approach.

Notably, further iterations of the form implemented some of the feedback provided by the monitors and contained simpler questions. However, the use of a structured reporting template clashed somewhat with how many of the monitors had originally thought about their work. The controversy around the monitoring methodology demonstrates that externally initiated early warning networks may struggle with aligning their data collection and analysis routines with the ad hoc and on-demand communication flows of information in which community-based monitors are embedded. They also indicate how quickly centralized systems may end up in tension with the initiative of individual volunteers in the network. The resistance of some of the monitors to the chosen analytical approach, sometimes excused by them, with a lack of comprehension or skill, makes visible that digital methods favour an epistemic approach. The information obtained by monitors had to be made sense of, interpreted, classified, and analysed first, before it could be acted upon. The apomediation of conflict becomes a necessity before peace can be envisaged. *Some* assessment of the conflict had to be undertaken before a response could take place. One conflict monitor described his surprise at this approach as follows:

> Initially, when we started having this training, I never got exactly the reason why Search for Common Ground wanted this. I was thinking they needed to do an immediate response. Like if a case happens today, I'm supposed to be reporting today, and Search for Common Ground takes Action . . . But then at the end of the day, we were told this is different. . . . they accumulate this (data), and they look for a long-term solution to the problem.[23]

The data collection arrangement thus invited the making of a gap between early warning and early response at both endpoints of the data transmission chain. On the one hand, field-based monitors would require sufficient information to collect the data; they had to adjust to the epistemic–analytical methodology enforced through the data collection tool, and it took them time to implement it. The structured collection approach, based on guidelines and templates, also meant that the responses they had to submit did not necessarily correspond with their learned or intuitive approach of making sense and responding to conflict. Moreover, the project's core team had to first aggregate and analyse the collected data before it could be used for a response. All this could contribute to the early warning–early response gap. Digitalization often promises to make things faster and to increase connectivity. In the case of localized conflict early warning, however, data analytics tend to break things up and slow things down by introducing a functional division between efforts to understand and act on conflict and violence, as well as through the enforcement of data analysis routines that may appear hardly intuitive for local monitors. Instead of simply emerging based on

[23] Participant, focus group discussion with conflict monitors, Juba, 21 November 2022.

situated knowledge and intuition, responses will, in this arrangement, be devised in accordance with a preconceived conflict analysis methodology. I will argue that this devising of responses also opens the door to all sorts of other intermediating actors to shape conflict prevention efforts.

The Co-Optation of Localized Designs

Drawing on the results of the conflict snapshots, SfCG and its partner organizations coordinated the responses by sharing its early warning reports with a selected group of stakeholders during so-called Early Warning and Early Response (EWER) meetings, which also aimed to involve officials from the state and national governments.[24] These efforts served the project's second objective: to 'foster vertical collaboration on atrocity prevention among local, regional, and national stakeholders', which also entailed creating or reinforcing linkages between community mechanisms and government institutions'.[25] According to one team member, the systematic collection and processing of conflict data through the Kobo Toolbox and the regular drafting of reports were meant to help share insights about local conflict dynamics with actors that could initiate a country-wide response involving relevant stakeholders at the state and national levels.[26] This ambition was also visible in the project's theory of change, which constructed a clear linkage between the monitoring and identification of local risks for conflict and the 'capacity' of 'key actors' to act on them.[27] Interestingly, these considerations ignored the question of the political will for action. However, at the beginning of the project, the team shared the reports only with a smaller number of stakeholders, such as local police, due to worries related to the 'sensitivity' of the data.[28]

The meetings mainly brought together local stakeholders to discuss a joint response, including influential community members trained by SfCG in peacebuilding and conflict prevention, such as youth leaders or elders, the conflict monitors, and governmental authorities, including the police.[29] The project staff also intended to use the EWER meetings to discuss and launch community-based initiatives to resolve the underlying causes of conflict, including through bigger 'Community Dialogue Forums' that involved up to forty influential community members, such as community representatives or women leaders. However, when attempting to obtain additional information about the reported incidents and discussing the adequate responses, the EWERs interestingly often relied on long-established social institutions. Many monitors attributed a particularly important role to community leaders and especially chiefs, because they would have the authority to 'investigate

24 Interview 33.
25 Search for Common Ground, 'Localized Atrocity Prevention', p. 4.
26 Interview 33.
27 Search for Common Ground, 'Localized Atrocity Prevention', p. 5.
28 Interview 62, staff, Search for Common Ground South Sudan, Juba, 2 December 2022.
29 Interview 33.

the . . . conflict parties' and support the monitors to obtain the 'full info(rmation)' because they would be able to 'access' it.[30]

This resurgent role of conventional authorities is partly the result of the South Sudanese context. Given that customary law is still practised widely across the country, the EWER meetings commonly tasked the traditional authorities to engage in an action that could help resolve conflicts and prevent a further escalation of violence.[31] Many of the conflict monitors I spoke to also conveyed a sense that the 'chiefs', 'sheikhs', or 'area leaders' in their community would be widely respected, including by individuals who may have committed acts of violence or crimes.[32] The reliance on such 'local' and 'traditional' authorities to prevent and resolve conflict is not uncommon in contexts such as South Sudan, where statutory governmental institutions are scarce. It is therefore unsurprising that they were also given an important role in the response element. However, some of the conflict monitors mentioned that while some of these authorities appreciated their efforts, others showed little support or even scepticism or mistrust towards their initiative.[33] Especially in situations where such leaders were usually associated with a specific community, they were ill-placed to deal with larger, inter-communal conflicts. Where this was the case, the community meetings could attempt to mobilize formal state organs, including the police, particularly when incidents allegedly involved a crime.

Enabling early action at the local level was also a key objective during the design and implementation of the mobile phone application for South Sudan's CEWERU. As was the case with the Atrocity Prevention Project, this required creative thinking about how to best involve local institutions. For instance, the runner-up to the winner of the application development challenge had built a specific button through which chiefs could get access to the data backend. The integration of this chief button corresponds with a particular vision of South Sudanese society, in which traditional authorities continue to play a key role in conflict resolution at the local level. Another designer had the idea of establishing a 'peer-to-peer' network of information exchange to strengthen information flows among different members of a local community. This was meant to help different members of the same community exchange early warning information and coordinate responses. In his view, communication, even between neighbouring families, was often lacking. If something was happening in the neighbourhood, it should be the neighbours who get to know about it first, and then after that, the information could be shared with the respective government officials. Building an application that could enable the exchange of information could enable a crowd-based response among them.[34] The application's name, 'AlertMe', still

[30] Participant, focus group discussion with conflict monitors, via messaging software, 19 September 2022.
[31] Participant, focus group discussion with conflict monitors, via messaging software, 19 September 2022.
[32] Participant, focus group discussion with conflict monitors, via messaging software, 19 September 2022.
[33] Participant, focus group discussion with conflict monitors, Juba, 21 November 2022.
[34] Interview 21, candidate of the UNDP Accelerator Early Warning Competition in South Sudan, Juba, 10 December 2021.

reflects this ambition since it suggests that the user, namely, the 'me', would obtain early warning information about security incidents. Indeed, the SSPRC's AlertMe was based on an open-source application with the same name, which, according to the developers, should 'make it easier for everyone to get involved in decisions made about the environment around them'.[35]

Some of the UNDP's project staff held comparable views, and they were particularly concerned with how to enable a timely and broad-based response among local communities, especially in rural areas where state institutions were spread so thinly. In their view, having responders based at the county level would be insufficient because the area they had to cover was vast and often difficult to reach. They thought there was a need to make sure that the early response capacity would be strengthened at the level where tensions could be defused most immediately, which was perceived to be at the 'grassroots' (UNDP South Sudan 2021). The application was meant to overcome this challenge, because members of the peace committees responding to violence and insecurity with the help of the application would be based at even lower administrative levels.[36] A member of the early warning network in Eastern Equatoria state, who had recently attended a training for the application, described this vision as follows: 'When I send a report . . . everybody who has this application AlertMe will access it. Then the solution [to the reported issue] will come in totality, and everybody can have a part in solving that particular problem'.[37] However, he also mentioned that the roll-out at the lower levels of government had been insufficient because not enough mobile phones and data had been distributed.

Indeed, in the first years after the launch of these initiatives, the vision of a tech-enabled grassroots response was never fully realized. This was partly due to the lack of resources made available during the rollout, with only a few mobile phones provided for each county. In consequence, local monitors were often spread too thinly to directly report on incidents, and they often had to rely on secondhand knowledge when submitting alerts. As in the other cases discussed above, the combination of digital data collection methods and limited grounding in conflict contexts led to a superficial engagement with the conflict context. As one civil society observer argued, the collected 'data keeps on losing the meaning from the original source': the reports might entail relatively basic information about security incidents, armed movements, etc. However, more detailed context information, such as about the intention or motivation of actors involved, could easily go missing.[38] According to the interviewee, sophisticated approaches such as the AlertMe application would make more sense if the relevant data were collected from 'relevant stakeholders', yet these were often located in rural areas where internet connectivity and literacy were extremely low.[39] This tendency to produce 'skimmed data' was aggravated by the input mask of the

[35] <https://github.com/CodeForAfrica/AlertME> accessed 15 November 2024.

[36] Interview 57.

[37] Interview 113, staff, CEWERU Eastern Equatoria State Technical Team, Torit, conducted by research consultant, 3 October 2024.

[38] Interview 58, Executive Director, Organization for Nonviolence and Development, Juba, 1 December 2022.

[39] Interview 58.

mobile application, which required a certain level of literacy to be submitted. Paradoxically, providing more context information via the application would factually require a higher level of literacy, as the example of the Atrocity Prevention Project discussed above demonstrates so well. However, it was exactly this kind of 'thick' and 'meaningful' information that would be necessary for a conflict-mitigating or de-escalating response, such as a mediation effort. Yet, such information could only be collected and assembled by 'intellectuals' who often lived in urban areas and were unable to support an immediate on-site response.[40]

Thus the application did not alter established social and political arrangements but instead accommodated them. This complicated the link between early warning and response, making it more difficult for local, on-site actors to be involved in the mitigative action. However, an even more decentralized response approach would have been at odds with the SSPRC's vision of a much more centralized mechanism. While the innovation challenge was portrayed as an approach to localize digital peacebuilding efforts, the design of the application could, in fact, result in the opposite because data submitted by local monitors would be directly saved in a centralized database accessible in a 'situation room' at the SSPRC, which was 'envisioned as a centre to receive early warning alerts, direct alerts to early responders, compile early warning and response data and monitor trends of conflict in the country', according to the UNDP (UNDP 2022). The acting Chairperson of the Peace Commission stressed that, thanks to the application, 'instead of the information going haywire everywhere, it has been controlled'.[41] Indeed, several staff at the commission reasoned that the application would help to reduce the risk that early warning communication via calls, radio communication, text messages, or handwritten notes could be eavesdropped by malevolent actors, including parties that cause insecurity and conflict in the first place. Alerts submitted via the application ended up directly in SSPRC's dashboard, instead of being potentially diverted.

The centralization of data flows formed the precondition for a more targeted distribution of early warning information. In this process, a range of conventional governmental actors gained access to the collected data, and many of these did not naturally fit into the picture of localized early warning, especially government agencies at the state and county levels. At the state level, the Governor's Office and a representative of the State Technical Team would receive relevant early warning data, whereas at the county level, the county commissioners heading the CEWERU obtained access. This meant that besides the technical members of the peace committees, the information was also shared with political appointees and governmental officials. The SSPRC's acting Chairperson argued that, since state governors and county commissioners also had 'security prerogatives', the information flows in the application should be designed in such a manner that early warning information could also be directly transmitted to them, through the assigning of user roles through

[40] Interview 58.

[41] Interview 44, Betty Oboy, Acting Chair Person, South Sudan Peace and Security Commission, Juba, 25 November 2022.

which they could gain access to a 'small dashboard' that would include early warning information relevant to their area of operation.[42] The influence of such conventional intermediaries on the seemingly decentralized, 'bottom-up' early warning and response initiatives, such as South Sudan's CEWERU, is thus clearly evident. They demonstrate that the promises of decentralized early warning and conflict prevention through networked connectivity can quickly evaporate once governmental actors become integral to the system, as I will demonstrate in the final section of this chapter.

How Early Warning Fails to Prevent and Sometimes Fuels Violence

Once conventional information brokers, such as state governors, county commissioners, or traditional authorities are integrated into early warning networks, chances are that they will use the obtained information in line with their own reasoning about conflict and insecurity, and their specific understandings of what is needed to mitigate them. In this section, I will therefore explore how the social and political contexts in which digital technologies are employed can shape the apomediation of conflict early warning data in such a way that it leads to what I describe as the co-optation of apomediated peacebuilding: socio-technical networks that were designed and implemented to support conflict prevention end up serving conflict stakeholders that pursue parochial interests that did not originally influence the design of these systems. I describe this as a process of co-optation because the originally conceived linkages between the identification (and thus construction) of the causes of conflict and the identification (and thus construction) of responses to conflict break down as the information about conflict is diverted and appropriated by powerful actors or institutions that pursue disparate and often narrow agendas.

The Atrocity Prevention Project discussed above was decisively designed primarily to facilitate conflict mitigation at the community level, while also establishing vertical linkages to conflict stakeholders at the state and national levels of conflict. The need for this approach was somewhat validated through the collected data once the conflict monitors had started submitting their reports. Many of the recorded incidents had been caused, or at least conditioned, by factors that did not exclusively stem from the local level, and thus couldn't be fully addressed there either. These included, for instance, cases of gang violence in some Juba neighbourhoods that had turned rampant due to high levels of youth unemployment, poverty, and ineffective policing.[43] They also involved cases of inter-communal violence about natural resources, including access to grazing land and water points, as well as cases of land grabbing. The data collected via the AlertMe application presented similar results. While the reported incidents, of course, happen in specific *localities*, many of them involve conflict stakeholders that are located *elsewhere*. This is particularly the case where

[42] Interview 44.

[43] Participant, focus group discussion with conflict monitors, Juba, 21 November 2022.

localized violence was perpetuated to maintain or gain access to valuable resources, since these types of struggles are often linked to national-level elite competition. However, in the case of the Atrocity Prevention Project, discussing such issues with the relevant stakeholders turned out to be a difficult task. For instance, the project had undertaken various efforts to engage local police to address the issue of gang violence in Juba. Yet, the police had noted that 'there was a lot of interference from other security agencies, which somehow tends to distract when they are following up' because 'some gangs are connected to some of the security agencies' or 'powerful individuals in these organizations that would make financial [gains] from their illegal activities'.[44] Several conflict monitors mentioned that they had been intimidated when trying to raise these incidents. According to one of them, this created a sense that, at the end of the day, 'you also had to take care of yourself because even if you are committed, you don't know who is hunting you from behind—somebody who wants to also get his way. This is the thing. It is that fear'.[45]

However, some early warning initiatives experience more than just a limited impact from the resilience of the institutional and structural drivers of conflict. They become outrightly permeated by them. This was the case with CEWERU's AlertMe application, where many actors that could be directly associated with incidents of insecurity, conflict, and violence documented by the application could directly or indirectly access the collected early warning data. For instance, the analysts operating the SSPRC's situation room shared information about security incidents with the army, the National Ministry of Defence, and the Ministry of the Interior. Moreover, the technical teams at the state level that had access to the data backend included representatives of civil society, but likewise the intelligence services, including military intelligence, the police and prison services, as well as the state executive. Reports would also be shared with the Office of the Governor, the State Peace Commission, and various other governmental entities.[46] This meant that the early warning data could be quickly shared across the different levels of the executive branch of government. In the view of the acting Peace Commissioner, the information needed to be shared to avoid further insecurity or escalation, because 'the army could also easily cause some insecurity somewhere' and this could only be mitigated from within the military hierarchy. At the same time, however, she made clear that the commission itself had no means to 'do' the response, but simply passed on the information to the 'relevant authorities'.[47] The privileging of the state's security apparatus in the early warning response was not only a preference of the SSPRC's leadership, but also corresponded with views held among some of its international and national partners. Indeed, staff at the UNDP were supportive of the idea that early warning information should be made available to the organized forces, at least for incidents that required a military response. While the CPRMs and peace committees were trained in conflict resolution, in their view, the violent nature of conflict in South Sudan often simply

[44] Interview 62.
[45] Interview 62.
[46] Interview 113.
[47] Interview 44, Chairperson, South Sudan Peace and Security Commission, Juba, 25 November 2022.

required an armed response.[48] This meant that, as one civil society member of the forum remarked, 'a few minor cases' could be addressed by a particular organization entrusted by the forum, which would use IGAD emergency funding to engage in dialogue activities. However, when armed violence was imminent, the forum would rely on the government to take military action.[49]

To coordinate the response, the SSPRC chaired monthly early warning meetings with a national steering committee in Juba. This involved representatives of government, including the military, intelligence agencies, the police, the Ministry of Defence, and the Ministry of Interior. It also included civil society groups, such as representatives from religious communities, women, and youth, as well as academia and international development and humanitarian organizations.[50] The meetings were intended to evaluate the information, differentiate actual incidents from rumours, and assess signals and alerts about imminent incidents. The participants of these meetings were also invited to share additional relevant information that they obtained through their own professional organizational reporting and professional networks.[51] In addition, the meetings commonly referred the issue to what the participants deemed to be the relevant authorities to respond, including state institutions at the national and sub-national levels. Members of the committee could also coordinate a joint response, including by utilizing CEWARN's rapid response fund facility, which supports civil society-led conflict prevention activities, as well as additional small grants from UNDP that could be provided to organizations located in the relevant localities. This approach was replicated in several state capitals, including Bor, Torit, and Rumbek, where the UNDP formed State Early Warning Forums composed of government and civil society representatives and headed by the state-level peace commissions, which met monthly.[52] Representatives of these committees could also quickly coordinate their action with the CEWERU members at the county level, which were headed by the County Commissioner and composed similarly, also containing representatives of the armed forces and the national intelligence services. In practice, this arrangement meant that governmental entities would play the leading role in the response. It also meant that the committee would assess and act on the early warning data days or weeks after an incident had occurred, while security services and the military could use the information to take immediate action. Given their hierarchical structure and vertical penetration of the country from the national level to the county level, they could often act quicker and more decisively to unfolding events than special mediation initiatives that first required coordination and funding.[53]

This constellation was prone to fuel further violence because it ultimately strengthened the capacity of state institutions to exercise force. In South Sudan, moreover, this state is controlled by an authoritarian regime that has historically responded

[48] Interview 57.
[49] Interview 63, Executive Director, Community Empowerment Organization, Juba, 2 December 2022.
[50] Interview 44.
[51] Interview 58.
[52] Interview 57, Interview 113.
[53] Interview 57, Interview 113.

to insurgencies and insecurity with the use of force, instead of pursuing a negotiated settlement, while centralizing much of its power in a tightly knit circle of elites that share power through controlling governmental institutions at the national level. When made available to such actors, early warning data can quickly lead to new suffering, grievances, and animosity. Several of the agencies that gained direct or indirect access to the data backend of the AlertMe application had been instrumental in maintaining the regime's authoritarian control, including by committing severe human rights abuses. A UN Security Council (UNSC) report published in 2019, for instance, attested that the National Security had 'expanded its unchecked security control to silence political and civic dissent', and that, alongside the military intelligence, had been involved in violent, arbitrary and extrajudicial activities to stifle political and armed opposition to the regime (UNSC 2020). This means that the early warning information collected and shared through the application could ultimately become diverted to serve a very different approach to dealing with conflict. Rhetorically, the application was heralded as an inclusive, bottom-up effort to conflict early warning and early response. However, it ended up supporting a repressive regime that many view as the source of conflict in South Sudan.

More concretely, several interviewees familiar with the initiative described how the specific kind of data collected through the application would end up in the hands of state actors who were prone to respond to violence and insecurity through the use of force. For instance, in 2022, early warning data received by the SSPRC from Ayod, a town in Jonglei state, were quickly shared with a member of the National Security Service (NSS).[54] For many years, Ayod has been a hotspot for insurgencies against Kiir's regime, especially by the so-called 'White Army', a militia group that mainly recruits from the local population identifying with the Nuer, and is led by the self-declared 'Nuer Prophet', Tut Makuach. The causes for this armed opposition are complex, but linked to South Sudan's elite struggle over power and to long-standing, ethnically framed grievances. When civil war broke out in 2023, the White Army aligned with the Sudan People's Liberation Movement-In Opposition (SPLM-IO), the main armed opposition group challenging Kiir's government. The peace process did not provide a lasting end to these antagonisms. In fact, the regime encourages other militia groups to continue the fight against this armed opposition, such as the Agwelek, which mainly recruits from the Shilluk population in neighbouring Upper Nile state. In 2022, tensions in the region came to a new boiling point, as the Agewelek clashed with White Army forces, supporters of the SPLM-IO, and other militias. The report submitted via the AlertMe application warned that Tut Makuach was in the course of mobilizing for another armed campaign, which indeed materialized in the following months, supported by elements of the SPLM-IO. The Governor of Jonglei state, formerly aligned with another opposition group and identifying with the Nuer ethnic group, sent a delegation to meet with Kueth, ostensibly to de-escalate the situation.[55] However, the response of Kiir's government was through the use of force (Small

[54] Interview 151, anonymous source.
[55] Interview 147, anonymous source.

Arms Survey, n.d.). In this case, the submitted report probably didn't make much of a difference regarding the government's preparedness and response. Instead, it simply vanished in the larger conflict context, basically providing intelligence to one party to the conflict. While the report provided information to the SSPRC, it's likely that the regime's security organs monitored the situation in Ayod through other means. Given its status as an opposition stronghold, the continued unrest in the preceding years, and the size of the mobilization, it wouldn't have gone unnoticed. However, the early warning system also clearly didn't lead to a mitigative response in line with what those affected by conflict in the area would have wanted to see.

However, this outcome is not simply the result of a failed tool but rather, of a more complex socio-technical arrangement in which the particular ways in which conflict and violence are commonly addressed by governmental actors in South Sudan play an important role. This becomes evident when we retrace a quasi-experiment that I accidentally came across while researching the roll-out and functioning of the AlertMe application. In the counties of Eastern Equatoria state, the system was fully operationalized during the research for this chapter. During that period, several monitors stated that, after reporting incidents related to armed violence, state- and county-level security organs had responded with disproportionate force. Amongst others, one monitor had reported that a group of youth had taken up arms in Pajok Payam in Magwi County, which has been traditionally a stronghold for military insurgencies associated with the armed opposition groups. In response, the State Governor dispatched a contingent of the army, which killed all insurgents in a gun battle and declared that they were 'rebels'.[56] Similar instances have been reported from other parts of South Sudan, where the application was used to report security incidents and armed violence.

Importantly, the government's response was comparable in places where the application had not become functional but where reporting took place through other means. For instance, in Rumbek, the application was rolled out, and several training sessions were conducted, in 2022 and 2023. However, it was ultimately not operationalized. Several of the monitors reported that, for reasons unknown to them, they were not provided with the necessary login credentials to access the application, and they lacked the mobile data credit to transmit the reports. While unable to submit reports via the application, several monitors resorted to reporting incidents directly to local government entities using text messaging services such as WhatsApp. While domestic violence incidents, such as rape, were reported to the relevant traditional authorities, events involving armed violence, such as cattle raiding or feuds between different clans or ethnic groups, were directly reported to the County Commissioner, the Governor's Office, or the Police Commissioner's Office at the state level. This ad hoc system thus employed a much simpler method, comparable to that used by the conflict monitors in Northern Ireland. It also 'missed' the gap—but with vastly different results. Several monitors reasoned that reporting incidents directly to the government was necessary because they controlled the security apparatus, and they

[56] Interview 140, anonymous source, conducted by research consultant, 17 September 2024.

praised the increase in digital connectivity because it would enable much quicker reporting and a faster response. Yet, in all these cases, the response involved the use of force and it often led to loss of life. In some cases, civil society-led mediation efforts initiated by organizations based in Juba aimed to bring conflict stakeholders together to initiate a dialogue, such as about how to manage land and natural resources contested by several communities. However, many monitors lamented that local civil society was virtually absent or too weak to take any mitigative action.[57]

The difficulty of keeping the early warning information intended to support conflict prevention separate from a violent regime waiting to exploit it demonstrates how such 'tools' have a political life. Development actors such as the UNDP put a prime emphasis on the national ownership of their projects, which necessarily also involved the ownership of data. The placement of the SSPRC under the Office of the President and the neat vertical linkages between levels of government gave the state's intelligence and military a comparative advantage over civil society initiatives, namely that they could take action faster. This privileged a security-focused, repressive approach to dealing with conflict and insecurity. The regime also attempted to contain civil society efforts that were designed to operate more interdependently, as was illustrated by the experience of the Atrocity Prevention Project, where security actors inserted themselves into the response mechanism. A comparison of these two initiatives also demonstrates that a data-centred approach to conflict prevention in contexts with limited digital access and literacy is likely to privilege a top-down, hard security response. The rich data needed for mediation, dialogue, and peacebuilding initiatives cannot be easily generated by the rural populations affected by violence and insecurity, while a repressive response by the security apparatus can content itself with the most basic data points, such as the place, time, and type of incident.

All this suggests that while the introduction of mobile applications for early warning enabled a quicker response, the socio-political context played a decisive role in shaping what kind of response would be pursued: this context includes both the broad acceptance of executive security organs as primary actors tasked with the response, as well as an established approach to dealing with conflict and violence that privileges the use of force over other less violent means. Without a change to these broader conditions, early warning data, whether sent via a specifically designed or off-the-shelf application, will likely continue to contribute to the reproduction of violence instead of mitigating or resolving it. This incongruence—between tech-solutionist claims about the kinds of peace applications such as AlertMe will help to build, and the kinds of peace they end up building—is a consequence of both their design *and* the digi-political playing field on which they operate. The UNDP project that supported the development process of the AlertMe application promised to make it ready-to-hand for peacebuilding approaches that leveraged the potential of South Sudan's youth entrepreneurs and contributed to strengthening participatory,

[57] This has been corroborated through two informal conversations with research participants familiar with the operation of the AlertMe app.

bottom-up responses to the country's continued insecurity. While the data were meant to be shared with a variety of actors that could use it for a diversity of mitigative measures, the application produced fairly skimmed data that were processed in the situation room of a commission that reports directly to the Office of the President, and were thus prone to serve the intelligence needs of a repressive regime. While meant to contribute to approaches to peacebuilding that enable vulnerable populations to play an active role in conflict mitigation, it ultimately became ready-to-hand for those who were more likely to seek an authoritarian peace.

Conclusion

Peacebuilding scholarship has for some time grappled with what is now commonly dubbed as the 'local turn', and the difficult question of whether—and how—localized approaches can make a difference (Mac Ginty and Richmond 2013). A particular concern in this debate is how technology could be used to view conflict and peace from the local communities affected by violence (Ginty and Ginty 2021). Efforts to employ digital technologies to enable effective conflict mitigation are no exception. Their localization, through the integration of local monitors into both early warning and response, has been viewed by the pioneers and designers of this system to counter the tendency of producing information that lacks context and is thus unable to support a comprehensive mitigation effort that puts the most vulnerable at the centre.

The localized early warning and response systems discussed in this chapter contribute to apomediated peacebuilding in various ways. They shape understandings of conflict through the aggregation and structuring of data, contributing to the skimming of conflict contexts through which the thick information necessary for comprehensive responses can get lost. They also organize and direct the flows of signals meant to help with detecting imminent conflict risk, and they may or may not widen the early warning–early response gap through things like conditional reporting masks that reduce the inclinations for immediate and intuitive responses. This chapter has first discussed how some of the pioneers of computer-assisted early warning and conflict prevention, such as Kumar Rupesinghe, were worried about the homogenizing effects of unified and centralized systems that would leave little space for generating information and knowledge stemming from the Global South. It has then demonstrated how initiatives on the African continent, including efforts to crowdsource and localize early warning and response, have not necessarily yielded much more encouraging outcomes. When aiming to understand the reasons behind this development, it helps to view these systems not as tools but as a network of socio-technical relations. The understandings of conflict that emerge in them and the mitigative actions they enable are conditioned by the particular relations between a diversity of agents involved in the apomediation and use of early warning information. This has become clear when comparing monitoring initiatives in South Sudan and Northern Ireland that used simple text messaging instead of sophisticated mobile applications, which showed that the type of mitigative actions depends on what kind of actors are involved in these efforts and with what intentions.

However, the datafication of mitigative action and the designing of complex, decentralized early warning and response systems complicate the picture further. The computational methods discussed in this chapter not only tend to 'skim' the conflict context, and thus produce impoverished understandings of conflict, but also often relegate local monitors to the role of mere information providers. This reduces the likelihood that these systems will contribute to immediate and context-sensitive responses. Nonetheless, the development of such systems is often accompanied by tech-solutionist claims, such as about how digital approaches enable crowdsourced responses or 'grassroots reporting' in the case of South Sudan's AlertMe application. The application also enabled the generation and sharing of early warning information which could enframe conflict mitigation measures in a variety of ways, fitting attributes such as youth-led, inclusive, participatory, or localized. However, the broader human–machine network in which the application was embedded could generate various mitigating measures, including diplomatic interventions by local peace committees or dialogue and reconciliation efforts, as well as the repression of armed violence by force. Importantly, the chapter showed that the kind of data that the application produced were, in practice, more ready-to-hand for repressive conflict mitigation efforts than they were for long-term transformative efforts.

This was, to a considerable degree, also the consequence of South Sudan's slanted digi-political playing field. Given the field's limited reach, depth, and flexibility, early warning initiatives need to invest heavily in establishing local monitoring capacities. This makes it difficult to achieve the ideal of 'crowdsouring' information from a networked, non-hierarchical pool of local monitors who would use it for direct response, and in ways that they see fit. There are many factors that get in the way of such horizontal, localized efforts to conflict mitigation. Limited digital connectivity reduces the number of people who could contribute, but more importantly, a lack of digital skills and necessary resources means that civil society volunteers will often struggle to provide information in ways that fit with formalized early warning methodologies. Moreover, South Sudan's slanted playing field considerably contributed to early warning that tended to have a regime-stabilizing effect, instead of enabling transformative responses. The AlertMe system relied on data from a relatively selected group of local monitors attached to the Peace Commission's CEWERU, and the submitted early warning information could be read by government officials at the state and local levels. This level of integration—or rather co-option—sits well with the governing regime's overall ambition to control state and civil affairs that could potentially undermine its power. Based on the empirical insights I was able to collect, it is difficult to tell to what degree this arrangement effectively *suppressed* early warning information that would incriminate the regime. However, it has become clear that the information could be used to violently *contain* potential threats to the regime's survival, such as localized armed resistance or insurgencies. This regime's efforts to control how early warning information would be used were likewise evident in the civil society-based early warning initiatives covered in this chapter.

The tendency of early warning to produce skimmed data, paired with South Sudan's slanted digi-political playing field, ultimately meant that a more transformative approach to conflict mitigation remained missing-from-hand. Sadly, Kumar Rupesinghe passed away in February 2022, so I wasn't able to ask him for his views on all that. Given his call to build early warning systems that would serve the vulnerable, he might have felt a sense of void—as do many who follow in his steps when associating data-driven conflict mitigation with inclusive and decentralized efforts that would ultimately empower those affected by conflict and violence.

5

Making Peace

AI-Enhanced Mediation and Dialogues

This chapter turns towards efforts to make peace between the parties in an armed conflict through mediation and the facilitation of dialogues, and explores how the increasing use of digital technologies affects such efforts.[1] Among the different peacebuilding activities discussed in this book, peace mediation is probably most strongly understood as a 'human-centred' activity, because mediators and conflict parties often insist that technologies are removed from the negotiation table to reduce the risk of information leakages. Such high-level mediations, commonly referred to as Track 1, are usually conducted by senior government officials or appointed mediators and their teams, such as United Nations (UN) Special Representatives of the Secretary-General (SRSGs), who are mandated to undertake efforts to support conflict parties in the search for a political settlement. They are often supported by a myriad of organizations that conduct complementary dialogue activities, such as informal dialogues and problem-solving efforts involving technical experts and civil society on Track 2, and activities involving ordinary citizens at the 'grassroots' level on Track 3 (Snodderly 2011). However, over the past few years, mediation actors have increasingly employed digital negotiation, facilitation, and dialogue methods, which have the potential to change peacemaking approaches and dynamics.

The desire to leverage the benefits of remote communication or computing power, to mention only two ways digital technologies enhance peacemaking efforts, is anything but new, as I have argued in the previous chapters of this book. Several factors are worth mentioning when attempting to explain the increased willingness of mediation actors to utilize digital technologies in support of their mandates. Among these is the realization that the uptake of digital tools by conflict parties and stakeholders has led to a tremendous growth of information that could influence peacemaking efforts. This includes content shared on social media and the many ideas, statements, comments, or petitions that are published online by advocacy groups or ordinary citizens to influence peace processes and their outcomes. The COVID-19 pandemic, which brought widespread lockdowns and travel bans, also led to a major 'digital disruption' that changed the 'script' of peace mediation (Aggestam and Hedling 2023). It forced many actors to seek remote and online solutions to maintain their peacemaking efforts and increased the willingness of mediation teams to experiment with

[1] Some parts of this chapter include material previously published in Andreas T. Hirblinger. 2022. 'When Mediators Need Machines (and Vice Versa): Towards a Research Agenda on Hybrid Peacemaking Intelligence'. *International Negotiation* 27: 1–32. https://doi.org/10.1163/15718069-bja10050.

Digital Peacebuilding?. Andreas T. Hirblinger, Oxford University Press. © Andreas T. Hirblinger (2026).
DOI: 10.1093/9780198986553.003.0006

technologies that helped with the facilitation of online meetings and remote intelligence gathering (Mandelik and Mohamed 2020). Moreover, mediation teams and mediation support actors have been confronted with an ever-growing marketplace of digital solutions that promise to make their work more efficient and impactful. The development and adoption of many such digital methods have been driven by a normative commitment to increase broad-based participation in peace processes and make them more inclusive. As I will discuss further below, applications leveraging artificial intelligence (AI) in particular promise to enable the large-scale and instant analysis of data and are increasingly embedded in participatory approaches.

Against this backdrop, this chapter draws on my previous work (see Hirblinger 2022) to describe how peacemaking efforts are shaped by what can be understood as 'hybrid intelligence', namely, peacemaking-relevant information and knowledge through a network of interdependent humans and machines. Such hybrid arrangements become necessary because human mediators are more and more dependent on intelligent machines to gather and analyse relevant information. On the other hand, despite the fast advances in AI, a scenario where peace mediation is entirely automated remains highly unlikely for multiple reasons, some related to technology and others to politics and ethics. Of course, we could think of many other information- and knowledge-generating processes studied in the previous chapters as similarly hybrid. However, the concept of hybrid intelligence is particularly useful when we want to take a close look at how the promises of automation that come with AI shape apomediation dynamics and efforts. In peace mediation, this uptake has been especially pronounced.

The chapter zooms in on the efforts by the United Nations Support Mission in Libya (UNSMIL) to maintain the Libyan peace process despite deadlocks and limited political will, high levels of insecurity, and the COVID-19 pandemic. Focusing on efforts to facilitate the Libyan Political Dialogue Forum (2020–21), it reconstructs the pragmatic blending of online and offline formats into hybrid arrangements. I also draw on participatory observations from similar contexts where the same digital methods were employed. The chapter describes how AI-enhanced human–machine networks apomediated understandings of conflict and offered specific pathways to peace, generating authoritative knowledge that was used in efforts to support a political settlement. As will become apparent, the uptake of digital methods can alter how mediators support conflict parties in their search for a political settlement, but not always in the ways we would expect. Employing the critical-reflexive lens introduced in Chapter 2, I argue that digitalization and new technological affordances have encouraged a tech-solutionist discourse, centred on inclusivity, that promises to transcend the track logic. Overcoming the hierarchical approach to peacemaking remains challenging because the track model is perpetuated by the sticky professional norms and bureaucratic routines that organize the apomediation of conflict and peace. Moreover, secrecy, exclusivity, and powerful hierarchies are reinforced by some of the affordances of online dialogue platforms and algorithmic approaches utilized to produce insights into mediation. I will demonstrate this by shedding light on efforts to employ digital technologies to broaden participation in peacemaking and strengthen the inclusivity of peace processes. Such digital inclusion efforts have been viewed by many as a solution to the failures of high-level peacemaking. However, as the case of the Libyan Political Dialogue Forum (LPDF) demonstrates, this ambition can quickly be curtailed by established international mechanisms and bureaucratic logics that privilege high-level politics and undermine the potential of digital methods to transcend politically enforced divisions and hierarchies.

The chapter thus illustrates how hybrid intelligence arrangements can disadvantage and curtail ambitions to enable a bottom-up human agency intended to initiate political change. I demonstrate how this is the case specifically in data-driven applications focused on documenting conflict stakeholders' preferences by adopting algorithmic methods that generate partly synthetic data. I argue that the tendency to employ the data outputs of hybrid intelligence arrangements as a means to predefined ends, paired with the reliance on statistical machine learning methods that generate data instead of fostering real-world human interaction, ultimately stifles any opportunity to diverge from the political status quo. This chapter thus suggests that the trends towards AI-enhanced peacemaking may leave a void for all those who intend to initiate and support forms of peacemaking that aim to transform conflicts by encouraging a change in power relations. However, drawing on insights from efforts to promote an intra-Palestinian dialogue that preceded the latest wave of violence in the Israel–Palestine conflict, I also point to uses of digital technology that promise, at least temporarily, to challenge existing structures of marginalization and seek to promote political renewal.

Hybrid Peacemaking Intelligence between Human and Machine Hubris

Peace mediation professionals have long been sceptical about the integration of digital technologies, especially AI, into peacemaking efforts. The first parts of this section reproduce some of my previously published work (Hirblinger 2022) to argue that we have nonetheless witnessed the rise of what I describe as 'hybrid peacemaking intelligence'. The predominant reasons for the relatively slow adoption of AI in peacemaking are concerns about the confidentiality of negotiations and, relatedly, risks stemming from the insecure transfer or storage of sensitive information. At the negotiation table, digital technologies have long been used only sparsely, if at all. The Iran nuclear deal was negotiated in a wood-panelled room in the United Nations Palais in Geneva, filled with heavy tables and leather armchairs. All technical equipment was removed to prevent information from leaking to the outside world and travelling through the channels of the early twenty-first-century social media ecosystem. While concerns with data security and the confidentiality of talks are often openly shared, there's also a less visible but likely even more influential reason for the cautious (and slow) adaptation of digital approaches, namely the persistent notion that mediation is human-centred (Eleiba 2020). In this view, good mediation outcomes rely on the facilitation of the process by eminent persons, in most cases senior or elderly men—preferably current or former statesmen or high-level officials—who assert their authority through charisma and wisdom. This is also reflected in the public representation of peace processes, of which we can get a first glance through a simple internet search. Almost all Google image searches for 'peace accord' and a given country name reveal pictures of senior male party representatives and mediators talking to each other, shaking hands, or signing an agreement.[2] Technology seemingly plays a role only at the margins, such as during the photo op. The representation of peace mediation as a human-centred process is further underpinned by the view that peace processes should remain an *art*, not turn into a *science*, because agreements cannot be achieved by creating objective, scientific knowledge or through engineered, mechanical solutions (Standfield 2020). Yet, such depictions of peacemaking create a false dichotomy between humans and machines, sustained by an outdated anthropocentric view of peacemaking that is more reflective of an ideal-typical view of professional culture rather than based on an assessment of actual peacemaking practice.

The work of peace mediation has traditionally been viewed as one of brokering information between the conflict parties, in an attempt to build trust and reduce uncertainty through providing information to conflict parties (Kydd 2003; Savun 2008). However, peace mediation is a prime example of how apomediation has affected efforts to make or build peace. The growth in process-relevant data and the emergence of apomediation means that it has become considerably more difficult for mediators to maintain their authority as providers of crucial information. For instance, most conflict parties and stakeholders employ social media in support of their interests and agendas, including influencing public opinion, generating political mobilization potential, and gaining tactical advantage on the battlefield (Rohwerder 2015). Moreover, peacemaking efforts increasingly produce large amounts of data, for instance, when mediators, mediation support actors, or civil society activists who attempt to influence the process launch interactive websites, online surveys, or messaging applications to gather information about the needs and interests of conflict parties and stakeholders (Hirblinger 2020). This means that mediators usually find themselves in what has been described as an 'ocean of information' (Varela 2021). These vast amounts of data cannot be analysed by human hand and eye, partly because of their size and partly because they are generated on social media platforms that influence how data can be retrieved and for what ends (Batrinca and Treleaven 2015). This means that the intermediary role of mediators is circumvented (or undermined) by the internet and social media networks, through which regular users retrieve and generate information. Rather than enjoying a privileged position, mediators and their teams today simply form part of the vast human–machine networks in which information about the conflict is apomediated.

Of course, one could assume that satellite images, video conferencing, or social media may enable mediators and their teams to be closer to what happens 'on the ground'. However, the opposite is very often the case because the vast amounts of

[2] I conducted such searches for Ireland, Colombia, Kenya, Mozambique, the Philippines, South Sudan, Sri Lanka, Sudan, Syria, and Yemen.

digital information about the conflict create distance. Since there is simply so much data being generated and collected, obtaining an overall picture and understanding patterns and trends becomes more challenging without automated data analysis. Therefore, an increasingly vocal group of advocates has argued that to stay connected to the peace processes, it is necessary to automate data analysis with the help of AI (Höne 2019). This adaptation process was further accelerated during the COVID-19 pandemic due to the widespread lockdowns and travel bans. This provided a unique opportunity to question and discard the human-centred orthodoxy of peacemaking because mediators often couldn't use their conventional approaches to be as close to the parties as they desired. Given the need to maintain processes, taking diplomatic and peacemaking efforts online became more acceptable, amounting, in the words of some practitioners, to a 'paradigm shift' that led to 'new opportunities for creative solutions' (Mandelik and Mohamed 2020). This entailed, inter alia, the use of videoconferencing software to organize working group meetings and, at times, high-level diplomatic gatherings. It also led to an uptake of data-collection exercises and interactive methodologies to involve ordinary citizens to compensate for the lack of in-person consultative processes. The large amounts of data created during such activities, in turn, strengthened the case for AI-enhanced data analytics.

While the lockdowns and travel bans have long been lifted, many AI-enhanced data-gathering and dialogue methods continue to be used. Indeed, the range and functionality of digital technologies employed in support of peace mediation, including those integrating different kinds of AI systems, have continued to grow and diversify. This is at least partly because of the decline in funding and resources in the sector, and partly because mediators and their teams find these methods convenient and effective. To start with, many ubiquitous applications that are likewise employed in other professional fields help with everyday tasks and increase efficiency, such as transcription, translation, or summarization programmes. Generative AI, for instance, is now commonly used in the peace mediation community to create bullet point overviews of meeting minutes and help with drafting and editing reports or documents. It comes in handy when developing materials for training sessions and simulations, such as negotiation scenarios.

Furthermore, various types of supervised machine learning techniques and generative AI models have been tested and employed to facilitate conflict analysis, including measuring the degree of polarization on social media or determining the positions of conflict parties (Hawke et al. forthcoming; Arana-Catania et al. 2021). Moreover, some emerging approaches use online interfaces or applications to directly communicate with conflict parties and stakeholders, enabling digital mediation processes that complement in-person efforts. This includes specialized applications for remote consultations, to facilitate what was referred to as 'digital focus groups' or 'large-scale digital dialogues' by the UN (Masood Alavi et al. 2022, 240), which I will discuss in greater detail below. The developers of these applications promise that, besides collecting insights about the needs, interests, or preferences of a large sample of participants, they enable 'deliberation' and 'building' consensus between representatives or constituents of conflict parties (Schirch 2024).

I will explore the specific tech-solutionist claims attached to these applications further below. However, before doing so, it's important to acknowledge that this adaptation process has sometimes been accompanied by a hubristic, tech-solutionist discourse, perpetuating the view that AI could simply automate peacemaking efforts and replace human mediation. If conflicts are increasingly automated, why not peacemaking? If (ro)bots wage war on social media and increasingly offline, why shouldn't they also resolve conflicts? Such thought experiments are at times inspired by the successful deployment of AI in other fields, reflected, for instance, in calls for a global data-driven and machine learning-powered system that could make predicting wars as straightforward as forecasting weather (Guo et al. 2018), or the idea that whole peace agreements that could be written with the support of machine learning tools (Rutherford 2021), which emerged as soon as observers noted that AI systems became capable of generating press releases, novels, and poetry (Lombardo 2019).

Today (or, at least, at the time of writing), the considerable advancements in Large Language Models (LLMs) such as ChatGPT mean that lay users can indeed task virtual assistants to draft parts of a peace agreement provision, with limited skills and knowledge necessary to tailor suitable prompts and instructions. Conventionally, AI research has differentiated between 'symbolic' systems (also known as 'classic' AI) that generate inferences based on existing knowledge, stored in a structured format, from rules or mathematical formulas, and 'connectionist' systems that operate based on training data, which enables inference of rules through patterns observed in the data (often referred to as 'machine learning'). Today, these two approaches are commonly combined in what has been called informed machine learning systems (von Rueden et al. 2020). Most prominent among them are LLMs, which have been trained on such large amounts of text data that they are able to reliably predict the sequence of words necessary to formulate a coherent answer to a prompt question. LLMs generate their results based on a measure of probability derived from generic training data. This means that Chatbot responses often appear rather broad-brushed, generic, and unattuned to specific contexts. However, when enriched with additional structured knowledge, provided with context-specific input data such as transcripts from negotiations or survey data, and employed in combination with the symbolic AI models, these systems seem increasingly capable of supporting mediation and dialogue efforts, as will become apparent in the course of this chapter.

Does this mean that the tech solutionists were, are, or will be justified? Are we possibly facing a scenario where AI applications may not only be ready-to-hand for peace mediators but may turn into the 'hand' themselves? As we will see below, such a scenario is not only improbable but also distracts from the critical task of scrutinizing how both human and machine agency condition efforts to settle armed conflicts. Much suggests that developing a 'broad' peacemaking AI will remain a considerable challenge for the foreseeable future. One reason is that peace processes (and peacemaking) do not follow defined and predictable rules or patterns, as is the case with traffic or weather systems. The heterogeneous, context-specific, and evolving nature of armed conflicts would require that the AI can reason in context (Shevlin et al. 2019)—yet it is precisely this context that is often very difficult to determine. Current digital applications that support peace mediation focus on much narrower tasks, or tasks that don't require much context knowledge. Martin Wählisch, who worked for many years with the UN Innovation Cell, predicts that:

> In the not-so-distant future, one could even imagine that personalized conflict resolution chatbots or virtual agents could serve a mediation function themselves, providing a safe and neutral space for parties to express their concerns and interests (Wählisch 2024)

Nonetheless, such applications will continue to rely strongly on human intelligence to produce meaningful outputs, for instance, the intelligence of conflict parties or stakeholders that share their insights, analysis, and interpretations of what is at stake in the negotiation. Crucially, the quality of the AI's output will primarily be determined by the quality of the human input they receive, that is, sufficient quality information about the relevant process and contexts, as well as suitable analytical categories that support their sensemaking efforts. I will discuss these two aspects in greater detail below.

A more likely scenario than a fully autonomous peacemaking AI is that AI systems will continue to be employed with a 'human in the loop'—in close collaboration with human experts that shape machine-generated outputs, amongst others, through the fine-tuning of models and the provision of quality input data, instructions, or prompts (Goodlad 2023; Wang 2019). Human-in-the-loop approaches are increasingly acknowledged as an essential safeguard to prevent unintended effects. Human mediation experts can stay 'in the loop' at various stages of AI development and deployment, closely supervising the design and selection of AI systems and the inference processes these systems will undertake. That way, they can exercise the necessary authority over the AI-generated information and analysis that is supposed to support the facilitation of a settlement between the conflict parties. However, there are other stumbling blocks in the way of automating peacemaking. As long as peace mediation is an activity that forms part of international politics and is something that is led by foreign governments or civil society organizations (CSOs) and mandated through decisions of the UN Security Council (UNSC) or comparable organs, it is difficult to imagine how human agency could be removed from it altogether.

Therefore, I have suggested thinking of peacemaking as a human-led practice that increasingly rests on 'hybrid intelligence' (Hirblinger 2022). This term acknowledges the growing influence of AI but stresses the continued relevance of human intelligence and agency in the production of the information relevant for the settlement of armed conflicts, suggesting that it is thus best to think of such information as a product of interdependent human and machine agency. Importantly, by focusing on the interdependency of human and machine agency in shaping peacemaking intelligence, the concept also invites us to critically examine how they influence one another. Hybrid intelligence systems can be understood as a specific kind of apomediation arrangement. They generate information about the causes of conflict and advance options for conflict resolution in decentred human–machine networks, just like the other applications studied in the previous chapters. These systems commonly include algorithms that organize, process, or distribute information, such as those used to help detect harmful speech or analyse social media data to map trending topics for conflict early warning. Many of the systems used in peace mediation perform similar tasks, such as analysing the sentiments, arguments, or positions of conflict parties or identifying statements that could bridge between the parties' positions.

A particular concern of this chapter is to demonstrate and problematize how computational concepts and algorithmic approaches have the power to challenge and influence the methodological orthodoxies that have long characterized peace mediation and how this affects transformative ambitions. Many applications tested in the years since the COVID-19 pandemic have operated with classical AI, which clusters algorithms that can be used to group statements together that contain a similar combination of words, such as 'k-means clustering'. Such methods have been experimented with and at times integrated into text-mining applications to help identify latent topics in a data set, for instance, of social media posts. The same method can also help identify different preferences among conflict parties or stakeholders. For example, an online platform called Poli.is has attracted a lot of interest in the peace mediation community because it promises to enhance public consultations on policy proposals, inter alia, by asking a large number of participants to provide written comments. The platform employs 'k-means clustering' to synthesize opinion groups based on the comments that the participants of an online consultation have submitted in response to a specific policy proposal. K-means clustering follows a highly complex procedure that translates a corpus of text items (these are often referred to as 'documents', but they could be short text responses to an open survey question or a social media post) into mathematical values and then organizes them into clusters that bear similarity. It first uses a statistical measure to evaluate the importance of a word in a text corpus, that is, the TF-IDF (Term Frequency-Inverse Document Frequency) that we already came across in Chapter 4. Using TF-IDF, the algorithm then translates text documents into numerical vectors and projects them into a vector space. This also allows statements to be compared mathematically and the distance between them to be measured. After that, k-means will cluster the various statements around several 'centroids' based on their TF-IDF vector distances, thus drawing statements with a similar TF-IDF pattern into one cluster (Wachowiak 2022).

This is only a cursory and likely incomplete summary of how the method works, yet it makes clear how hybrid intelligence systems introduce modes of 'inference' that are very different from how mediation professionals usually go about the work of identifying overlaps, similarities, or differences between what conflict parties want. Conventional participatory methods, such as employing sticky notes that are commonly used in Track 2 dialogues and other settings where collaborative ideation is necessary (Close et al. 2020) do not rely on a mathematical calculation of similarity between words, but human thought processes and, importantly, dialogue between the participants. Instead of simply presenting a result, such conventional methods utilize the process of getting to the result to do the work of conflict transformation, as will become clear further below. The results of automated clustering exercises can intuitively make 'sense' to human analysts—but they won't know why. This is because k-means clustering operates through unsupervised learning, which means that the method is not entirely transparent because the inference process is often difficult to interpret (Hu et al. 2024). This makes the outputs of hybrid intelligence systems potentially delicate because mediators or facilitators who rely on algorithmic methods won't be able to explain to the negotiation parties and other stakeholders how these results were produced.

Unfortunately, the reason for why AI-based methods are taken up is often not because they are seemingly fit for the task at hand. Many computational methods are integrated into digital peacemaking approaches, less based on an assessment of what is required for a dialogue, but rather because they are simply available, promise to deliver useful results, and are ready to be employed. While they may be novel to peace mediators and their teams, they have often been developed to suit other types of clients and their needs, such as market research teams that aim to test product prototypes or to understand how popular and trending specific products are and why. This is the case with sentiment analysis tools that are commonly employed to analyse product reviews, for instance. While they have made their way into conflict analysis toolkits employed in support of peace processes, their value for mediators and their teams is often contested—perhaps except for the use case of measuring the popular sentiment around a peace process or mediation. Introducing such methods and committing financial and human resources to make them work also contributes to altering how peace mediation is understood and practised as a professional activity. For instance, it is not unlikely that the uptake of sentiment analysis will make mediation teams care more about their public appearance and strategies to 'sell' a deal to the public instead of investing in methods that can enable transformative change.

There are several other examples where hybrid intelligence arrangements lead to a clash between how humans and machines make sense of the conflict, and where the limited transparency of AI methods complicates any effort to seek a common understanding. These differences may be less visible because AI methods are often described in terms familiar to peace mediators. Yet, a closer look reveals that humans and machines don't have the same perspective on the world they try to make sense of. For instance, a team of machine learning experts at the universities of Warwick and Oxford suggested that one could measure the distance between conflict party 'positions' employing a popular Natural Language Processing (NLP) model developed by Google, often referred to as BERT, which stands for 'bidirectional encoder representations from transformers'. BERT is an LLM, which can be employed to generate a statistical understanding of language use by assigning a 'linguistic position' to words in an abstract multidimensional space, whereas words with a similar meaning are given positions with greater proximity. The team used this method on a data set of notes from a dialogue process, based on the assumption that 'the model would be able to extract party differences by measuring the distance of the language used by one party from that of another party'. (Arana-Catania et al. 2021, 14). The team recorded relative success in operationalizing this model, producing, inter alia, a visualization of the quantitative distance between conflict parties in relation to specific topics such as 'Constitution' and 'Demobilization'. Yet, these results missed any qualitative illustration of what these quantitative values meant in practice.

Mediation experts familiar with the intricacies of language in mediation contexts may rightly object that the meaning of certain terms is highly context-dependent. For instance, one party arguing for a 'decentralized state' may appear diametrically opposed to a second party arguing for a 'centralized state', whereas a third party arguing for 'federalism' may be closer to the first party. However, anyone who has closely followed the federalism debate in South Sudan, for instance, will know that the actual fault line there ran between advocates of a federal state and those preferring a decentralized state, which provided a means of retaining most powers in the national government (Johnson 2014). Similar intricacies of language use in negotiations seem to have become clear to the research team. However, given the limited transparency of BERT, it remains impossible to reconstruct how the model produces its outputs and to compare its inference process to the thinking of human analysts. Unsurprisingly, the team concluded that machine learning tools 'cannot replace human analysis' and that the introduction of such tools would require 'adaptation on the part of users' as well as a 'need to adapt the tools' (Arana-Catania et al. 2021, 23).

A comparable set of challenges emerges when LLMs are employed in support of peace mediation, even though in a slightly different order. Models such as Open AI's ChatGPT can be easily customized and tailored to fit a specific task at hand, including by attaching additional knowledge and instructions to the model, such as guidance on particular methods for conflict analysis. This means that the concept that the model tries to detect in the data will resemble those created by human mediators. For instance, in 2024, a team at an organization by the name of Frontline Associates led by Claude Bruderlein spearheaded the development of customized ChatGPTs that implement methods such as the 'Iceberg Model', which analyses the positions, interests, and needs of conflict parties. To this end, the GPT was tasked to employ Bruderlein's own handbook that explicates the method, which was attached as knowledge (Ma et al. 2024). At least two challenges with this approach are worth pointing out. First, getting suitable data to analyse, given that the parties' actual interests and needs are often not to be found in the public domain. Yet, chatbots tend to be programmed to always provide an answer, even when their knowledge base is scarce, and they often tend to do so with prescriptive confidence. In contrast, many mediation initiatives conduct time- and resource-intensive efforts to organize participatory data collection exercises that can generate these kinds of insights.

A second fundamental challenge pertains to the data used to train the foundational model, which tend to be sourced from publicly available (mainly online) text sources. However, many armed conflicts take place in digitally less connected parts of the world, and the particular ways of speaking, thinking, and interacting that are common to populations in those places may not yet be sufficiently represented on the internet. As the training data shape the quality and appearance of the output, this risks erasing much of the linguistic and cultural heterogeneity with which conflict parties and stakeholders usually voice their positions, interests, or needs.

This problem is manifest particularly in 'zero-shot' prompting approaches, where users task the chatbot to answer the question without any specific training examples, which means that the model will rely only on its pre-existing knowledge and semantic understanding of the issue (IBM 2024). Such approaches thus implicitly assume that AI can be employed to make sense of conflicts across a diversity of contexts through one single model of the world (Amoore et al. 2024). Yet, we know that armed conflicts are not only strongly heterogeneous, but more crucially, that dealing with the differing views on the conflict, its causes, and its dynamics that conflict parties hold is often key to resolving it.

Despite these and other challenges, recent innovations in AI have created widespread optimism about how it could help overcome some of the key challenges that peace mediation efforts commonly face: they operate under enormous time pressure, with limited resources, and they need to make sense of, and accommodate, a vast number of interests and needs. The remainder of this chapter will thus explore how the increasing use of AI and the rise of hybrid peacemaking were sustained by the hope that increased connectivity and processing power could make peace processes more inclusive, more efficient, and more sustainable. However, I will shed light on various hurdles that stand in the way of these objectives, including ingrained institutional cultures and practices that favour high-level deals, the tension between human and machine conceptions relevant to peace, and the ways in which human mediators approach, value, and use data.

Digital Inclusion and the 'Tracking' of Apomediation

While peace mediation has seen a considerable uptake in digital innovations over the past decade, it has also been characterized by limited successes and the insight that the majority of negotiated settlements will break down at some point (Bosetti et al. 2017). One factor that was widely discussed as influencing the sustainability of settlements is whether they were negotiated exclusively in high-level processes involving elites or whether they involved a broader section of society. While high-level mediation efforts have often resulted in exclusive elite deals, recent research suggests that sustainable and lasting agreements require broad-based participation (Krause et al. 2018; Nilsson 2012). This research was accompanied by the rise of a distinctly normative call for more 'inclusivity' through participatory efforts that could ensure that the 'interests' and 'needs' of all relevant stakeholders are taken into account, as was noted in the UN Guidance for Effective Mediation (United Nations 2012), amongst others. This trend towards inclusive peacemaking was partly a response to studies that found that high-level elite deals enjoyed limited legitimacy and thus often didn't last very long, and partly, a normative drive towards strengthening the participation and protection of marginalized groups and women (Hirblinger and Landau 2020). In this context, technocratic questions about the right 'design' and 'sequencing' of peace processes increasingly gained traction. This entailed maintaining a discussion of how the outcomes of mediation activities at Tracks 2 and 3 could be 'transferred' to high-level negotiations, assert pressure on the conflict parties, and make possible more lasting agreements (Cuhadar-Gürkaynak 2009).

One widely acknowledged feature of digital communication is that it easily transcends places, spaces, and many of the institutional, structural, and functional divisions established by states and society. Digital technologies, the internet, and social media have often transgressed modern forms of social and political organization, such as the nation-state, but they have also helped reproduce them. Apomediation has the potential to overcome existing hierarchies and undermine established authority, as I have discussed in Chapter 2. It is, therefore, hardly surprising that digital methods also became viewed as a potential solution to peacemaking efforts focused exclusively on high-level politics and imposing solutions from above. For instance, several early contributions to the digital peacebuilding scholarship discussed how digital technologies could enable 'bottom-up' approaches that could provide viable additions—if not alternatives—to the 'top-down' approaches of international organizations (Tellidis and Kappler 2016; Tellidis 2020). Even earlier, Welch et al. discussed how the 'web' could enable 'alternative spaces' for peacemaking and provide opportunities for empowerment (Welch et al. 2015).

However, much suggests that this emancipatory potential should not be taken for granted. In my earlier applied research on how digital technologies can be utilized to foster digital inclusion, I discussed the risk that digital methods could be employed in an extractive manner, especially when data about the preferences, interests, or needs of conflict populations are generated without their active participation. As an essential safeguard against this risk, I suggested that such data should be provided intentionally by those who are affected by, and have a stake in, the conflict. More specifically, I drew on the works of Albert Hirschmann, Judith Butler, and Aletta Norval to argue that digital inclusion should bring the 'voice' of stakeholders into the peace process. The data provided through digital methods should contain an account of the participants that reflects their 'attempt to change an objectionable state of affairs' (Hirblinger 2020, 41; see also Norval 2009). If understood in this particular way, digital inclusion could form part of a political project that contributes to conflict transformation; one that is aimed at changing how the parties in conflict relate to one another and the political institutions through which these relations are governed. By extension, digital inclusion could thus break with conventional approaches, including the tendency to organize peace processes along hierarchical tracks. Digital technologies were increasingly viewed as a means to support and catalyse such efforts, such as by empowering marginalized stakeholders who didn't have access to the negotiation table.

Admittedly, the most successful sales pitches for digital approaches had much more practical, and arguably more modest, overtones. A common counterargument against the proponents of inclusive and bottom-up approaches is that broad-based participation takes considerable time and resources, which could further reduce the likelihood of successful settlements. In response to this shortcoming, the use of digital technologies in inclusion efforts seemingly provided a solution to efficiency concerns by automating the collection and analysis of data. This vision is best expressed in a paper entitled 'Faster Peace Through Inclusivity', in which the authors argued that AI-enabled large-scale digital consultations would resolve the 'tension between inclusivity and efficiency' (Bilich et al. 2019). This paper was meant to provide a proof-of-concept for an application that became widely used across the peace mediation community in the years after it was published, initially through a partnership between the UN Innovation Cell and the company that had developed the AI-enabled platform called Remesh. The thinking underpinning the article exemplifies how narratives around AI as a productivity enhancer have encouraged a tech-solutionist discourse, centred on inclusivity, that promises to transcend the track logic and make peacemaking both more efficient and more effective. Using Remesh's platform, a mediation team can communicate with up to 1,000 participants via a mobile-device-friendly web platform. Combining multiple-choice polling questions with open-ended questions, the team can ask the participants about their views, demands, or proposals concerning a specific issue. The participants' responses are then shared with other participants and ranked by them, and the outputs of this exercise are used to assess the level of agreement among the participants. Based on the collected data, the platform's algorithms can produce a ranked order of preferred proposals. They can also associate specific proposals with certain sub-demographies across the focus group, that is, groups with specific identity markers, such as gender, religion, or geographic origin (Masood Alavi et al. 2022; Bilich et al. 2019). While commonly referred to as enabling 'digital dialogues', the Remesh platform is not specifically geared towards horizontal conversations between the various participants. Instead, it essentially aims to identify statements that are most representative of various sub-groups across the participants.

As I will argue in the case of the LPDF below, the desire for a 'fast' and 'inclusive' peace did not materialize, mainly because of the blending of digital approaches with conventional approaches to peacemaking that are underpinned by professional norms, bureaucratic routines, and international mandating practices that ultimately reproduce a 'tracked' peacemaking model. The primacy of the track logic also influences how mediation teams produce and utilize peacemaking-relevant information and knowledge. Put differently, when employed in support of peace mediation, digital technologies do not enable a relational and decentred apomediation of conflict and peace. Instead, mediators and their teams employ them to reassert themselves as influential intermediaries and brokers of information, and this ultimately reproduces a tracked peace process.

Hybrid Peacemaking in the Libyan Political Dialogue Forum

These dynamics are evident in practice. I will explore them in greater detail by shedding light on international support for the LPDF, which was held from November 2020 to May 2021, during a time when in-person meetings were difficult to organize due to the travel bans and lockdowns that resulted from the COVID-19 pandemic. In 2011, soon after the fall of Muammar al-Gaddafi, the UNSC mandated the United Nations Support Mission to Libya (UNSMIL) to support the political transition process in Libya. Further mandates also tasked it with undertaking 'inclusive political dialogue, promot[ing] national reconciliation, and embark[ing] on a constitution-making and electoral process'—all the while, the country descended deeper into armed violence as various factions with competing visions for Libya's future aimed to dominate the process through competing transitional institutions (USIP 2022; Asseburg et al. 2018). In 2016, UNSMIL's mandate was revised to support the implementation of the Skhirat Agreement (also referred to as the Libyan Political Agreement), which aimed to establish a unified transitional government composed of representatives of two rival parliaments aligned to the conflict parties: the House of Representatives (HoR), aligned with the Libyan National Army commanded by General Haftar, the General National Congress (GNC), aligned with the Libya Dawnn Coalition, and independent stakeholders, such as from political parties and civil society. However, the process was characterized by multiple setbacks, including a failed attempt to organize a National Dialogue in 2019. At the beginning of the COVID-19 pandemic in January 2020, the conclusions of a high-level meeting in Germany, the so-called Berlin Conference on Libya, urged all parties to 'engage in the intra-Libyan political process under UNSMIL auspices' (Arab League et al. 2020). The LPDF was a direct outcome of the Berlin Conference and endorsed by UNSC Resolution 2510 of 2020 (UNSMIL 2021), tasking it to agree on a unified transitional executive authority and to develop a roadmap for national elections, which members of the UNSMIL mediation team viewed as the ultimate goal.[3]

The efforts to form a transitional and unified government, facilitated by UNSMIL, in the first two years of the pandemic took on a distinctly hybrid character. Travel bans and lockdowns initially caused considerable delays to the process. After an in-person consultative meeting facilitated by the HD Center with key stakeholders in September 2020 in Switzerland, UNSMIL initiated the LPDF with the aim of facilitating a broad-based process (UNSMIL 2020). Acting Special Representative of the UN, Secretary-General Stephanie Williams, who led UNSMIL's mediation, announced that the process would be held according to a 'hybrid' formula, combining 'face-to-face meetings' and 'virtual consultations', with the objective of generating consensus on a unified governance framework leading to the holding of national elections in the *shortest* timeframe possible. The virtual components of this process initially included online conversations via videoconferencing platforms, such as Zoom, with specific constituencies, such as women, youth, and mayors of Libya's municipalities, to broaden the political process. Each of these groups designated a rapporteur, who conveyed the stakeholder group's recommendations to members of the LPDF ahead of an in-person meeting held in Tunis in mid-November 2020. More importantly, the virtual sessions also entailed the deployment of the AI-powered platform for digital focus groups referred to by the name of the parent company, Remesh, whose designers had earlier promised it would bring 'faster peace through inclusivity'.

[3] Interview 106, former Political Affairs Officer, UNSMIL, via videotelephony software, 26 June 2023.

Five rounds of 'digital dialogues' were held between October 2020 and January 2021. Through these, UNSMIL initially aimed to broaden participation in the process. Responding to criticism about the composition of the LPDF, the team worked with a local polling company to obtain a quality sample of Libya's different demographic groups to generate a representative picture of the participants' views (Masood Alavi et al. 2022, 240). However, these digital consultations were also turned into an efficient vehicle for testing and validating many assumptions that the mediation already held (Masood Alavi et al. 2022, 241). Moreover, the mission decided to broadcast the dialogues live on television in order to increase their visibility and thus strengthen the legitimacy of the process. Remesh's application was employed at a critical moment in the process, that is, in the months leading up to the creation of a unified transitional government (the Government of National Unity (GNU)), when the mediation aimed to increase the political pressure on the parties to the dialogue process. The series of digital dialogues touched on several topics, including the impact of the civil war, economic issues, the distribution of oil revenues, human rights, and future elections (Irwin, n.d.). The final digital dialogue held by the LPDF before the formation of the GNU was used by the mediation to develop a list of questions to be answered by the members of the newly formed transitional government.

According to several experts involved in the process, through the combination of digital dialogues and media outreach, 'the election of the GNU by the LPDF appeared to provide some sort of legitimacy' (Masood Alavi et al. 2022, 241). However, it remains unclear how exactly it did so. In fact, while the GNU was elected to office in February 2021 during an in-person meeting in Geneva, the constitutional process stagnated in the months after, and the GNU lost the confidence of several key parties before the end of the year. The envisaged presidential and parliamentary elections aimed at establishing a permanent unified national government were postponed and never rescheduled due to continued disagreements over the electoral framework, among other issues (USIP 2022). While it's clear that the digital dialogues were employed to engineer a process that not only appeared broad-based and legitimate, but also fast and efficient, the peace it was meant to bring to Libya has remained just as elusive in the years that followed.

Notably, the digital inclusion efforts via videoconferencing and the Remesh platform were designed in such a manner as to fit a process design that was established in advance. UN high-level mediation efforts are mandated by UNSC Resolutions, and these clearly define their scope of activities. In the case of the LPDF, the UNSC endorsed the outcomes of the Berlin Conference through Resolution 2510 and requested the Secretary-General to task UNSMIL in line with the operationalization paper produced by the conference (United Nations 2020b). The mission leadership interpreted these as prescribing that the 'formal' negotiations would unfold along several complementary thematic tracks, focusing on the political and economic, military and security, and human rights and humanitarian aspects of the process.[4] Williams

[4] Interview 110, Stephanie Williams, former Acting Special Representative of the Secretary General (ASRSG) to Libya, via videotelephony software, 31 August 2023.

described the digital inclusion efforts as 'sub-tracks', a term that suggests a hierarchical relationship between the high-level process and these broad-based activities. Yet, she spoke of these activities as all contributing to Track 1, arguing that they were advertised, public, and at times even widely broadcast.[5] Indeed, the digital consultations were a means to make sure that 'we were somehow . . . picking up the public sentiment', 'just, kind of, check our assumptions', and 'making sure we weren't operating in a bubble', according to Williams[6] To this end, she briefed the participants of the political track regularly about the results of the digital consultations.

This is telling of an approach to digital inclusion that is not at all meant to provide a bottom-up alternative to the formal elite-centred process because it clearly plays a subordinate role by 'supplying' legitimacy to high-level conversations. Crucially, the Remesh application was mainly employed to confirm established assumptions and generate outputs that would fit a preconceived process, with the ultimate goal of establishing a transitional government and holding elections, rather than opening up to an alternative political process. According to one mediation team member, the objective was to 'listen to the people' and to understand their opinions, demands, and priorities, and then bring 'that together and put it on the table for the LPDF members . . . to put pressure on them'.[7] For this, the mediation team used a report and a slideshow summarizing the most relevant demands of the participants, as well as the statistical distribution of demands across the focus group population.

Moreover, the use of online platforms provided a suitable means for the mediation to control and moderate public participation in the process. According to Williams, 'physical meetings involving Libyans . . . tend to be like, it's like bees to honey especially if they think it's a political process that might yield a new executive. You know, they show up in the 1000s, around the hotel, and they try and pedal influence.' A staff member tasked with organizing and moderating the first preparatory Zoom meeting in advance of the formal launch of the LPDF reported that it was a 'tense and stressful experience', mainly because of the large number of online participants (seventy-five), but also because of the many technical issues that the participants faced. Once underway, the facilitator of the sessions took a highly structured approach, employing a written 'script' with a fixed order of speakers, which made the moderator feel 'a bit more in control'. However, as the facilitator noted, this left little room for the human dimension of the dialogue:

> I started with a more or less friendly tone, though, in the beginning, it was a bit rigid, very formal. And you appear in that context like more of . . . not military, but more of . . . there is no room for humanity, it's like robotic.[8]

These remarks reflect the fact that online platforms tend to allow for a more interventionist approach than would be used during in-person dialogue meetings. For

[5] Interview 110.
[6] Interview 110.
[7] Interview 108, staff member, public information officer, UNSMIL, via videotelephony software, 6 July 2023.
[8] Interview 108.

instance, the participants were asked to raise their virtual hand or write in the chat if they wanted to give a statement, and they would only be able to speak after they were unmuted by the facilitator.[9]

While struggling with, and managing, the potentially overwhelming demand for participation, the online process arguably also allowed the participants to be more outspoken. In particular, the opportunity to contribute anonymously meant that certain voices could be represented more prominently, and this was in line with UNSMIL's mandate of an open political dialogue that would consider minority voices. As Williams put it:

> If you were going to [meet] in eastern Libya, you couldn't be critical of Mr. Haftar because that would result in a knock on your door. For women . . ., showing up physically would often have required the consent of . . . a blood-male relative. . . . so I do think the anonymity of the digital process, gave some, some Libyans the assurance that they could more freely express themselves.[10]

While Williams thus thought that the digital platforms would enable a more 'genuine' conversation, the online encounters unfolded along preconceived lines and were largely viewed by the mediation as a venue to prepare for in-person meetings of the formal process. For instance, the online meetings held with youth representatives were conducted largely to prepare for the in-person meeting in Tunis in November 2020, which was viewed as most influential in shaping the political track of the process.[11] Several additional online sessions were held to facilitate 'mini-tracks' or 'committee' meetings, for instance, by the 'legal committee', to prepare for a second in-person meeting held in Geneva in February 2021.[12] As one team member put it, the digital tools they employed were first and foremost 'enablers' for the process.[13]

Overall, the online meetings and digital consultations were primarily employed as vehicles to prepare and support the high-level LPDF process that was meant to form a transitional government. While ostensibly they were efforts to foster digital inclusion, they also reinforced a tracked peacemaking approach geared towards concluding a settlement between Libya's political leadership. The UNSMIL mediation played a key role in this because it functioned as an intermediary between a large number of digitally connected citizens and the high-level process to which they transferred relevant information. By moderating online meetings, admitting, muting, and unmuting the dialogue participants, the mediation could assert unprecedented control over a usually much messier and chaotic process. Moreover, the affordances of Remesh's digital consultation allowed the UNSMIL mediation to take control of the participatory knowledge-generation processes by setting survey questions, structuring the consultative process, and presenting the results of the consultations to the members of

[9] Interview 108.
[10] Interview 110.
[11] Interview 108.
[12] Interview 108.
[13] Interview 108.

the LPDF. This meant that the mediation did more than structure the flow of information. It also shaped how causes of conflict and possible solutions to conflict were perceived, namely against the pre-defined goal of establishing a transitional government and conducting elections. Williams understood the focus group discussions as an

> [i]nformal poll that you conduct online to answer questions such as do you want elections? Yes or no?. What types of elections do you want? Do you want simultaneous presidential and parliamentary elections? Or, when asking the participants about their view on the economic situation in the country, was there anything the participants had to say about lack of transparency, corruption, services?[14]

Williams' recollection of the process suggests that the mediation used the broad-based process to collect information not only in support of a preconceived process that was meant to result in elections, but also based on established problem statements that conditioned what could be said about the underlying causes of conflict in Libya. The use of digital tools to enable the hybrid facilitation of the LPDF thus demonstrates that digitalization does not necessarily lead to apomediation dynamics in which peacemaking knowledge is constructed in decentred and relational human–machine networks. Indeed, intermediaries such as mediation teams can use digital approaches to reassert themselves powerfully, potentially controlling the flow of information even better and with more authority than in an offline setting. However, in the case of the LPDF, this approach meant that thousands of collected inputs were employed to prop up a political process that soon after began to stagnate and ultimately failed.

How Data Drive Peacemaking

The case of the LPDF illustrates how mediators use the data output of digital inclusion efforts for pre-defined ends. In fact, the data that AI systems generate are commonly described as an 'output': a final product that a human user can utilize. Hybrid intelligence systems generate these data with the active engagement of humans, including data scientists and domain experts fine-tuning the model or mediation team members formulating queries and participants providing the input, to mention only a few options. I will argue in the following section that such interactions are commonly geared towards the means of producing a particular data output, instead of enabling a more open-ended process that could contribute to conflict transformation. The increasing use of digital technologies and AI for peacemaking thus not only makes peacemaking more data-driven but also more driven by data. These data are commonly employed for preconceived goals that the mediation derives from its mandate. This means mediators and their teams tend to establish and control apomediation

[14] Interview 110.

dynamics that are best suited for an already designed process, instead of using apo-mediation to define goals in a decentred and participatory manner. In the case of the LPDF, for instance, the mediation employed digital meetings and dialogues to generate an analytical output that could inform and influence the high-level process intended to form a unity government and pave the way to the elections. The data for this output were produced through computational methods that integrated the participants' inputs into a rigid data generation pipeline. As I will argue in the following two sections, this approach can result in lost opportunities to facilitate conflict transformation during the data creation process itself.

While I didn't have the chance to closely observe the deployment of the Remesh platform during the LPDF, I am developing my argument based on the participatory observation of another Remesh deployment in a comparable context in 2023.[15] That said, the following discussion should not be read as an evaluation of the Remesh platform or that particular process. Such an effort would have limited value, as such technologies undergo regular technical updates and improvements that change their specific functionalities. My aims are also much more foundational, namely, to demonstrate how powerful algorithmic techniques organize the interaction between conflict-affected populations and the mediation, and how this affects the prospects of conflict transformation. Given the limited space, I also won't discuss or review all features of the platform. Rather, I will focus on selected aspects to make a more significant point, namely, that AI-enhanced approaches risk reducing human–machine interactions in peace processes to a means of a pre-defined end. This, in turn, reduces the human agency of conflict-affected populations to providing factual data about the world instead of contributing to changing it.

Computational approaches commonly direct the mediation towards observable data. These data are usually treated in a 'sincere' manner, namely as 'evidence' that approximates the 'world as it is'. This approach to data rests on a metatheoretical foundation that we can describe as a variant of 'empiricism', which aims at producing information that corresponds as much as possible with an observable reality (for a more detailed discussion, see Hirblinger 2023; Hirblinger, Wählisch et al. 2024). One conviction underpinning such evidence-based approaches is that gaining as accurate a picture as possible of the conflict context and the needs and interests of conflict stakeholders and affected populations, and inserting this empirical information into high-level negotiations, can result in more legitimate negotiation outcomes. In the context of the Remesh deployment that I observed via my web browser, the participants were informed, for instance, that the results of the dialogue would be used to feed the 'citizens' viewpoints in(to) the national and international efforts' by the mediation. As I will discuss in the remainder of this section, this goal of using the collected data as a means to a pre-determined end informed the entire set-up of the digital dialogue, precluding other forms of dialogical interaction.

A strengthened role for empiricist methodologies risks that other forms of information will be less valued, including the tacit (and informal) knowledge of mediators,

[15] For confidentiality and ethical reasons, I cannot share additional details about the case.

conflict parties, and stakeholders, or intuitive knowledge that often cannot easily be written down, categorized, and quantified. This also entails knowledge relevant to the conflict context that is not machine-readable, such as about informal political arrangements, cultural habits or preferences, and informal practices. This is a common issue with machine learning approaches employed to recognize pre-trained entities, which are most successful in detecting and analysing visible and easily categorizable objects such as weapons (Yadav et al. 2023), but struggle much more with picking up community-specific language that could be used as an indicator for polarization between groups (Hawke et al. 2024). However, it also affects AI enhancements that employ LLMs because these are trained on an observable archive of online text sources. The UN Innovation Cell, for instance, which employed the Remesh platform across various Middle East and North Africa (MENA) countries, also supported the development of NLP modules for various Arabic dialects. This enabled AI systems to process natural language inputs from a diversity of national and sub-national groups in the MENA region (American University of Beirut 2023). However, such context-specific modules were not employed in the case that I observed. While the dialogue was held in Arabic, the facilitation team suggested employing the web browser's automated translation to read the results in English. The automatic translation often produced text snippets that didn't amount to clear and unambiguous text. This made it difficult to perceive what the participants wanted to express.

Many fully automated text-mining methods face challenges when tasked to produce quality outputs, such as summaries of conversations or clusterings of topics, due to factors such as noisy and unstructured data, domain-specific terminology, and a vast vocabulary. Unsurprisingly, the most important analytical processes I observed on the Ramesh platform relied on direct participant inputs that were not very different from standard survey methods, including multiple choice and rating or ranking exercises that provide insights about the participants' preferences. I will discuss these in greater detail below. At this point, it's important to note that these methods aim to generate unambiguous information that can be easily translated into numerical data and used to make sincere determinations about the participants' preferences. Indeed, one of the perceived benefits of survey methodologies is that they can generate results that are fine-grained and context-sensitive. This helps with identifying preferences held by a majority of the population, but just as much to study sub-populations that may have diverging preferences, which could spoil the peacemaking effort or an agreement that results from it. While collecting the subjective preferences of each participant, such methods are ultimately geared towards producing an apparently objective and quantifiable representation of preferences across the group.

However, negotiations and dialogues are fundamentally shaped by their path-dependent character and by uncertainty (Hirblinger, Wählisch et al. 2024). This pertains not only to the process outcome but also to the process itself *and*, most importantly, to the participants' positions. In fact, the participants' views on specific topics are hardly ever cast in stone, even though it might seem like it. Instead, they change depending on the context and the process, and indeed, they often *have* to change so that a settlement can be achieved. Mediation is commonly understood as an effort by a third party to help the conflict parties achieve a mutually acceptable agreement (United Nations 2012). For proponents of conflict transformation, this requires facilitating a process that encourages the antagonistic parties to recognize their mutual positions as equally valid and to reflect on their own interests, needs, and beliefs vis-à-vis those of others. Only when such preferences are not set in stone can the transformative change towards peaceful relations become a possibility. In the digital dialogue I observed, the facilitators could shape the dialogue dynamic somewhat, for instance, by pacing the sharing of questions with the participants or by asking follow-up questions. However, the whole exercise was meant to identify the participants' preferences as if they were given rather than changeable, and the results of the consultations were meant to document them in an objective manner. The session was structured through a sequence of questions clearly meant to provide actionable insights for the mediation, such as what groups should be represented in a forthcoming process, how the process should be launched, and which agenda items should be prioritized.

The effort to establish with as much precision as possible what conflict-affected populations want or need tends to care little about other information that commonly informs dialogue efforts. This includes information that derives its value not from revealing the world *as is*, but from expressing a will for political action and change. I have suggested thinking of this kind of information as enabling a 'subjunctive' attitude or mood, one that is concerned less with how the world is and more with how it *could be* (Hirblinger 2023). Indeed, when observing the dashboard of the digital consultation, something else emerged on my screen, despite the structured design of the process and the strong data processing features of the platform: the participants' desire to be part of a political process. For instance, while some participants responded to the fact-focused questions in line with the facilitator's expectations, other participants seemed intent on using the session to initiate a conversation with others that could mobilize for joint action. This was telling of the fact that they had ambitions other than simply providing data according to a rigid data collection process. Instead, their answers expressed the desire to create political momentum for joint action and enter an actual conversation with others about how this could happen. For instance, in responding to the question 'what are the most important issues that should be covered by any ceasefire agreement?', the answer which received most approval did provide a lengthy set of suggestions, including facilitating humanitarian aid and urgent financial and psychological assistance, but most importantly, it started out with a call to action seemingly directed at the other participants: 'First ... on the level of advocacy ... can we see how we [can] support the implementation of the recently signed agreement?'[16]

However, the platform did not provide the opportunity to respond to the participant's call for action. Instead, it treated all submitted inputs according to

[16] The participant mentioned a particular agreement. I have removed its name to ensure the process can't be identified.

pre-defined analytical steps to identify responses and statements that would be most representative of the whole group. To this end, participants were asked to evaluate other participants' responses through two complementary exercises: a simple yes/no vote to determine whether they agree with the statement, and an exercise where participants are presented with two statements and have to select the one they agree with more. The platform translates the participants' evaluation of each response into numerical values called 'utility scores'. The results of these exercises are visible to the facilitating team via a ranking interface that displays the percentage of agreement for each response across all the participants, and this result can also be broken down according to the demographic categories that the participants had initially provided, such as gender, political affiliation, or geographic origin. The call for action to support the Jeddah agreement received an approval rating of 78 per cent. However, I was left wondering what this percentage really meant and what conversation would have unfolded if the participants had been gathered in a single meeting room.

The Generation of Voice

The platform goes further in its effort to transform each of the human-generated data snippets into an aggregated result that claims to represent the group's preferences. Given the potentially large number of participants and questions, asking each participant to evaluate every single response by another participant is unfeasible. The time needed for this assessment grows exponentially with each new participant and recorded response: if ten participants submitted ten responses, each participant would have to evaluate ninety responses—yet if one hundred participants submitted ten responses, each would have to evaluate 990 responses. To overcome this issue, the platform integrates a machine learning model that uses the recorded utility score to 'fill in utility scores for all thoughts for all participants' (Remesh 2024). Critically, this means that the overall percentages made visible to the mediation team are, to a large extent, not generated through actual human input but synthetically produced. While the final data output is thus partly based on human input, a considerable proportion is composed of synthetic data. While the results are meant to accurately represent the participants' preferences, the 'voice' it contains is partly authentic and partly generated through a hybrid data generation process that aims to generate insights into what conflict-affected populations need and want. More specifically, to generate the missing values, the model undertakes what Remesh refers to as a 'matrix completion process', which arranges all observations about the participants and their responses in a data table (i.e. a matrix) to predict and fill in missing values based on the patterns in the matrix. This approach rests on the assumption that respondents with a similar profile will also likely have a similar preference for an item to which they have not provided a response. According to an information leaflet produced by the company, 'though we can't make explicit decisions about how every participant feels about a response . . . we can make an educated inference based on other data in the matrix' (Remesh, n.d.).

Importantly, this process necessarily reduces the natural heterogeneity of the participant profiles because it fills in the missing data based on a predicted value. Imagine two participants who have both voiced a preference for a secular state, a decentralized state, and a government led by civilians instead of a military junta. We also know that both participants identify as men, are between forty and fifty years old, and have the same level of education. One of them also voiced a preference for enforcing a women's quota in parliament, whereas the other has not provided this information. Would you be able to say with certainty that this person would also vote for that women's quota? At least, we should be curious to find out instead of taking the most likely scenario for granted. What's at stake here is well represented in Chimamanda Ngozi Adichie's concerns with the 'danger of a single story': the risk that we are too quick with reducing complexity for the sake of the (political or personal) convenience of sticking to dominant narratives, and a lack of knowledge of others (Adichie 2009; Shumpert 2018). Yet, challenging these narratives is exactly what is required to advance a more nuanced understanding of the world. The work of transformative peacemaking entails, inter alia, augmenting a diversity and complexity of views and narratives, which helps conflict-affected populations understand that matters aren't black and white and that social and political cleavages often rest on simplified and static narratives (Deiana and Riley 2024; Little and Maddison 2017). Put differently, societies that can accommodate and handle a complexity of identities and interests are also less prone to exhibiting violent antagonisms (Lehti and Romashov 2022, 1292). However, through the 'matrix completion process', which forms part of an algorithmic recommender system called 'collaborative filtering', authentic human interaction and preference for complexity are sacrificed to make sure large amounts of data can be processed at scale and speed; a need that Remesh shares with other private sector companies that aim to understand a population's preferences to obtain recommendations on what products to advertise to a large number of customers:

> Given that we can have hundreds of participants online at a time, it is impossible to collect voting data for each participant on all responses during the voting period. This is where collaborative filtering comes into play to rank and structure the responses—the same model adopted by Netflix or Amazon to make predictions for consumers.

According to IBM:

> Collaborative filtering is a type of recommender system. It groups users based on similar behavior, recommending new items according to group characteristics. Collaborative filtering is an information retrieval method that recommends items to users based on how other users with similar preferences and behavior have interacted with that item. In other words, collaborative filtering algorithms group users based on behavior and use general group characteristics to recommend items to a target user. Collaborative recommender systems operate on the principle that

> similar users (behavior-wise) share similar interests and similar tastes. (Murel and Kavlakoglu 2024)

The interaction with the participants, and the data collection from them, is thus chiefly geared towards matching individual users with specific user groups that ostensibly share a certain preference. Instead of assuming that the participants' views could diverge, the approach assumes they are the same.

Much suggests that the needs of mediation teams tend to differ considerably from those of streaming services or online shops: they rarely have already finished 'products' in stock that simply need to be sold. Instead, the need to create them in the first place. Those who use streaming platforms will also know that much of the content their recommender systems offer may not meet their needs. You will hopefully switch that new show off again after the first five minutes, realizing that it isn't what you really wanted to watch. In dialogue settings, the participants may share and agree to single preferences, such as about the system of government, the distribution of natural resources, or the role of religion in the state. However, these preferences do not amount to a full-fledged proposal for political settlement, and more importantly, they say very little about what steps and change processes are necessary along the pathway to this settlement. While certain aspects of an algorithmically recommended TV show that don't resonate with you won't ruin your evening, having one or two critical 'gaps' in your peace agreement matrix filled with the wrong preferences could well lead to you wanting to spoil the process. Moreover, unlike the next sci-fi-crime series that doesn't quite resonate with your preferences, once signed, peace agreements tend to be around for longer and they have immense consequences for anyone governed by them. Simply put, the costs of 'switching off' an agreement are much higher than pressing a button on your TV remote.

Furthermore, a concern with how AI methods could help with clustering positions and generating consensus risked ignoring the human—and political—aspects of making peace, while at the same time fundamentally reshaping how the participants interact with one another during peacemaking efforts. Indeed, the example of the Remesh platform captures well how the use of machine learning models to generate synthetic data can affect how digital inclusion efforts unfold. To start with, the interaction between the participants that I observed was highly structured. During the live session, participants were bound to answer each question within a set timeframe to allow for the process to unfold synchronously, that is, they would first record their preferences and subsequently evaluate somebody else's preferences. After the live session had ended, the participants could finish these tasks at their own pace, making the process comparable to a normal survey experience. Importantly, both options structured the interaction so that it was difficult for the participants to relate to one another through a direct communicative exchange, as would be the case, for instance, in a simple text messaging chat.

This meant that the opportunities for an actual 'dialogue' were virtually non-existent. Instead of enabling a natural language conversation among peers, the platform placed the participants into an indirect relationship that seemed strangely impersonal—and hierarchical. Instead of equal participants in a dialogue, the users were turned into 'judges' of each other's statements—and they did so without access to the demographic information that the platform had acquired at the beginning of the process. Put differently, they didn't really encounter each other, only their stripped-down statements. Relatedly, and probably most crucially, the platform treated the participants' preferences as pre-given and unchangeable. In the process I observed, users submitted their views on a particular matter only once, and these views were then treated as objective and stable data, which is necessary to process them algorithmically. However, this approach cares little about the fact that conflict party and stakeholder preferences tend to be conditioned by the conflict context itself, and that hardened—and seemingly mutually exclusive—preferences can be both a cause and consequence of violent conflict. Crucially, this approach left little space for many of the conventional dialogue and mediation methods that help, for instance, with humanizing the other, generating empathy and understanding, or challenging harmful convictions and beliefs on which the participants' preferences may rest.

Yet, this kind of work is often necessary for transformative peacemaking. In the search for a political settlement, dialogue facilitators will not only look for a compromise position where the interests and needs of conflict parties and stakeholders overlap. This would mean taking the status quo for granted, which leaves little room for manoeuvre when positions are hardened and seemingly irreconcilable. Instead, transformative approaches to peace mediation highlight the need to work with (and on) the participants' positions. The aim of such efforts is not a convergence of positions and preferences but rather, to find a constellation where they can be acknowledged and accommodated. For instance, in their seminal work, Bush and Folger argued that mediators must encourage the dialogue participants to better understand their own situation and needs and those of others. This recognition of conflict party preferences can then provide the groundwork for mutually acceptable agreements (Folger and Bush 2007). Likewise, agonistic approaches to peace would encourage the participants to enter into dialogue through which they can recognize, acknowledge, or explore a plurality of preferences, views, and interests in an effort to make them more acceptable and to critically rethink their own (Lehti 2024; Suransky and Alma 2017; Strömbom et al. 2022).

The process I observed did not provide an opportunity for the participants to reflect on their preferences. In fact, the platform interface stood in the way of an actual dialogue, that is, a communicative exchange in which the participants have the opportunity not only to share their points of view but can attempt to persuade other participants of the merit of theirs. This is not least because each individual user was only exposed to a fraction of statements. Crucially, these statements are neither necessarily those that have the most sway among the participants nor are they statements that the participant who reviews them may be most opposed to. While the participants will thus take note of the preferences of others, they do so to evaluate them against their own. In the process, their own and the others' positions turn into seemingly stable data snippets with an attributed value, namely, the 'utility score'. If, how, and to what degree this evaluation process influenced the participants' views is ultimately impossible to determine because the participants will submit their own preferences before they get to see those of others. While the mediation team had access to the aggregated data that might be telling, for instance, about which preferences are popular and among whom, these insights were not shared with the participants. That said, it's not unthinkable that mediation teams may decide to do so after the exercise is completed or use it in combination with other facilitation methods that enable the participants to engage with each other's voice.

For scholars or activists who associate digital inclusion with the opportunity to empower groups and transform the relationships between people in conflict, algorithmic approaches that focus on producing reliable representative data by extracting information and synthetically generating participant profiles will likely leave a sense of void. It is important to stress that Remesh does not purport to support the particular view on conflict transformation that I have introduced at the beginning of the book or any of its components, such as encouraging respect for difference or empowerment. Thus, this section should not be read as a critique of the Remesh platform or the particular dialogue against my theoretical framework. However, digital dialogue applications are widely viewed as technical solutions to enhance the inclusivity of peace processes, and inclusivity has been viewed as a vehicle for empowerment and bottom-up approaches that could change the dynamics of peacemaking, as I pointed out above. This section thus demonstrates what can happen if algorithmic methods start driving digital peacemaking efforts, namely that the participants' voice may be reduced to rather static inputs. When used in a hierarchical track framework, such methods risk reducing the dialogue process to a process of algorithmic computation aimed at producing data that are meant to influence high-level negotiations, instead of a process that can generate broad-based change by focusing on the interactions between the participants.

The Limits of Evidence

Could we imagine employing hybrid intelligence differently? In such a way that it wouldn't simply be used to produce data for pre-defined ends, but as an end in itself? Put differently, could the production and exchange of understandings of conflict and peace form part of the effort of terminating it? I think so. This, I will suggest in the remainder of this chapter, would require that mediators, facilitators, and dialogue participants view data information, which is generated and employed through digital methods, with less sincerity. A *sincere* approach towards digital peacemaking, which takes the information contained in data as evidence about how the world 'really' is, often stands in the way of the transformative effort of changing how conflict parties and stakeholders view the conflict. Since these perspectives are closely related to their interests and needs, it's crucial to find ways to use data and information as vehicles to change them, instead of merely proclaiming them as facts. I'd like to suggest that one way of pursuing this goal is to explore how digital methods can be employed in a *subjunctive* manner, that is, to encourage an engagement with how the world could be.

In my previous research (Hirblinger 2023), I demonstrated that change processes can benefit from digital methods that playfully operate between the world *as is* (such as a world filled with violence and antagonism) and a world that should or could be (such as a world of non-violent adversity). I also pointed to several potential subjunctive affordances of digital peacebuilding technologies that can facilitate this movement towards possible (and possibly more peaceful) worlds. Among these are the ability to unlock (or challenge) existing social structures and invoke new ones. I will focus on this affordance in greater detail in the final section of this chapter to demonstrate how hybrid peacemaking intelligence could indeed help transform conflict.

To demonstrate the need to challenge the world *as is*, I'd like to return once more to the topic of inclusivity. One of the key challenges of digital peacemaking efforts is that they often struggle to ensure that the interests and needs of all relevant stakeholders are considered. Paradoxically, while the discourse around digital inclusion has portrayed digital approaches as a means for strengthening participation, the limited digital access and digital literacy that shape many conflict-affected contexts commonly mean that large parts of the population cannot usually participate in such efforts, or at least not as easily as others. Poor connectivity, narrow bandwidths, and low levels of digital literacy all shape the playing field on which conflict and peace are apomediated. Undeniably, these forums of material marginalization are in most cases also closely linked to structures of inequality and marginalization, and are actively exploited and at times reproduced by the conflict parties. For instance, in South Sudan, political leaders aiming to perpetuate violence and armed conflict build their support on a large loyalist following of rural populations that have limited education, limited access to information, and thus a limited capacity to question their leaders' intentions and motives. Ordinary people also often remain politically and economically dependent on their leaders' protection, such as through intricate patron–client relationships through which some of the privilege and wealth accumulated by political elites is trickled down to their loyal followers. In other contexts, such as that of Sri Lanka and Palestine, power imbalances between the conflict parties play a major role. In the case of Palestine, discussed further below, the voice of ordinary Palestinians is also considerably curtailed, amongst other things, by Israel's strategies of military occupation, the annexation of land, and the surveillance of communication.

When aiming to enhance the inclusivity of peace processes with digital means, many mediation initiatives and support actors have looked for ways to counter such constraining factors temporarily. For instance, the UNSMIL team facilitating the digital dialogues in Libya worked with other UN actors, such as the UNDP, to increase digital access. Moreover, they also undertook efforts to ensure that the participants would feel safe when participating, including through close collaboration with the content management teams of social media that would delete harmful speech directed at dialogue participants.[17] This means that certain factors that could constraint the participants' voice were temporarily mitigated, so that they wouldn't reduce the quality of the data sample. The logic behind this thinking resembles that of

[17] Interview 108.

a positivist science, which imagines that the phenomenon under study should remain, as much as possible, insulated from context factors that could influence the measurement. This thinking is also visible in the methodological discussions that surround these applications. For instance, Remesh's platform, when paired with a systematic respondent sampling effort, promises to create a representative sample, even though digital access and freedom of expression are limited. This approach promises to generate 'objective' data, but it won't have a direct effect on the digi-political playing field itself.

Where and when such efforts are successful, hybrid peace mediation efforts that enable the participation of a larger part of the ordinary population could increase the pressure on high-level negotiations to initiate a long-term change, such as through peace agreement provisions asking for socio-economic development or the safeguarding of human rights. There are examples where broad-based processes achieved such objectives. For instance, the outcome document of the Kenyan National Dialogue concluded in July 2008 asked for the 'increasing availability' of 'appropriate technologies to create an enabling environment for poor communities to take part in wealth creation'. Furthermore, in January 2024, the outcome of the Yemen National Dialogue called for constitutional provisions to criminalize the possession of 'electronic interception devices' or their use to conduct 'surveillance of citizens'. The situation is different where the outcomes of digital inclusion efforts are meant to feed into high-level processes, such as was the case with the LPDF, simply because they fail so often.

Importantly, there are ways in which the participants' direct engagement with the context factors of dialogues, including those stemming from the digi-political playing field, can nourish conflict transformation efforts instead of merely constraining them. This means going beyond simply providing internet access, data bundles, or digital skills crash courses to marginalized participants so that they can contribute their data. What's required is to direct attention to how these context conditions stand in the way of peace because they nourish hatred and enmity. An exclusively sincere engagement with this harsh reality will often have a disempowering effect, given that tremendous challenges usually stand in the way of a comprehensive transformation of the relations that reproduce conflict and violence. However, information about these constraints can be employed with a different objective, namely, not only to sincerely engage with the world as it is but with how it *could be*. Using information and data to create such a subjunctive attitude may create momentum that initiates resistance to the social, political, and material relations that constitute the very status quo that keeps parties in an antagonistic deadlock. Therefore, the last section of this chapter will discuss how digital technologies may be used in ways that enable both sincere and subjunctive attitudes towards the conflict.

Data and Hope

I will attempt to illustrate this alternative approach to digital peacemaking by shedding light on dialogue efforts aimed at supporting peacemaking in Palestine in the years before the most recent escalation of violence. In the case I will discuss, no AI-driven technologies were used. Yet, it is informative for how digital methods and data can support conflict transformation in ways that enable an engagement with the relations that underpin armed conflicts. At the time of writing, Palestinian lands and Palestinian life lay in shatters in the rubble of Gaza, following the relentless Israeli military campaign in response to Hamas attacks of 7 October 2023. Before this latest wave of violence, the Israel–Palestine conflict was viewed as one of the most protracted conflicts globally, yet several organizations continued efforts aimed at resolving it. Among them was a dialogue project by the CMI-Martti Ahtisaari Peace Foundation implemented in 2019–21 with a Palestinian partner organization, Masarat - the Palestinian Center for Policy Research & Strategic Studies. The project conducted intra-Palestinian dialogue activities in preparation for an eventual resumption of the peace process between Israel and Palestine. During this time, there was no political momentum for high-level talks involving Israeli representatives, which meant that this effort differed from the LPDF process discussed above in that it had a much more moderate goal of empowering Palestinians in their quest for peace. While not directly engaging with Israeli counterparts, this project aimed to support conflict transformation by working with one side of the conflict, assisting a diverse group of Palestinian stakeholders in finding a joint posture and vision with which future negotiations with Israel could be approached.

In the decades before the project, Israel's policy towards Palestine had increasingly been one of colonial domination through a combination of military occupation of East Jerusalem and the West Bank, the expansion of Israeli settlement activities, and the siege of Gaza. All of these actions undermined peacemaking efforts and reduced the ability of the Palestinians to sustain their livelihoods, thus playing into the hands of those who preferred to maintain the status quo characterized by deep and often lethal antagonisms. The prospects of a revived peace process were further diminished by the fact that politics in Palestine had been characterized by more than a decade of internal division, splitting the territory between the Palestinian Authority (PA) in the West Bank and a de facto government led by Hamas in the Gaza Strip. The absence of a unified Palestinian leadership, mistrust between the two main political actors, Fatah and Hamas, and the disintegration and waning legitimacy of Palestine's governing institutions further reduced the chances of a revived Israeli–Palestinian process and the conclusion of a peace agreement. CMI's project in Palestine operated in the context of a peace process that the implementing team described as 'stuck' or at a 'dead end.'[18] Due to the many failed diplomatic efforts since the 1993 Oslo Accords, the possibility of a conflict settlement through the two-state solution seemed difficult to achieve. As a project team member put it, 'we keep hearing that they [the Palestinian partners] don't see any opportunity for the actual negotiations to even start in the near future because of the [intra-Palestinian] division and the different, opposing viewpoints between Israeli[s] and Palestinians.'[19]

[18] Interview 151, Project Manager, CMI-Marthhi Athissari Peace Foundation, via videotelephony software, 13 October 2021.
[19] Interview 151.

In this context, CMI's project facilitated a future-oriented dialogue process among key Palestinian stakeholders, including political parties, the private sector, unions, civil society leaders, and academics. The project aimed to enable a joint, participatory scenario-building process to increase Palestinian resilience in the face of the unresolved conflict and to develop policy recommendations on issues such as the secure reintegration of refugees and national unity. Due to the COVID-19 pandemic, most of these efforts were carried out remotely and with the help of several digital methods through which the dialogue participants collected, analysed, and discussed data about the current state of affairs. In January 2020, the project team convened a workshop to establish a group of experts of key Palestinian stakeholders and researchers called the National Strategic Dialogue (NSD) Advisory Group. After that, the team conducted a multifaceted participatory dialogue process involving regular online meetings, supported by background studies of available household data and online surveys to gather expert assessments. The project aimed to build several scenarios that described Palestine's future and develop recommendations for how to respond to them. This process was, in many ways, characterized by the interplay of sincere and subjunctive uses of data and information, as will be discussed in the following.

To start with, the project was clearly driven by a sincere desire to know and plan with as much precision and reliability as possible. For instance, the team aimed to identify key drivers shaping future developments using the so-called PESTEL framework, widely used to identify macro-environmental factors relevant to strategic planning and for managing adaptation processes, specifically, factors related to the political, economic, social, technological, environmental, and legal dimensions of the conflict. Moreover, some team members argued that it was essential to underpin the scenario exercise with measurable indicators that would objectively tell where Palestinian society was headed.[20] To this end, the team employed a combination of scientific research and expert consultations to explore key drivers that would influence the future of Palestinian society (Masarat 2021, 14), including factors such as poverty, economic growth, and the availability of basic services. According to some team members, grounding the scenario-building effort in 'empirical realities' was necessary to take the dialogue effort beyond what they viewed as merely personal perspectives, narratives, and beliefs.[21] On the other hand, some team members thought that using too much evidence could be detrimental to a project meant to identify ways of moving beyond the current status quo. As one team member put it, 'if you are stuck in what the world looks like now, then it might be difficult to imagine different kinds of futures'.[22]

Crucially, the sincere approach to data and information was guided by an overarching perspective on what was necessary, from the standpoint of the Palestinian participants, to prepare the ground for peace. This perspective became encapsulated

[20] Interview 155, Head, Middle East and North Africa, CMI-Marthhi Athissari Peace Foundation, via videotelephony software, 15 November 2021.
[21] Interview 154, Associate Professor, Al Quds University, via videotelephony software, 10 November 2021.
[22] Interview 155.

in how they approached the project's objective of strengthening Palestinian resilience. Notably, resilience was operationalized in a particular way. The Arabic variant of the term, الصمود (Somoud), is best translated as 'steadfastness' or 'perseverance'. According to CMI's partners, it 'came to denote an ideological approach and political strategy coined in the wake of the 1967 Six-Day War and has defined Palestinian resistance to the occupation ever since' (Masarat 2021, 13). 'Somoud' was necessary because, in the words of one team member:

> We all know that the Middle East conflict will not be resolved within the next five years. Nobody believes that the Palestinians will gain control over their land in the foreseeable future. Therefore, the goal of this project was to provide the Palestinians with the ability to endure, to at least keep the land they have left over—for that, they need steadfastness.[23]

This localized interpretation of the term 'resilience' centred around the need for an attitude in which Palestinian society would stand united to be prepared for, confront, and manage internal and external pressures related to the conflict, which would also be a prerequisite for any future negotiation with Israel.[24] The project was thus meant to encourage Palestinian society to be concerned not only with the past and the present but also with the future and to bring 'hope to the people'.[25]

Somoud guided the use of the joint scenario-building process, through which the participants discussed alternative chains of events that led to different outcomes to identify how they could influence future developments.[26] This effort had a strong subjunctive trait. In fact, the team's approach to the scenario planning was somewhat unconventional because it organized two scenarios—one being a continuation of the status quo, and one being a worst-case scenario—around a third one, the *preferred* scenario.[27] The latter was meant to guide Palestinian society towards a version of the future in which Palestinians could realize life in dignity. As one team member remarked, this was a more 'emotional' rather than 'analytical' approach to developing scenarios.[28] The inherently normative dimension of the framework meant that the move towards the future would be in line with the core values grounded in Somoud. This scenario would, amongst other things, include the desirable goal of Palestinian unity, in which cohesion between the various parts of the population would enable them to strive towards a more peaceful future. At the same time, the scenario planning exercise itself was meant to be a demonstration of the feasibility of collaboration across intra-Palestinian fault lines; it was meant to enable a 'united discussion'.[29]

[23] Interview 153, Senior Adviser, CMI- Marthhi Athissari Peace Foundation, via videotelephony software, 18 October 2021.
[24] Interview 151.
[25] Interview 152, Analyst, Inclus, via videotelephony software, 14 October 2021.
[26] Interview 154.
[27] Interview 153.
[28] Interview 155.
[29] Interview 151.

The large amounts of data collected along the dimensions of the PESTEL framework, however, didn't make it easy to find a joint perspective. The complex picture that emerged through this sincere quest to analyse the current situation had to be translated with an engagement with how Palestinian society could be resilient in the future. To this end, the dialogue facilitators employed a participatory mapping exercise[30] meant to identify the most important priorities for achieving the preferred scenario in order to enable Samoud. The method combined a participatory ranking via an online interface with live data visualizations employed to facilitate the participants' discussion.[31] The prioritization exercise combined several sincere and subjunctive affordances. To begin with, it asked the participants to engage with the world *as is* by identifying concrete drivers that affected Palestinian resilience, again using the PESTEL framework. On the other hand, the exercise also enabled a discussion of how Palestinian resilience *could* be realized, by identifying priorities for action. To this end, the participants were asked to rank the list of identified drivers according to two dimensions, namely the degree to which each identified driver would impact Palestinian resilience, and the degree to which the participants thought they could influence the drivers. The scatter plot in Figure 5.1 below shows the results of this ranking exercise.

This method enabled the participants to draw a boundary between what should have priority and what shouldn't, and thus to move away from a paralysing complexity towards a jointly constructed, more manageable outlook. When used in the dialogue sessions, this visualization was not meant to describe objective facts but to stimulate the participants' discussion around the joint goal of Palestinian resilience. Drawing on the data collected through desk research, consultations, online surveys, and dialogue sessions, the team crafted the final project report and a policy roadmap, outlining the necessary steps to achieve the preferred scenario of Palestinian resilience, along with the list of identified drivers influencing each dimension of the PESTEL framework. While the exercise was driven by a subjunctive desire to explore how a more resilient future could be achieved, the series of dialogues thus also produced actionable results.

The intra-Palestinian dialogue effort differs considerably from the other two examples of digital dialogues discussed above, mainly because it didn't involve Israeli representatives and because it wasn't geared towards contributing to a high-level peacemaking effort. That said, it illustrates well how digital methods can be employed for ends other than establishing simply what conflict parties and stakeholders want or need. The scenario-building and digital facilitation methods, such as the prioritization exercise, weren't primarily meant to generate representative and objective facts. They were employed to work with the diverse representatives of one side of the conflict to overcome their internal divisions and, more importantly, to foster a posture and identify concrete steps that could help Palestinians endure the status quo in

[30] This method was originally developed by the Finish company Inclus.
[31] I have discussed the use of this method in other dialogue efforts elsewhere, in a joint article with CMI staff, see Hirblinger, Brummer et al. (2024).

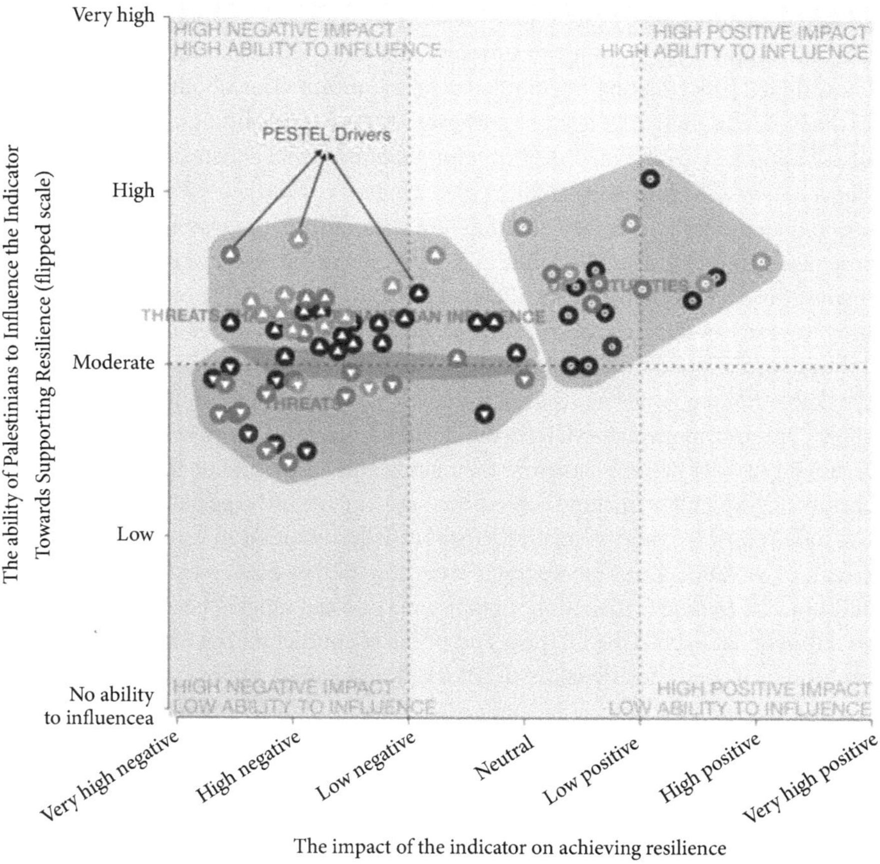

Figure 5.1 Issue-mapping exercise to assess PESTEL indicators in Palestine/Software applied from Inclus

such a way that a solution to the Israel–Palestine conflict could become possible in the future.

The data collected during the dialogue process and many concrete steps identified through it have been overtaken by the latest wave of violence. However, the vision of Samoud persists, and so may a sense that Palestinian unity is necessary to maintain it. CMI's project thus demonstrates the benefits of digital peacemaking approaches that take the context in which they operate as more of a set of intervening variables that need to be controlled, and treat data as more than just a mere tool to generate evidence about the world.

It also shows how digital peacemaking can become much more than employing specific tools when it engages with context-specific, historically and culturally grown understandings of what it means to struggle for peaceful co-existence. At the same time, the harsh reality and the global context of the Israel–Palestine conflict were impossible to change through the project. Against this backdrop, sincere data-driven efforts that simply collect information about the status quo and the conflict stakeholders needs can quickly become disempowering, because they don't go beyond simply documenting an unbearable reality. That said, the case exemplifies how digital peacemaking can employ data and information about this context in ways that foster a subjunctive attitude may provide a more fruitful, and strangely more 'realistic', approach to conflict transformation than those that equate data merely with evidence. Such uses of digital technologies are less concerned with using data to describe the conflict as it *is*. Instead, they use data to pave the way to explore what a more peaceful future *could* look like and what's needed to achieve it.

Against the backdrop of Israel's military actions in Gaza since October 2023, which destroyed many of the prerequisites for sustaining Palestinian life in the territory, and the genocidal policies of the government of Israel that led to mass starvation and unspeakable human suffering, the subjunctive invocation of Samoud appears both strangely apt and inadequate. Given the declared objective by members of Israel's government to evict Palestinians from their land, or at least facilitate their 'voluntary' departure, it's difficult to imagine how a peace agreement, and especially one entailing a two-state solution, could be realized without Palestinian steadfastness. The latest outbreak of violence also demonstrates well how digital peacemaking efforts that merely aim to generate evidence about the status quo can quickly become irrelevant. Even if timely, data about the interests and needs of conflict parties and stakeholders holds little promise if positions are deadlocked and parties choose aggression over a negotiated settlement. In the face of a stalled peace process characterized by the unwillingness of Israel's government to engage with the demands and needs of the Palestinian side, the project's effort to strengthen Palestinian resilience was a much more adequate goal. Hamas' 7 October attack, which demonstrated the group's desire to annihilate its enemy, can clearly be read as belated proof of the need to explore ways in which Palestinians can endure the consequences of this unresolved and costly conflict, without resorting to the use of force. Nonetheless, the subjunctive uses of digital technology documented above can also create a sense of void, just as the many others I have described in this book. The unprecedented wave of violence since October 2023 demonstrates painfully and mercilessly that what's required to resolve the Israeli–Palestinian conflict has largely remained out of sight. Something is missing-from-hand in the socio-technical relations needed to end and transform this conflict, and it's clearly more than current digital innovations are able to provide.

Conclusion

Contemporary peace mediation and dialogue efforts to end armed conflicts widely use digital technologies, and they increasingly integrate AI, despite the persistent notion that peacemaking is human-centred. The consequence is that such technologies have a growing impact on peacemaking approaches and dynamics. Yet, peacemaking is to no lesser extent influenced by *claims* about technology and society. Importantly, the tech-solutionist ambitions that emerge in narratives about which social (and political) problems technologies could solve are commonly conditioned by the social and political environment in which digital methods are employed. This environment includes organizational cultures, existing social networks and the skillset of mediation teams, or political mandates. Therefore, I have suggested thinking of contemporary peacemaking as a practice that relies on hybrid intelligence, which encourages us to explore the *interdependency* between human and machine agency. Like other human–machine networks, hybrid intelligence apomediates understandings of conflict and peace. However, the integration of AI means that algorithms will have a bigger impact on these dynamics because they autonomously generate knowledge. Moreover, some AI applications also constitute a 'black box' that is impenetrable to human intelligence, which means that mediators, facilitators, and dialogue participants won't be able to reconstruct how specific outputs relevant to peacemaking were derived.

This chapter has demonstrated that AI-enhanced digital methods have been viewed as a ready-to-hand tool to foster the inclusivity of peace processes, make peacemaking efforts more efficient, and achieve peace faster. Many international mediation attempts launched over the past two decades have achieved very few tangible and lasting outcomes. Against this backdrop, AI's data processing capacities, paired with the instant and far-reaching connectivity of mobile internet networks, promise to overcome the limitations of conventional approaches, which were often either too exclusive or too cumbersome to deliver political settlements that last. However, I demonstrated that the hybrid arrangements likely left a sense of void among those who had hoped that these digital technologies could make processes such as the LPDF more inclusive and more effective. While digital methods can enable a more broad-based and inclusive process, this form of inclusion can ultimately be disempowering if its outputs are fed into stagnating, or failing, high-level processes. One could, of course, as many observers did, argue that the failure of the LPDF to establish a stable transitional government and pave the way to democratic elections was primarily due to the unwillingness of the main conflict parties to commit to the process. This serves as an important reminder that digital 'tools' are always part of socio-technical arrangements and that they are always employed in a social, and more importantly, political context. Digital peacemaking approaches could enable a more radical break with the established and widely accepted ways of making peace. Part of this promise is associated with the fact that recent improvements in digital connectivity in many conflict-affected contexts enable new forms of networked organization that could transcend existing hierarchies and undermine authority—an idea that is at the core of the notion of apomediation. Yet, in Libya, such potentials were not realized, largely because the UNSMIL mediation used various digital tools towards pre-defined ends and integrated them into the tracked logic of the process.

Moreover, the specific design of digital dialogue platforms greatly affects the kind of dialogue dynamics they enable. In the case of the employment of the Remesh platform described above, the participants' ability to freely express themselves and shape the dialogue effort was clearly limited, as the platform valued their inputs only insofar as they could help generate an aggregated picture of the population's interests and needs. The use of algorithmic approaches to 'fill the gaps' in this picture points to the risk that

AI-enhanced dialogue may generate voice that contains synthetic data, thus potentially replacing other methods that would explore the plurality of views and opinions, especially of stakeholders that may hold unknown views. The gearing of all data gathering to serve this process not only restricted the kinds and the degree of interaction between the participants but ultimately created a simplified representation of the population that is ultimately less diverse. This way of using AI somewhat mirrors a behavioural science paradigm where human needs and interests are reduced to observable phenomena that can be documented and quantified to serve as evidence, instead of something that has an inherently subjective and emergent quality that can be valorized in a transformative peacemaking effort.

The chapter thus points to the risk that AI is employed *in lieu* of human interactions that are key to seeking understanding, accommodating differences, and achieving political settlements. Such applications of hybrid intelligence thus leave a sense of void for those who would like to see technology support transformative efforts that go beyond an approach to inclusion that treats the interests and needs of conflict parties and stakeholders as pre-given and static. The last section of this chapter made some suggestions for how such an ideal could be approached, namely by thinking of hybrid peacemaking intelligence not as a means to sincerely produce evidence that serves a preconceived process. Instead, it would be used to enable a subjunctive attitude concerned with how a more peaceful political order *could* be. There are many aspects to this, and the illustrative examples from the intra-Palestinian dialogue discussed in the last section of this chapter can't do justice to them. However, they at least indicate the contours of such an approach. One part of this effort would entail treating the context factors that shape digital peacemaking, including the properties of the digi-political playing field as potentially intervening variables that need to be isolated, but to let them come into the conversation and into the data picture. Another part would entail data and digital methods to nourish a sense of how existing social structures and power relations could be challenged, resisted, or at least endured, instead of merely documenting them as pre-givens. Such subjunctive approaches to digital peacebuilding could produce information and knowledge meant to destabilize and upend existing patterns of antagonistic conflict. This is where transformative peacemaking could begin.

6

Building Peace

Everyday Online Interactions for Political Change

The last chapter of this book explores how digital technologies enable new approaches for addressing the deep-seated societal cleavages that underpin many armed conflicts. Once the guns have gone silent and a peace agreement has been concluded, societies need to sustainably transition from war to peace. The peacebuilding efforts addressing this need are often subsumed under the term reconciliation, which is conventionally understood as an organized process aimed at transforming the relationship between antagonistic groups and building trust between them. This entails, for example, formalized and institutionalized efforts, for instance, by appointed courts that carry out transitional justice, or truth commissions tasked with conducting public inquiries to collect evidence about atrocities. Reconciliation helps to build a joint narrative and memory of past conflicts that is also suggestive of how the former antagonists could approach future ones. Moreover, to support lasting conflict transformation, such efforts will also facilitate a change of relationship between the conflict parties, which entails 'the removal of the negation of the other as a central component of one's own identity' (Kelman 2004). This requires viewing fellow citizens who identify with other (ethnic, religious, etc.) population groups no longer as antagonists that need to be subdued, expelled, or annihilated, but as agonists that form part of the same political community (Strömbom and Bramsen 2022). Evidently, such reconciliation efforts thus also have a preventative effect, as they aim to make the reactivation of enmity more difficult.

Reconciliation efforts are often highly political because they are seen as an opportunity to build the foundation of the post-war political order. International organizations working within the liberal peacebuilding paradigm have often intertwined their support for reconciliation with support for transitional justice and the promotion of human rights (Gibson 2004; Skaar 2012). However, reconciliation processes can also be exploited by victorious regimes to give legitimacy to their view on the conflict, as was, for instance, the case in Rwanda (Clark 2014). Moreover, they can also make marginalized experiences and perspectives of certain groups visible and inscribe them into public memory. This requires that ordinary citizens affected by war and violence can play an active role. Moreover, reconciliation may involve what Roger Mac Ginty (2021) describes as 'small acts of everyday peace' that create a 'space for civility, reciprocity and solidarity' to help former enemies practise collaborative behaviour that challenges prevailing group relations. Such an approach to reconciliation is often more realistic, feasible, and inclusive than formal transitional justice or truth-telling efforts (Mitchell 2023). This chapter builds on this insight. It discusses

Digital Peacebuilding?. Andreas T. Hirblinger, Oxford University Press. © Andreas T. Hirblinger (2026).
DOI: 10.1093/9780198986553.003.0007

how everyday online interactions apomediate understandings of conflict and peace in ways that enable a change in relationships between former enemies. In doing so, the chapter contributes to the growing literature interested in the agency of ordinary citizens in conflict transformation and in the role of everyday digital practices in global politics (Randazzo 2016; Eggeling and Versloot 2023; Adler-Nissen and Eggeling 2022).

Exploring everyday online interactions and their potential to contribute to conflict transformation is particularly relevant because of the current dearth of initiatives and mandates by international organizations. The United Nations (UN), for instance, had been strongly involved in setting up reconciliation mechanisms in many postwar places in the course of the 1990s and early 2000s. Past transitional justice and truth-seeking efforts have often been broadcast via the radio and television, and they have commonly entailed the production and distribution of a large number of publications and much audiovisual content (Price and Stremlau 2012). To consider how online resources and social media could be utilized in such efforts is thus anything but far-fetched, especially if we are interested in exploring how antagonisms could be overcome through everyday practices of ordinary citizens, who obtain information about the conflict via the internet and social media. However, we know very little about the effects of digitalization on the broader societal efforts to terminate and transform conflict. This is somewhat surprising because civil society initiatives have employed digital technologies, especially internet-based platforms, for some time to support the long-term transformation of agonistic relationships. A growing number of digital peacebuilding initiatives aim to support organized reconciliation efforts, such as through online memorialization, advocacy campaigns, and the establishment of inter-communal online networks. Yet, as I will explore in the first part of this chapter, these initiatives often have limited reach—and impact—because they tend to be viewed as 'nerdy' and elitist. More importantly, the internet and social media provide virtually unlimited opportunities for ordinary users to contribute to societal change processes that help overcome grievances and enmity and thus reduce the risk that conflict will (re-)occur. This makes it worthwhile to explore how everyday digital practices of producing, sharing, and consuming content online and on social media provide alternatives to organized reconciliation efforts, especially given their role in the re- or de-construction of collective identities and political institutions (Milan 2015).

Many readers might rightfully object that thinking about the internet and social media as a place for overcoming antagonisms, hate, and grievances seems somewhat naïve because contemporary armed conflicts are frequently accompanied—and fuelled by—online polarization. There are many factors that contribute to such divisions and some stem from the same politics that also make armed violence and war more likely. In particular, political regimes with authoritarian leanings may exploit online and social media platforms to maintain or deepen divisions through harmful online narratives and hate speech as part of what I suggest describing as digital divide-and-rule tactics. DD&R has two main components: it creates antagonisms along ethnic or religious lines that make it more likely that online polarization becomes deadly. Importantly, not all polarization stands in the way of peaceful co-existence. Indeed, agonistic politics require a healthy level of adversity, which coincides with different narratives and beliefs (Marichal and Neve 2019). However, DD&R creates antagonisms by fostering divisions that justify oppression and violence. Furthermore, DD&R also curbs online freedoms to stifle political dissent, mobilization for change, and collaborative interactions between the divided groups (Jones 2022; Yilmaz et al. 2022; Garbe et al. 2023; Hattotuwa and Wickremesinhe 2022; Bradshaw and Howard 2019).

Instead of taking the influence of these professional provocateurs as given, I argue that it is pertinent to explore how everyday online reconciliation efforts could help diminish it. Therefore, the chapter examines efforts to counter DD&R through everyday online practices and explores whether digitalization offers new opportunities and methods through which peacebuilding actors can promote conflict transformation. I will look at particularly how everyday online and social media interactions can help counter DD&R, by contributing to leaderless social movements that bypass conventional social institutions and enable popular protests across societal divisions (Castells 2015, 2013). In certain past cases, these movements have had the potential to topple authoritarian regimes and support transitions to more peaceful societies (Dudouet 2014; Lambach et al. 2020). However, we know very little about how they use digital spaces for everyday reconciliation.

When exploring how everyday, unplanned, and often undirected online interactions can contribute to reconciliation, I employ the concept of apomediation introduced in Chapter 2 to shed light on how human–machine, peer-to-peer networks (or in short, apomediaries) condition what can be said about the causes of conflict and options for overcoming it. More specifically, the chapter explores under what conditions and how apomediation may contribute to relationships between ordinary citizens that are less defined by enmity and more by adversity, and therefore make non-violent forms of political struggle more likely. To this end, the chapter investigates the interplay of relation-making and meaning-making in apomediaries, that is, the creation of connections between specific users and platforms on the one hand and the creation of particular understandings of conflict and peace on the other. I argue that online spaces in conflict-affected contexts exhibit at least three types of apomediaries relevant to peacebuilding, distinguishable by their relational organization and the meanings of peace and conflict they produce as they respond to DD&R tactics. First, most peacebuilding contexts tend to be characterized by *programmatic* apomediaries that organize around specific topics that form part of peacebuilding actors' technical repertoire and policy. Second, contexts with a slanted playing field also tend to see *reformative* apomediaries that organize vertically to communicate popular grievances against incumbent regimes. Third, most peacebuilding contexts are characterized by efforts to foster social and political integration across societal cleavages through *integrative* apomediaries that organize horizontally, aiming to foster communication exchange across societal, cultural, or political fault lines.

The chapter develops and explores this typology by comparing three empirical case studies of recent peacebuilding efforts in South Sudan, Sri Lanka, and Northern

Ireland. With that, I will demonstrate that the prevalence of each of these apomediaries is conditioned by the slantedness and adaptability of the digi-political playing field. How much an incumbent regime slants this playing field in political terms and the degree to which digitalization affords internet users the capacity to adapt to such constraints shape the users' ability to build relations and share alternative understandings of conflict and peace. The relationship between varying digi-political configurations and the prevalence of certain apomediaries can be captured in a conceptual-explanatory typology that provides insights into when and how everyday online interactions can effectively counter DD&R. I describe this constellation as an instance of *transformative* apomediation, characterized by a productive fusing of programmatic, integrative, and, most importantly, reformative apomediaries.

Programmatic, Reformative, and Integrative Apomediation

This section explores online reconciliation as a decentred process that manifests in everyday interactions of producing and sharing information online, especially on social media. I suggest that these dynamics are shaped by several apomediaries that advance distinct understandings of conflict and peace. While such understandings emerge first and foremost through human cognition and communication, digital technologies play an increasing role in producing, sharing, and consuming them. The previous chapters of this book shed light on the interplay of relation-making and meaning-making involved in this process. This chapter makes clear that these dynamics are not limited to specialized applications, such as for analysing hate speech on social media or localized conflict early warning. They are also shaped by how ordinary internet and social media users relate to one another. These relations that shape meaning-making are, to some degree, produced by hardware, such as fibreglass cables or Very Small Aperture Terminal (VSAT) stations. However, they also result from software that provides distribution channels, such as webpages, podcasts, or newsletters, or through social media connections or groups that deliver information to internet users based on their browsing profile. It goes without saying that these relations are created and sustained through human agency and organization. Indeed, as will become evident below, social organizations and structures, including professional and civil society networks, play a considerable role in shaping the apomediaries relevant for reconciliation. What's also important is that the many users who take part in the production and sharing of information are commonly influenced by their assumptions about both the societies in which they operate and the technologies they employ. Limited digital connectivity or literacy, for instance, manifests not only as a material fact but also as one that is assumed, declared, or at least taken as a given. This makes it important to consider too how the claims held by human actors within the apomediation network shape apomediation dynamics.

I suggest differentiating between three ideal types of apomediaries that condition how everyday conflict transformation efforts play out online and on social media.

With Max Weber, I understand ideal types as abstract concepts with distinctive characteristics that fit together logically, reflecting the meaning that social actors bestow upon their actions (Hekman 1983). The three apomediaries I identify differ in the interplay of relation-making and meaning-making. 'Programmatic' apomediaries organize in *centred* relations around specific technical subjects, 'reformative' apomediaries organize in *vertical* relations to express popular grievances from the bottom up, and 'integrative' apomediaries aim to bridge between antagonistic segments of the population in *horizontal* relations. The three apomediaries should not be understood as directing communication patterns in a definitive way but as initiating or encouraging one pattern and not another. Furthermore, while per the logic of ideal-typical comparison, any apomediation effort should be attributable to one of the three ideal types, these do not constitute mutually exclusive forms of social organization. Put differently, social media users can interact with, and contribute to, several apomediaries at a time. Indeed, it will become clear that this is of critical importance for transformative apomediation dynamics.

The first type of apomediary commonly found in conflict-affected contexts constructs and conveys narratives of peace and conflict aligned with professional peacebuilding projects and programmes. The practices of meaning-making and relation-making in these *programmatic* apomediaries tend to be *centred* around their themes. This type is nourished mainly by information produced by international, intergovernmental, and regional organizations, bilateral development organizations, national civil society, community-based organizations, think tanks, and academic institutions, and, of course, the employees who work for them. Experts, volunteers, and interested citizens likewise contribute to programmatic apomediation. Programmatic apomediaries largely further issue-based discourse on conflict and peace, informed by the peacebuilding organizations' mandates and programmes, established thematic priorities, and the repertoire of international norms governing peacebuilding interventions. Thus, programmatic apomediation also commonly maintains and reproduces the established canon of assumptions, claims, and knowledge, legitimizing peacebuilding projects and rallying political and financial support. Furthermore, issue-based and problem-oriented debates can also take place in domestic politics, which means that politicians and policymakers associated with the conflict parties likewise contribute to programmatic apomediation. In fact, they may contribute to the bulk of programmatic apomediation in contexts characterized by a relative absence of professionalized peacebuilding organizations.

The centring of information exchange on specific topics means that programmatic apomediation often only involves a selective community of experts and the informed public. Professional peacebuilders often engage in programmatic apomediation to foster public attention for their topics of concern as part of their 'strategic communication' efforts. They may, to this end, also implement dedicated outreach projects to transmit more complex and comprehensive knowledge, for instance, via social media, mailing lists, or specialized applications, websites, and webinars. Where organizations aim to promote long-term societal change, such as promoting human rights and gender equality, they also commonly link up with national civil society and community-based organizations, or create decentralized networks of volunteers and social media influencers tasked to promote and multiply content. However, these relations are exclusively organized to generate attention and interest in the specific subject matter. This approach is well captured in the words of one expert, 'we use Twitter, but essentially, it's on a single send-out message. We don't start having conversations with people who connect with us.'[1]

The second apomediary can be described as *reformative*. Social media is also commonly a space where social movements, protest movements, and the political opposition articulate their grievances with incumbent regimes and formulate political alternatives. In contexts affected by armed conflict, such online practices are usually not directly associated with peacebuilding as a distinct social or political activity. However, in settings with slanted playing fields, they often address many underlying causes of conflict. The reformative apomediary is entertained by members of the organized political opposition, including political parties and activist groups, and may also involve parts of the organized civil society, such as trade unions, or human rights or religious groups. In cases of widespread popular protest, reformative apomediaries have the potential to integrate large numbers of ordinary citizens to mobilize for protests that articulate popular grievances.

Linking the causes of conflict to the governing regime, and presenting solutions to conflict in terms of people-driven change, requires communicating popular grievances from the bottom up. Therefore, relation-making and meaning-making in reformative apomediaries are organized in a *vertical* manner. Reformative apomediation enables this through the articulation of popular demands and grievances on social media with the aim of challenging the power of incumbent regimes. These articulations may not directly reference 'peace' or 'peacebuilding'. However, the views on the conflict circulating in reformative apomediaries may well inform formal peacemaking efforts, such as when demands for specific agreement provisions or political reforms are voiced. Reformative apomediation efforts are often supported and sustained by organized civil society actors, including political parties and unions, that maintain close communication links with their members and encourage the articulation of demand from the bottom up. These networks tend to contain a broad cross-section of the population organized alongside specific identity markers, interests, or professions. Reformative networks also commonly maintain strong connections to diaspora groups and exiles who can express grievances against incumbent governments with a far lower risk of repercussions. During political crises and widespread popular discontent, reformative apomediation may unfold and grow organically, leading to widespread, heterogeneous, and decentralized user networks within which content is produced and shared without the involvement of organized actors.

The first two apomediaries discussed above admittedly tend to do very little to establish online relations that cut across the societal divisions commonly produced by DD&R. This means that their claims about peace and conflict risk being

[1] Interview, Director of Communications, International Organization, South Sudan, 28 November 2022.

overshadowed by divisionist narratives or distributed along conflict-defining social cleavages. However, some peacebuilding actors focus their online engagement on establishing cross-cutting apomediaries in a deliberate attempt to enable communication and collaboration across divided groups. The work of such *integrative* apomediaries can have a lasting effect on how ordinary internet and social media users relate to one another, how they make sense of the conflict, and how they can contribute to reconciliation efforts. When relating across group divisions, integrative apomediaries thus commonly rely on cross-cutting relations. Furthermore, peacebuilding organizations may maintain their own integrative social media presences that are inclusive and stimulate information exchange across groups. Integrative apomediaries thus organize *horizontally* to counter divisive political designs, both in terms of offline organization (such as along political parties and religious institutions), social demography (such as regarding geography and language), and in terms of online platform organization (such as exclusive messaging groups) and platform design (such as by algorithmic recommender systems).

In terms of meaning-making, integrative apomediaries encourage conversations and information exchange across societal divisions, for instance, by training social media influencers to produce cross-cutting content or hosting online activities that intentionally bring together a diverse group of participants. They also produce and circulate social media content promoting inclusivity, diversity, and tolerance, including through visual artefacts such as online memes or video content. Moreover, their content may showcase examples of successful horizontal integration and peaceful co-existence, which are framed as solutions to conflict, while practices of division, essentialization, and Othering are called out as drivers or causes of conflict. At times, the meaning-making undertaken by such apomediaries may be intellectually demanding and thus less inclusive, for instance, in the context of online memorialization projects, which activate scholarly expertise concerned with historical and deconstructive methods. At other times, integrative apomediation may amount to not much more than ad hoc, amiable information exchanges between users identifying with different groups that stand in contrast to the divisive rhetoric and practices of incumbent regimes.

The Power of Digital Critical Mass

Admittedly, the description of these three ideal types of apomediation—and their role in enabling long-term change towards more peaceful societies—has remained somewhat abstract so far. In the remainder of the chapter, I will thus put the typology to work, using it to explain how apomediation dynamics condition the possibility for conflict transformation. To this end, I will compare apomediation dynamics in Northern Ireland, South Sudan, and Sri Lanka to discuss how the relative presence or absence of the three types of apomediaries shapes (digital) peacebuilding dynamics and how this varies with the characteristics of the digi-political playing field. While comparing apomediation dynamics in Northern Ireland, South Sudan, and Sri Lanka at distinct critical junctures, I will advance what Elman (2005, 296) has described as an 'explanatory typology' that may be helpful for other cases as well.[2] It builds strongly on my discussion of how the digi-political playing field shapes apomediation dynamics in Chapter 2, which I have already demonstrated in the preceding chapters. However, the impact of the degree of political slantedness and digital adaptability becomes even more visible when we study everyday and public online practices.

In contexts such as South Sudan, where regimes attempt to slant the playing field through high levels of *suppression*, such as through internet filtering, online discussions that relate incumbent regimes to the causes of conflict or call for political change as a peacebuilding strategy will gain limited traction. Furthermore, high levels of *containment* will mean that where apomediaries may advance alternative understandings of conflict and peace, they will have difficulty translating these into offline action, such as in the form of public protests or civil society activities. South Sudan's playing field also has limited digital *reach*. Internet access will be mostly restricted to urban elites and those living abroad, and apomediation is limited to a minority of the population that often benefits the status quo, except for some parts of the diaspora. While rudimentary internet access may exist in some rural areas, only a small percentage of the rural population uses social media to exchange written content, and they usually form part of the administration or civil society. Images and videos circulate to a limited extent, but given the restricted depth of the field, they often reproduce stereotypical conflict narratives. Therefore, South Sudan's playing field overall has a weak adaptability. In contrast, cases with more extensive reach due to dispersed and widespread internet usage, and greater *depth* and *flexibility* due to higher levels of literacy and digital skills, may build connections that involve political opposition and socio-economically marginalized groups, enabling more varied and frictional meaning-making, as in Sri Lanka. I will argue that such levels of digital adaptability are critical for apomediation dynamics that can counter DD&R.

As illustrated in Figure 6.1, we can make sense of the impact of the playing field on the apomediation of conflict and peace by first differentiating between contexts where the regime attempts to slant the playing field and contexts where this is not the case. We can then contrast playing fields based on their relative digital adaptability. Based on that, we can identify three possible dynamics of apomediated peacebuilding, characterized by variation in the prevalence of the programmatic, reformative, and integrative apomediaries, which I suggest calling *digressive*, *regressive*, and *progressive* apomediation. I will briefly summarize each of these possible outcomes before discussing them in greater detail in the empirical case studies.

To start with, all three countries were characterized by considerable online polarization, but Northern Ireland lacked the slanted digi-political playing field that coincides with forms of authoritarianism that were prone to fuel armed violence in Sri Lanka and especially South Sudan. Northern Ireland's playing field was very

[2] Such typologies are composed of a 'multidimensional conceptual classification based on explicitly stated theory' and containing descriptive and classificatory elements 'in a way that incorporates their theoretical focus'.

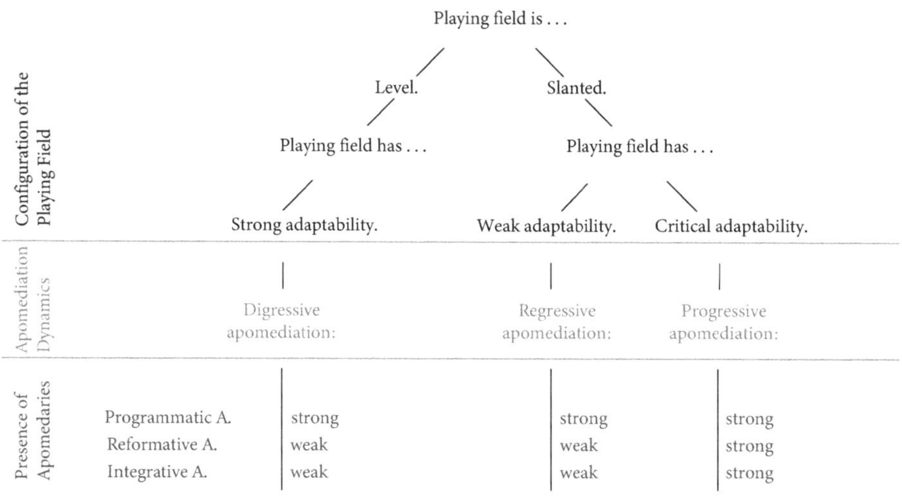

Figure 6.1 Explanatory typology depicting the relationship between the configuration of the digi-political playing field, apomediation dynamics, and presence of apomediaries

adaptable, but this capacity isn't needed to counter DD&R. That said, I will describe Northern Ireland as a case with *digressive* apomediation dynamics because they produce forms of unproductive adversity. Digressive apomediation occurs in cases with a relatively level playing field and is dominated by programmatic apomediaries that are polarized across antagonistic population groups. I call these dynamics digressive because the issue-centred information exchange breaks down into smaller online communities dominated by group-centred narratives. This derails conversations that could otherwise help transform group relations. What would be needed to counter these polarizing dynamics are integrative apomediaries that increase the exchanges between the different exchanges and help develop common understandings of the conflict and options for its resolution. However, at least in Northern Ireland, the few organized integrative efforts that exist seem to have had limited influence on public debates, which remain deeply entrenched in hurting inter-group dynamics. At the same time, reformative apomediation tends to be weak. Since citizens can vote their governments out of office, the need for political mobilization from the bottom up is understandably rather limited. Overall, online relations among Northern Ireland's communities are polarized in ways that reproduce the status quo. Yet, they tend, by and large, to miss an authoritarian incumbent willing to exploit lethal antagonisms to stay in power.

In contrast, in cases where regimes slant the playing field, polarized understandings of conflict and peace are instrumentally employed to divide populations and oppress them. In such cases, as I will illustrate when discussing South Sudan, everyday online interactions may be shaped by both programmatic and integrative apomediaries that produce subject-specific conversations and promote interactions between users who identify with antagonistic population groups. Programmatic apomediaries will advance technocratic understandings of conflict and peace, while integrative apomediaries may improve group relations. However, neither poses a fundamental challenge to DD&R. Reformative apomediaries remain relatively weak, as online interactions that construct a relationship between the incumbent regime's politics and the underlying causes of conflict are quickly repressed. I will describe these dynamics as a case of *regressive* apomediation dynamics because they reproduce the context conditions for armed violence.

How, then, can everyday online interactions enable conflict transformation in cases where regimes slant the playing field and risks of large-scale armed violence remain high? A detailed reconstruction of apomediation dynamics during the mass protests in 2022 in Sri Lanka suggests that online interactions may better realize their peacebuilding potential if a regime's efforts to slant the playing field is confronted by a critical mass of (digitally) literate citizens that can adapt to the regime's DD&R tactics, such as online censorship and internet blackouts. This critical adaptability can sustain what I suggest describing as *transformative* apomediation dynamics, which are characterized through a strong presence and concerted effort of programmatic, integrative, and, most importantly, reformative apomediaries. A broad-based online protest movement that formulates popular grievances with the regime, and links them to the causes of conflict, provides the necessary catalyst for these dynamics. In such cases, a sufficient number of internet users perceive that they can exploit the depth and flexibility of the playing field to create synergies between the three apomediaries that enable political change with the potential to counter DD&R effectively.

Discussing the preconditions of conflict transformation, Lederach and Thapa (2018, 110) argued that what's necessary is a 'critical yeast', that is, 'a group of improbable relationships crossing social divides, just large enough in number and unique enough in its formation to create a sustainable, new space of interaction'. What's crucial is not the number of people involved, but the quality of the platform that enables exponential growth. For Lederach, 'it is a question of the quality of relational spaces, intersections, and interactions that affect a social process beyond the numbers involved' (Lederach 2005, 117). The same holds for digital relations. However, I will show that in authoritarian peacebuilding contexts, the size of the network matters as well. The term 'mass' captures this need better—its double meaning points to the necessity of a large number of connections, as well as the understanding of mass we might recall from physics class, where it tends to be defined in terms of resistance to the application of a force to an object.[3] Figuratively speaking, we can think of the mass of digital connections as being 'critical' when they are numerous and strong enough to resist the slantedness of the playing field, formulate a critique of the political status quo, and translate it into political change. As I will demonstrate, mobilizing a digital critical mass matters because it protects from regime repression. The more

[3] This definition is admittedly incomplete. In physics, the resistance is that of a body of matter to acceleration, the mass of that body is a measure of the amount of matter it contains, the bigger the mass the bigger the resistance.

nodes there are that can grow and strengthen reformative, integrative, and programmatic apomediaries, the broader their effects will be felt across society, and the less likely it is that a regime can escape the alternative understandings of conflict that they advance and the political changes towards a more peaceful future they demand. In the following, I will illustrate these three scenarios in greater detail.

Digressive Apomediation: Negotiating Northern Ireland's Status Post-Brexit

Northern Ireland is an exemplary case for how, in contexts with a comparably level and highly adaptable playing field, online polarization nonetheless shapes understandings of conflict and peace. Digital and general literacy and digital access are among the highest in the world, and there are no known hurdles imposed by the governments of the United Kingdom (UK) or Northern Ireland on access to the internet, nor is there any noticeable censorship or filtering of online content. Northern Ireland's peace process was formally concluded with the Good Friday Agreement of 1998. However, social and political cleavages between the Unionists and Republicans remain, and a power-sharing agreement between the political parties has contributed to freezing instead of overcoming them. Following Brexit, the UK's withdrawal from the European Union (EU) in January 2020, disagreements between the EU, the UK, and the former conflict parties have led to new frictions, especially regarding the status of Northern Ireland in the European Single Market (ESM) and the necessity of a sea border between Northern Ireland and the UK mainland (Gormley-Heenan and Aughey 2017; Murphy 2021). The tensions between the two communities are also visible in polarized online interactions, sustained by divisive social media influencers who reproduce and foster the conflict cleavage. However, ordinary social media users identifying with the two communities in conflict likewise contribute.

Thus, Northern Ireland somewhat serves as a 'counter case' in the comparative analysis presented in this chapter. It demonstrates that societal and political cleavages can persist in contexts that lack autocratic leaders who fuel and exploit group divisionism to maintain their grip on power. More importantly, the Northern Irish case demonstrates how both programmatic and integrative apomediation efforts alone may be insufficient for transforming the conflict and promoting reconciliation between groups. Likely because of the nature of armed conflict in Northern Ireland, and the strong involvement of the UK government, international organizations provide limited support for efforts to reduce community tensions and work towards group relations that are no longer antagonistic. However, a few exemptions exist, such as the EU's funding for several peace programmes since the 1990s, which supported long-term conflict transformation through community-driven initiatives (SEUPB 2021). Interestingly, other cases with a similar digi-political playing field and more recent histories of tensions or violence, such as the United States or Spain, also lack the type of professional peacebuilding sector that can be found in contexts with less level and less adaptive playing fields. This may be because international peacebuilding actors tend to prioritize support to countries that also meet other criteria for development assistance, that is, countries that are usually associated with the Global South. Besides common factors such as high levels of poverty and comparably weak democratic institutions, these countries also commonly share a history of international intervention and colonial rule.

At the same time, many Northern Irish civil society institutions, such as religious organizations or sports clubs, are deeply divided across sectarian lines. In consequence, actors who would promote integrative apomediation barely exist. During my research, I found only a few examples of local-level cross-community social media initiatives, such as mitigating communal violence or reporting harmful speech (The Bytes Project 2022; FactCheckNI 2023). However, there are also traces of more ambitious efforts to leverage the internet for peacebuilding, including programmatic apomediation efforts to collect evidence about the conflict relevant to academic research and education. Especially worthwhile mentioning are the now rather ancient-looking static HTML pages making up the Conflict Archive on the Internet (CAIN) hosted on the web servers of Ulster University, which contain a wide range of information on the Northern Ireland conflict (CAIN 2024). However, the challenges of keeping CAIN online, well documented on the archive's pages, clearly point to the fact that at the time of research, there was a limited public demand, and with that, limited financial support to launch initiatives that could have an impact on current politics and society.

Given the absence of a strong involvement in subject-specific and technical debates about policy, programmatic apomediation in Northern Ireland tends to blend with online debates in the formal political arena and is shaped by the everyday political online interactions between government officials, political party representatives, interest groups, and journalists, to mention some of the most important actors. While the Northern Irish tend to be critical of their political leaders and the lack of political progress, momentum for far-reaching political changes is relatively low, at least of the kind that would pose a fundamental challenge to how politics in Northern Ireland are organized. Therefore, the only way in which Northern Ireland can currently be governed is through a power-sharing arrangement that mandates that members of both the Republican and the Unionist communities must be represented in government (Doyle and Connolly 2019).

After Brexit, however, this power-sharing agreement became dysfunctional because Unionists perceived that the Northern Ireland Protocol, which formed part of the withdrawal agreement between the UK and EU (European Council of the European Union 2024), undermined the institutional and political framework established by the Good Friday Agreement. Consequently, the Democratic Unionist Party (DUP) withdrew from the power-sharing institutions in February 2022, in protest against the Protocol and especially the trade agreement provisions, effectively collapsing them (Centre on Constitutional Change 2024). This hiatus lasted for two years, until a separate agreement between the DUP and the UK government was reached in January 2024, which paved the way for the formation of a new Executive (McCormack 2024). However, in contrast to South Sudan's transitional power-sharing arrangement discussed below, Northern Irish voters could hold their government officials accountable by voting them out of office. This may explain why there was little resistance and no widespread protests against the DUP's actions. Indeed, the DUP maintained a relatively steady level of votes in the 2023 local elections, which suggests that its decision to withdraw from the power-sharing arrangement enjoyed popular support (O'Leary and Pow 2023). Unsurprisingly, reformative apomediation gained no observable traction during this period. Indeed, the instances of public 'disorder', observed by the authorities around Easter 2021, seemed to be directed against the Protocol itself and thus followed the pattern of Northern Ireland's divisionist politics (HM Government 2021).

The policy discussions around the Northern Ireland Protocol between the EU and the UK on social media in 2021 and 2022 remained strongly polarized, characterized by community-centred framings of key issues. Observers familiar with the context compared the online interactions around the Northern Ireland Protocol with earlier moments of high-level contentious politics, where online spaces were mainly used to frame the issue in terms of an attack against group identities, which reinforces group divisions. A long-term expert in online communication in Northern Ireland, described it as follows:

> Activists . . . are once more using online platforms to articulate their opposition to the 'Irish Sea Border', and also the Belfast/Good Friday Agreement itself . . . Meanwhile, Twitter hashtags such as #brexitriots are used to both dehumanize loyalists and to score political points about the negative consequences of the UK's decision to leave the European Union. (Reilly 2022)

Efforts to promote evidence- or fact-based political debates were quickly swallowed by these dynamics. As the director of one non-governmental organization (NGO) that aims to help improve community relations through fact-checking put it, 'we can post the most glorious fact check, or information about something, and it doesn't matter because it doesn't save the narrative of a particular agenda'. Their organization would be quickly accused of bias when presenting fact-based information about peacebuilding-related topics such as the cost of Brexit. In their experience:

> [B]oth 'camps' are so organized, with their teams online and how they deal with us, how they deal with people who share our work . . . one day they love you and share your work and the next day (we) are the IRA or something—you know what I mean.[4]

In other words, subject-centred social media conversations about Brexit, the Northern Ireland Protocol, and trade relations quickly became polarized. Paradoxically, integrative apomediaries also seemed ill-positioned to change these dynamics. Most peacebuilding efforts that aimed to strengthen inter-communal linkages through collaborations and exchanges between people identifying with the Unionist and

[4] Interview, Director, Northern Irish Civil Society Organization, via telephony software, 27 June 2023.

Republican groups were largely confined to local communities and specific demographic groups, such as youth. For instance, the majority of measures by the Peace IV Programme funded by the EU, which were implemented in the years leading up to Brexit and after, utilized digital technologies and social media to promote digital skills and capacities in schools or youth clubs, focusing on marginalized and disempowered youth (Special EU Programmes Body 2019). While fostering cross-community ties was an explicit objective of some of these efforts, many focused on improving highly localized community relations, such as among the segregated neighbourhoods of Belfast's interface areas or among various ethnic communities (Belfast City Council, n.d.).

Maybe the participants of such measures didn't feel prepared to intervene in high-level contentious politics. They might also not have felt compelled or empowered to do so. I observed a similar tendency in South Sudan and Sri Lanka. Digital capacity-building initiatives focusing on improving cross-community relations can appear oddly apolitical. I will explore the reasons for this in greater detail below. For now, I suggest describing the apomediation dynamics in contexts such as that of Northern Ireland as *digressive*: programmatic apomediaries produce and share issue-related content, yet this content tends to be framed in a community-specific narrative that reproduces polarized understandings of conflict and peace. This means that apomediation touches on peace-related topics but ultimately diverts from the question of how to sustain peaceful relations between the divided groups by framing the causes of conflict through community-specific, antagonistic narratives.

Regressive Apomediation: Debating South Sudan's Protracted Peace Process

South Sudan resembles Northern Ireland in how apomediation tends to be shaped by community-specific framings of the conflict. In many other ways, it is a starkly different case. The country's social media landscape is characterized by intense polarization between the followers of respective political, military, and ethnic groups, as well as rampant harmful speech, which remains largely unmonitored and unregulated due to linguistic diversity, lack of national oversight, and little content management by social media platforms. The widespread sense of ethnic fragmentation and the risks of renewed or increased ethnic conflict have been employed to justify delays in elections and strengthen democratic institutions (Kuol 2020), which indicates how Kiir's regime benefits from ethnic divisionism. Unlike Northern Ireland, the country is characterized by very low levels of digital access, with broadband mobile internet connectivity confined to the biggest urban centres and limited or no coverage in most rural areas. Digital and general literacy are among the lowest in the world, which strongly reduces the adaptability of the playing field. Since the outbreak of a civil war, South Sudan has also seen increasing political oppression that slants the playing field, including internet filtering, social media censorship, and the shutdown of internet connections in areas of armed insurgency (Amnesty International 2021a; Okello 2022).

South Sudan's authoritarian regime officially embraces a political vision of 'unity in diversity'. Yet, its political hardliners, as well as informal influence groups, clandestinely promote ethnic divisionism, which also manifests online and on social media. Unsurprisingly, like Northern Ireland, programmatic apomediation involving South Sudanese stakeholders quickly becomes polarized. As a South Sudanese social media activist put it, many discussions, including on seemingly technical topics such as environmental protection, 'appeal to ethnic identity and group dynamics . . . and compliance is very much required'.[5] In contrast to Northern Ireland, however, programmatic apomediation is also strongly shaped by professional peacebuilding, humanitarian, and development actors, which have played a sizeable role in the state- and peacebuilding process since the country's independence in 2011. The institutions created by the peace agreements signed in 2016 and 2018, the diplomatic community, and a pool of national experts and political elites, including those based in the diaspora, also commonly create and share content regarding the peace agreement implementation and respective hurdles to building peace in South Sudan.

However, many political elites who hold government positions do not have an official social media presence that they could use to share information about their work or to facilitate engagements with citizens or constituents—and if they do, they use it only sporadically or inconsistently.[6] A South Sudanese journalist explained the reasoning behind this as follows: 'Let's say when you are a member of the Parliament, and you've been there for ten years, and there's no development in the place [from which you hail], and you are not a part of any solutions, if you are involved in social media, maybe you will get a lot of comments. Some people will go on your neck!'[7] Given the lack of democratic oversight, the continued postponement of elections, and the fact that the regime successfully managed its hold on power through a combination of violence, intimidation, and clientelism, engaging with ordinary South Sudanese about government policy seemed both unnecessary and cumbersome for those in power. Programmatic apomediation nonetheless develops some traction, largely because international organizations and national civil society share content relevant to their mandate and projects, relying on their expertise, technical equipment, and infrastructures to create, consume, and share content.

Crucially, though, many communications specialists working for some of the major international organizations assumed that the content they create and share has very limited direct reach among the population. Given the material and tangible realities of poor connectivity and digital literacy, this view is not at all surprising and it is also echoed by local experts such as journalists.[8] However, it is important to

[5] Interview, Member, Board of Directors, South Sudanese Civil Society Organisation, Juba, South Sudan, 15 June 2023.

[6] Interview 65, Civil Society Activist, Juba, South Sudan, 3 December 2022.

[7] Interview 53, Journalist, Juba, South Sudan, 30 November 2022.

[8] Interview 53.

note that the portrayal of South Sudanese society as digitally disconnected and illiterate also justifies the decoupling of programmatic apomediation from large parts of the public. Many professional peacebuilding organizations aim to foster information exchange, first and foremost among technical experts working in international organizations. For instance, a spokesperson of the United Nations Mission in South Sudan (UNMISS), the international peace operation that supports conflict prevention and peacebuilding efforts in the country, described the purpose of their social media and web presence as follows:

> We put things up for the record. But we don't think that we're particularly hitting the general population and the communities in South Sudan via those channels. However, we think . . . that we're hitting the international community.[9]

Such established assumptions about the field at times also prevent peacebuilding professionals from trying to engage a broader section of the population in apomediation. For instance, as one communications expert at the commission tasked with overseeing and supporting the implementation of the peace agreement, the Reconstituted Joint Monitoring and Evaluation Commission (R-JMEC),[10] put it, 'when we travel out of Juba, you do have to ask yourself, how many people here are tweeting? Next to none, you would imagine . . . So, we have to valorize and place emphasis on those things we think are going to work.'[11] In the case of R-JMEC, this involved mainly in-person meetings with community representatives. In his view, the audience they intended to reach via the internet and social media to inform about the peace agreement implementation entailed some South Sudanese, but more importantly, diplomats, journalists, and staff working for international and regional organizations. While it has been difficult to quantify the exact reach of South Sudanese by international peacebuilding organizations on social media, it's clear that their approach to communicating about peacebuilding and process-relevant topics was not meant to enable a broad-based and interactive exchange with the country's population. Other communications staff of international and bilateral organizations took a similar approach, largely spreading information and sharing updates about their activities on social media, without engaging in back-and-forth social media exchanges with South Sudanese social media users who were largely perceived as poorly connected, poorly educated, and polarized.[12]

This is despite the fact that a comparably small but growing group of educated and connected South Sudanese regularly use social media to learn about and discuss

[9] Interview, Spokesperson, International Organization, Juba, South Sudan, 28 November 2022.

[10] R-JMEC is short for the Reconstituted Joint Monitoring and Evaluation Commission. JMEC was first established as part of the peace agreement reached in October 2015. However, the peace process and power sharing agreement broke down in the year after. R-JMEC was created in November 2018 to oversee the implementation of the revitalized peace agreement.

[11] Interview, Director of Communications, International Organization, Juba, South Sudan, 28 November 2022.

[12] Interview 105, three Staff Members, International Development Organization, via telephony software, 21 June 2023.

peace-relevant issues, ranging from questions of the sharing of natural resources, the integration of the armed forces, to gender equality or climate change. These conversations also include politically sensitive topics such as financial mismanagement and corruption, to which ordinary South Sudanese contribute. Some observers and journalists facilitating these exchanges argued that many contributors perceived that contributing online was safer than engaging in such topics in person on the street or in the market, given that contributors can somewhat conceal their identity.[13] Nonetheless, the programmatic apomediation efforts that could support the termination and transformation of conflict in South Sudan are characterized by ill-defined, yet commonly acknowledged, red lines. The information shared online and on social media tends to focus almost exclusively on factors that influence the prospects of peace in South Sudan but that do not threaten to undermine the power of the incumbent regime. International actors involved in programmatic apomediation are arguably relatively well protected from the effects of a slanted playing field due to their status. However, they still depend to some degree on the acceptance of the incumbent regime to implement their mandates. Indeed, Kiir's regime has not held back from intimidating or expelling diplomats, international technical experts, or journalists who have publicly shared their criticism, and access to online news pages and the webpages that convey critical content is regularly blocked. Therefore, online content posted by international organizations tended to remain strongly focused on specific technical subjects.[14] It also prioritized 'positive' messaging focused on how specific efforts may help build peace rather than on the causes of conflict (Leib 2023).

National civil society organizations (CSOs) and ordinary South Sudanese are even more vulnerable to direct intimidation and control, which is exemplified in the long list of instances of harassment, intimidation, arbitrary detention, and even targeted killings of civil society actors who have criticized Kiir's regime for its role in perpetuating armed conflict in South Sudan. For instance, while online discussions about corruption or poor service delivery are common and also often agitated, contributors based in the country often stop short of attributing the causes of conflict to those in power. Social media users who directly criticize government officials by name risk repressive reactions, as demonstrated by the large number of attacks on, and arbitrary arrests and disappearances of, regime critics.[15] For instance, several observers mentioned that critical voices would commonly be 'followed' by members of the National Security Service (NSS), or discouraged from posting critical content by friends and family members who had received 'well-meaning' warnings from government employees and risked more serious repercussions, including physical assaults and arbitrary detentions.[16] The regime also commonly blocked critical internet and social media resources, such as independent media houses, and human rights and

[13] Interview 41, two Journalists, Juba, South Sudan, 24 November 2022.

[14] Interview 41, Staff of Diplomatic Delegation, Juba, South Sudan, 24 November 2022; Interview 51, Staff Member, International Peacebuilding Organisation, Juba, South Sudan, 29 November 2022.

[15] Interview, Civil Society Activist, via telephony software, 26 February 2025; Interview 104, Board Member, Civil Society Organization, via telephony software, 15 June 2023.

[16] Interview 53; Interview 52, Journalist, Juba, South Sudan, 30 November 2022.

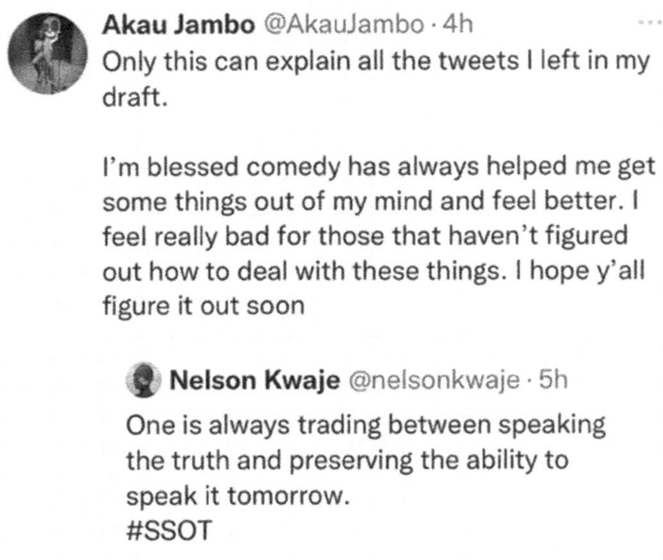

Figure 6.2 Social media conversation about self-censorship between a South Sudanese comedian and a civil society activist specializing in social media, 5 February 2023, reproduced with permission of the authors

pro-democracy organizations. One journalist, who was involved in efforts to report on developments in the peace process and translate them to the wider public via social media, described the problem succinctly when he stated that 'we could use the media, including social media, to translate the peace process and to inform [the public]. But it [the effort] was attacked by the government. There's no freedom of speech, and that's what makes it difficult.'[17] The issue for ordinary social media users who spoke out was the same: '[T]he people who are speaking on social media face a lot of issues, restrictions, and also harassment and other things because the government knows that social media is a strong tool for mobilization and informing society.' This doesn't mean that South Sudanese social media is free of content that blames the country's governing elite for war and violence. However, speaking out comes with risks. As one civil society activist put it in a Tweet (see Figure 6.2), 'One is always trading between speaking the truth and preserving the ability to speak tomorrow.'

Moreover, where criticism of the regime emerges, online interactions quickly become polarized. For instance, when Kiir's government announced its first extension of the transitional period and a delay of national elections in August 2022, civil society activists launched several WhatsApp groups to discuss this step as unlawful and counterproductive for peace, arguing that a leadership change was necessary to

[17] Interview 53.

end the suffering of the population. Observers of the social media discussions surrounding the extension noted that there were not only opponents but also supporters of the proposal and, more importantly, that these camps were divided along ethnic lines.[18] The social media exchanges about the extension proposal and whether it was good or bad for the prospects for peace in South Sudan were thus shaped by the societal cleavages that Kiir's regime exploited to sustain its grasp on power. Furthermore, the regime successfully managed to quell the online opposition to the proposal by securing the buy-in of some major voices in civil society. The discussion on social media quickly died down after the head of one of South Sudan's major CSOs, CEPO, was invited to speak at a meeting in which the extension was formally declared, officially abandoned his critical stance, and embraced the extension.[19] While it was difficult to reconstruct the exact tactics employed by the regime to stifle dissent, the examples illustrate well how offline and online dynamics work together in introducing a divisionist logic into technical policy discussions, and in slanting the playing field to suppress oppositional voices.

Content created and circulated by professional peacebuilding organizations can also quickly become divisive. For instance, as a project officer of one organization implementing peace-support projects recalled, 'we conducted a training in one of our partner counties, and we posted about it online. And somebody posted and asked, why do you always go to this county? South Sudan is such a big country, and it needs more support, and you are doing very little.'[20] The fact that seemingly factual information could be quickly weaponized in polarized online debates may explain why some peacebuilding actors refrained from engaging in social media exchanges altogether. The Director of Communications at R-JMEC explained, 'we don't go down the route of getting into Twitter debates with people you can't identify'. While he acknowledged that there was a 'number' of active Twitter (now known as X) users in South Sudan, he argued that the platform wouldn't be suitable for their mission because 'it's Twitter. People have strong opinions. They fire it back and forth.' While maintaining a Twitter account, R-JMEC didn't use it to enable two-way communication, but simply to 'fire messages out'. Instead, the commission preferred other means of (digital) communication that offered better control, including publishing regular updates on its website or sharing updates via conventional media such as the radio. Conversations with ordinary South Sudanese, however, were limited to some sporadic exchanges. As the Director put it, 'if an individual in the diaspora, a South Sudanese person, wanted to get involved, an email might be more effective, but Twitter quickly descends, so we stay away'. Comparable considerations informed the thinking of many other peacebuilding professionals I spoke to, who, when engaging with ordinary South Sudanese on social media, either chose a top-down approach or a platform that allowed for strong content moderation.

[18] Interview 41.
[19] Interview 53.
[20] Interview 105.

Importantly, integrative apomediation efforts that could counter these tendencies seemed surprisingly weak, given that the country has seen myriad attempts to foster dialogue between its warring parties at the national and local levels and across the country. There are a few CSOs, including religious actors, that could cut across the political and ethnic cleavages that sustain South Sudan's decade-old conflict. Still, their authority to do so has regularly been undermined by allegations of ethnic clientelism (Wilson 2019). Like in Northern Ireland, some South Sudanese peacebuilding organizations have also aimed to build volunteer networks that include participants from many different parts of the country, such as those helping monitor harmful speech discussed in Chapter 3 or supporting localized conflict early warning, discussed in Chapter 4. For example, it is a common practice to create WhatsApp groups among the participants of these activities, which often include users from a diversity of locations and communities. While promoting cross-community engagement and understanding commonly forms part of these efforts, their ability to effectively counteract online and social media polarization remains limited because these networks tend to be established to implement their narrow technical mandates defined in the projects.

South Sudan has also seen some attempts at reformative apomediation. However, these collective efforts to use the internet and social media to communicate popular grievances with the regime that are shared across the population were quickly curtailed. This is well illustrated in the experiences of a group of civil society activists, the People's Coalition for Civic Action (PCCA), who attempted to use social media to mobilize for peaceful protests in 2021, just after the ten-year anniversary of South Sudan's independence. Launching accounts on Twitter and other social media platforms, the coalition aimed to mobilize for a 'day of awakening' on 30 August 2021. The PCCA comprised several CSOs, a think tank, and individual activists that launched a public campaign with the slogan 'We have had enough'. The coalition blamed Kiir's regime and the leadership of the ruling party, the Sudan People's Liberation Movement (SPLM), for failing to govern the country effectively and demanded political change. It also used various social media platforms and messaging applications to mobilize for the protest and coordinate events in the run-up to the action day.

The PCCA's campaign and messaging constructed a direct link between the absence of political leadership and the abuse of public offices for individual enrichment, with the continued economic hardship and, crucially, war and violence. In a post on 3 August 2021 and directed to the 'leaders of South Sudan', it stated that 'you have achieved nothing but total destructions (sic), wars, sufferings (sic), tribalism, corruption, poverty and underdevelopment for the people of South Sudan' (The Civil Action SSD [@CivilActionSSD] 2021b). Its spokesperson was cited in an online newspaper arguing that South Sudan's 'dysfunctional government cannot deliver peace, security and stability which the country desperately needs' (Sudan Tribune 2021). In a video message posted on X (formerly Twitter) a few days before the day of action, the PCCA stated, 'if you live in the village in terror because of communal violence, cattle raiding, war, child abduction & flooding, this is your chance to speak out' (The Civil Action SSD [@CivilActionSSD] 2021a). This suggests that the PCCA also tried to mobilize ordinary South Sudanese outside the well-educated and connected urban population. Indeed, one of its members had attempted to lay the ground for a broad-based movement several years before, for instance, by inviting ordinary citizens to monthly collective action events that were formally branded as an environmental effort to tidy Juba's streets. The accompanying hashtag #NadafaLe-Beledna ('Clean our Country'), the social media commentary published alongside it that linked to hashtags and accounts relevant to the peace process, such as @JMEC-southsudan, and the symbolism employed—including a raised fist—were intended to create ambiguity as to what exactly the volunteers wanted to clean up.[21] As Michael Wani, the campaign organizer, stated in an interview re-published on social media: '[A]s young people, it's time to clean up this country, and we have a responsibility to bring about peace and to bring about co-existence among ourselves' (Deutsche Welle 2018). The PCCA's messaging, in comparison, was even less veiled, calling directly for a change in government.

However, when attempting to take the PCCA's calls for transformative change beyond the capital, Juba, the Coalition was faced with challenges such as limited connectivity and low literacy rates. According to one of the activists, reaching rural populations would have required travelling around the country and arranging in-person meetings. However, given the regime's tight control that extended down to local administrations tasked with welcoming and introducing visitors, a comprehensive outreach was difficult to implement.[22] Nonetheless, the PCCA's campaign generated considerable attention in the capital and the regime responded swiftly to it, including by spreading disinformation on social media, threats of arrest against possible protesters, a multi-day curfew, and the heavy deployment of the police and military in Juba, as well as an internet shutdown on the planned day of action (Reuters 2021). The crackdown particularly targeted the ringleaders of the PCCA, including through the freezing of their bank accounts (Amnesty International 2021b). Using virtual private networks (VPNs) and other methods to hide their digital traces, most activists managed to go into hiding and escape the hands of the regime for several weeks. Fearing for their liberty, safety, and even their lives, many subsequently left South Sudan and sought refuge abroad.

The experiences of the PCCA make clear that South Sudan's strongly slanted and limitedly adaptable playing field makes it difficult, if not impossible, to establish or maintain reformative apomediation dynamics that could connect a broad range of ordinary internet users to articulate popular grievances, identify the incumbent regime as an underlying cause for the continued armed violence in the country, and mobilize for peaceful change. While the PCCA's efforts were bold, the reformative apomediary that it mobilized quickly faltered and fell short of its goal of initiating a political change process that would undermine the power of the country's governing elite and bring peace. Overall, I thus suggest describing these apomediation dynamics as *regressive* because they not only divert attention from important technical subjects

[21] Interview 149.
[22] Interview 149.

and the detrimental effects of authoritarian rule but also strengthen the incumbent regime's hold on power by reproducing divisionism, and with it, some of the main factors that make violence possible.

Transformative Apomediation: Economic Crisis, Mass Protests and Political Change in Sri Lanka

In terms of its digi-political playing field, Sri Lanka sits somewhat in between the two other cases discussed above. The country's average population has considerably more internet access, literacy, and digital literacy than in South Sudan, but not as much as in Northern Ireland. And while Sri Lanka has seen substantial efforts to slant the playing field, its governments have repeatedly been voted out of office. Since its independence in 1948, the country has experienced various cycles of violence. The armed resistance and war against Sri Lanka's government led by the Liberation Tigers of Tamil Eelam (LTTE) was originally triggered and then sustained by grievances of the Tamil minority population, which demanded a stronger devolution of power or independence from the Sinhala-dominated state. However, political reform efforts, such as through the 13th Constitutional Amendment that I discussed in Chapter 3, led to little tangible change. The war ended with a military victory of the Sinhala-nationalist government under Mahinda Rajapaksa over the LTTE and other separatist groups in May 2009. In the decade after, the Rajakpaksas aimed to extend and consolidate their power over the Sri Lankan state, relying on a combination of Sinhala-nationalist populism and coercive control to keep the political opposition in check. The country's transition from war to relative peace and stability has thus often been described as a form of authoritarian peacemaking, leaving little civic space for organized peacebuilding activities that could help overcome the continued discrimination against minority ethnic groups (Heathershaw and Owen 2019).

However, the Rajapaksas' grip over the post-war Sri Lankan state was not total, thanks to a strong democratic political opposition and well-established democratic institutions. This first manifested in January 2015 when the regime was voted out of office. Some observers attributed the success of the political opposition to an increase in online connectivity and social media use in the years since the end of the war, which made mobilizing resistance easier. The ousting of the regime led to a brief period of political liberalization, during which many social media users felt less constrained to voice their political opinions and views online. The success in voting the regime out of power also emboldened many Sri Lankans to speak out more freely. This development did not reverse once the Rajapaksas regained power a few years later.[23] Therefore, compared to South Sudan, many more ordinary citizens viewed social media and other online spaces as an important venue to express their desire for change towards a more peaceful political order and to mobilize for it.

[23] Interview 67, Journalist, Colombo, Sri Lanka, 3 February 2023.

The need for this arose when the Rajapaksa regime returned to power. The Easter Sunday terror attacks in April 2019 had a strong polarizing effect on Sri Lankan society, which also manifested online and on social media. In the immediate aftermath of the attacks, the government briefly blocked access to the major social media platforms to suppress the sharing of disinformation and inflammatory content (O'Brien and Wright 2019). However, the effects of the attacks were long-lasting and provided a fertile ground for Sinhala-nationalist agitators aligned with the regime, who aimed to further deepen the societal fault lines through polarizing online and social media content. Many social media posts blamed and insulted Tamil and Muslim politicians, created hatred against religious and sexual minorities, and promoted Sinhala-nationalist ideology, which had been used as a political weapon already in the years before. This stirred many voters back into the nationalist camp and returned the Rajapaksa regime into government in November 2019.

At the beginning of the research for this book in 2020, the regime continued to maintain its political influence by perpetuating a Sinhala-nationalist ideology that spread hatred against the Tamil population as well as Muslim and Christian communities. These efforts had a strong online and social media element to them. Social media influencers, often with close links to Buddhist extremist groups and the regime, widely used hate speech and disinformation to stimulate animosities. A prime example of this is the widely known Buddhist monk Gnanasara Thero, head of the hardline Bodu Bala Sena, known as the 'Buddhist Power Force', which is one of Sri Lanka's leading ultra-nationalist Sinhalese Buddhist organizations. Thero played an instrumental role in stoking anti-Muslim sentiments before and after the attacks. For instance, in May 2019, he spread unfounded allegations about Muslim restaurant owners putting 'sterilization medicine' into their food to reduce the fertility of the majority Sinhala population, and called for the stoning of Muslims. This inflammatory rhetoric spread quickly on social media (University of Minnesota Law School Human Rights Clinic, n.d.). Critical observers highlight that such divisionist rhetoric played an instrumental role in politics. In the words of one journalist working for an independent online news platform, 'the politicians are always playing us, they create these divisions and keep them going for their own personal benefits so that they can rob the country'.[24]

Like South Sudan, Sri Lanka has witnessed considerable international support for the peace process and efforts to promote long-term conflict transformation and reconciliation. In the brief interim period between 2015 and 2018, international efforts to promote long-term reconciliation and support truth-seeking and transitional justice initiatives picked up. Among others, Sri Lanka co-sponsored a Resolution at the UN Human Rights Council, committing to a comprehensive approach to reconciliation and transitional justice (Asia Justice and Rights et al. 2022). It also established key institutions and policies, such as an Office of National Unity and Reconciliation (ONUR) and a National Policy on Reconciliation (Fernando 2020). However, these efforts were quickly curtailed again after the Rajapaksas re-established their

[24] Interview 67.

power, reducing the scope of formal peacebuilding and reconciliation activities. Consequently, institutions like the ONUR became dormant (Institute for Participatory Interaction in Development 2021).

Nonetheless, several international and national organizations continued supporting various peacebuilding initiatives, albeit falling short of the ambition of enabling a nationwide and government-supported reconciliation process. These initiatives were diverse in terms of their activities and objectives, ranging from empowering marginalized communities through rural development projects to strengthening women's participation in politics and society, to promoting and memorialization through online mapping (Hettiarachchi 2018) or digital museums and art exhibitions (International Centre for Ethnic Studies, n.d.). Where these organized activities had an online component, they contributed to a diverse effort that shed light on various topics and concerns relevant to a lasting transition from war to peace, thus contributing to a programmatic apomediation of conflict and peace. For instance, the projects focused on promoting a culture of remembrance commonly associated the causes of conflict with the fact that the Sinhala majority population had limited awareness of the past and that violence and political oppression of minority groups had, to a large degree, been carried out clandestinely. From this, it followed that sharing testimonies of violence would contribute to conflict transformation by promoting empathy and mutual recognition of grievances, thus contributing to a re-humanization of former antagonists (GIZ, n.d.). However, many of these projects shared the common challenge of struggling to do more than 'preaching to the converted' by reaching beyond the well-educated middle and upper classes.[25] This made it difficult to achieve broad-based social and political change.

However, many of Sri Lanka's programmatic apomediation efforts were also sustained by a vibrant civil society that didn't view its mission primarily as contributing to peacebuilding, yet it expressed a diversity of views on the country's tense intercommunal relations and what to do about them. Comparable to Northern Ireland, Sri Lanka also has a long-established culture of democratic debate and an even more diverse landscape of political parties and CSOs, despite the risk of 'ethnocratic' dominance by the Sinhala majority (DeVotta 2021; Hensman 2022). Thus, many politicians and civil society voices engaged in issue-based debates relevant to conflict and peaceful co-existence, and they would commonly extend to social media. However, like in Northern Ireland, these debates tended to be polarized along political party, ethnic, and religious lines.

Some peacebuilding initiatives aimed to bridge these divisions and thus strengthened the potential for integrative apomediation. Particularly impactful were those that first facilitated offline encounters between members of different communities and subsequently established communication infrastructures that could help maintain contact between them. For instance, Sri Lanka Unites, a youth-led CSO

[25] Interview 22, Staff Member, Civil Society Organization, Colombo, Sri Lanka, 9 January 2022; Interview 24, Staff Member, Civil Society Organization, Colombo, Sri Lanka, 13 January 2022; Interview 28, Staff Member, Civil Society Organization, Colombo, Sri Lanka, 22 January 2022.

founded in 2007, has a long track record of facilitating exchanges between youth by building partnerships between schools. Most schools in Sri Lanka only use one language of instruction, which institutionalizes ethnic divisions by nourishing in-group thinking that commonly reproduces negative stereotypes about other groups. This creates a fertile breeding ground for bigotry and hatred. In the words of the founder of Sri Lanka Unites, 'that's how we've had cycles of violence every 15 to 18 years because you are able to brainwash enough of your people to assume that their patriotic duty is to hurt the Other!'.[26] To counter this trend, the organization viewed student exchanges as a straightforward way to enable cross-community exchanges. It started facilitating such exchanges in 2009, by organizing an annual conference, bringing together representatives of several hundred schools. The conference participants were encouraged to create dedicated school clubs that offered extracurricular activities, as well as partnerships with clubs located in different communities. Importantly, since the rapid uptake of social media in the past decade, the organization has also used in-person exchanges between students to create lasting online connections, encouraging the participants to connect via social media platforms after they had met in person. Through this approach, the organization reached many thousands of students every year and across all parts of the country.

The assumption behind this effort was that having more diverse groups of social media contacts would encourage the users to reflect more strongly on what they are sharing or posting—and it created some sort of 'check system', in the vision of its founder: 'If I met you and now you are putting racist stuff against me, I can ask you what happened.'[27] Several other CSOs similarly used offline engagements to work towards a stronger integration between different ethnic and religious communities on social media (Helvetas, n.d.). It is difficult to determine how much of an impact such efforts to facilitate cross-community social media connections had. For the sake of comparison, however, it suffices to say that this kind of civil society activism was unthinkable in South Sudan. This is arguably due to many factors, including the fact that Sri Lanka has a much more vibrant and well-capacitated civil society. Yet, the characteristics of the digi-political playing field, including higher levels of general and digital literacy, as well as fewer hurdles for organizing such meetings, were also more favourable in Sri Lanka. South Sudan has also seen some school exchanges, but on a much smaller scale, and arguably with less impact. For instance, the UN mission, UNMISS, facilitated a two-month cultural exchange between six secondary schools in March 2023, to strengthen social cohesion (UNMISS 2023). While it is difficult to determine the degree to which this effort translated into online relations, South Sudan's low levels of connectivity clearly made it more difficult than in Sri Lanka.

In other ways, the playing fields in Sri Lanka and South Sudan were more similar. Certain invisible red lines that curtailed apomediation on social media also remained in Sri Lanka, shaping to what degree and how civil society groups would associate the causes of conflict with the incumbent regime. In the decade following the victory over

[26] Interview 66, Prashan de Visser, Founder of Sri Lanka Unites, Colombo, Sri Lanka, 3 February 2023.
[27] Interview 66.

the LTTE, the Rajapaksa regime demonstrated repeatedly that it did not shy away from intimidating and punishing political opponents and critical civil society voices that challenged government policies, including through beatings, arbitrary arrests, surveillance, and death threats (Human Rights Watch 2020). As one prominent civil rights activist put it, 'there was self-censorship and a concern that, if you didn't toe the line, to a certain extent, you were going to get it in the neck'.[28] The red line was drawn around topics such as war crimes and human rights abuses, but also allegations of political mismanagement and corruption—issues which, for many, were closely linked to Sri Lanka's decade-old conflict. In particular, journalists or activists calling out the regime's human rights record and involvement in war crimes commonly faced severe repercussions (Human Rights Watch 2021). Therefore, many civil society representatives tended to exercise restraint in directly criticizing government officials and members of the regime for their divisionist politics and involvement in the war. Critical voices among the Tamil minority, as well as civil society activists, lawyers, and journalists, would commonly be censored through the application of the Prevention of Terrorism Act (PTA), which allows arrests without warrant for unspecified 'unlawful activities', including speech-related offences (Lewis 2020, 19). Calls for reconciliation and solidarity with the Tamil and Muslim populations were effectively stifled through this legislation, as well as through internet filtering and social media censorship. Like in South Sudan, this meant that programmatic apomediation efforts would ultimately remain somewhat toothless when it came to the need to discuss the regime's role in Sri Lanka's protracted transition from war to peace.

All this was challenged in 2022 when Sri Lanka was hit by a harsh economic crisis, characterized, among others, by high inflation and severe energy shortages that disrupted the country's economy, service delivery, and everyday life. Many voices in the opposition and civil society quickly attributed the unfolding crisis to the regime's economic and fiscal mismanagement. Faced with worsening living conditions, thousands of Sri Lankans joined country-wide mass protests calling for the resignation of the Rajapaksa regime. These protests, known as the 'Aragalaya'—the Sinhalese word for 'struggle'—led to the tumultuous overthrow of the government in July 2022. The protests were primarily driven by a disgruntled middle class that perceived the impacts of the economic crisis as an existential threat to its way of life. Many observers noted that social media played an instrumental role in the movement's efforts to communicate its grievances and coordinate the protests. The movement effort also gained traction outside urban centres, reaching rural populations via chat groups and social media posts. Unlike South Sudan, Sri Lanka's high levels of connectivity thus enabled a mass mobilization across large parts of the country—a mobilization that proved to be much more potent than the PCCA initiative discussed above.

Even more relevant for understanding the protest movement's success were the comparably high levels of literacy and education among Sri Lanka's social media users, which not only played a role in enabling the sharing of insights and analysis

[28] Interview, Executive Director, National Civil Society Organization, Colombo, Sri Lanka, 9 February 2022.

about the causes of the crisis across a large cross-section of society, but also constituted an intrinsic factor in mobilizing against the regime. High levels of education and literacy gave rise to a vibrant political culture, which manifests, for instance, in the participation of ordinary citizens in local government affairs.[29] At the same time, preserving access to quality education also became a declared goal of the protest movement, especially among student activists with a pro-democracy and social justice agenda (Dulanjana 2024). The protest movement benefited from this mutually reinforcing relationship between high levels of education and political activism when it employed online and social media platforms to articulate criticism of the regime's financial and economic mismanagement (Gamage 2023; Centre for Policy Alternatives 2023).

Within months, the protests quickly threatened the survival of the regime, culminating in the storming of the presidential residence in July, the subsequent escape of President Gotabaya Rajapaksa from the country, followed by his resignation on 14 July, and the creation of an interim government under Ranil Wickremesinghe. This outcome was, to a considerable degree, due to the sheer size of the protest movement: an estimated one-third of the Sri Lankan population took part in the protest at some point (Centre for Policy Alternatives 2023, 9). Not all protesters engaged online, and different degrees of digital literacy conditioned the degree and depth of engagement.[30] That said, the level of online engagement, especially in coordinating protest activities, was much higher than during the PCCA action in South Sudan. This meant that the number of organizers the regime had to intimidate, prosecute, and arrest was much higher. By September 2022, more than 4,000 people had been arrested due to their role in the Aragalaya; however, this was mainly for acts with heightened visibility, such as forcibly entering government buildings and not just for engaging in the protest (Warakapitiya 2022). Thus, many protesters felt that the sheer size of the movement would provide a degree of protection.

Seeking cover and anonymity among the millions of co-protesters, the well-equipped and networked protest movement formed what I suggest describing as a digital critical mass, which changed the cost-benefit calculation for many activists. That is to say, taking part in connected, decentralized mass action was perceived as providing sufficient protection to balance potential risks stemming from the regime's authoritarian tactics. At the same time, quietly accepting the status quo became increasingly unbearable, as one observer put it: 'When your basic needs are not met, what is there for you to lose? If I lose my job? If I can't feed my kids?'[31] Importantly, social media served as a key vehicle to organize and coordinate protest action, allowing the creation of a leaderless protest movement that coordinated offline action in Facebook and WhatsApp groups, TikTok, and Telegram channels. Such platforms were also widely used by individual protestors to livestream events, voice grievances, and share content. Some of their messages were elevated and made visible with the

[29] Interview 67.
[30] Interview 86, Senior Lecturer, University of Colombo, Colombo, Sri Lanka, 9 February 2023.
[31] Interview 68, Journalist, Colombo, Sir Lanka, 3 February 2023.

help of social media influencers who had previously exclusively generated entertainment content but now posted about topics such as democracy and civil rights.[32] Furthermore, many civil society activists also employed sophisticated tactics and tools to circumvent the online surveillance, censorship, signal jamming, and internet blackouts the government put in place to stifle the protest movement.

The protests were largely driven by grievances about the impact of the economic crisis on the population, such as a lack of fuel for transportation or cooking. This prompted calls for regime change through slogans and hashtags such as #GotaGoHome. The mass protest was dominated by the Sinhala majority, primarily motivated by the economic crisis, and aimed at ending the financial mismanagement, corruption, and nepotism associated with the Rajapaksa regime. In contrast, many people identifying with minority groups and located in the North were reluctant to join the street protests due to fatigue associated with long-term deprivation, fears of more political oppression, or a sense that the struggle wasn't theirs (Ellis-Petersen and Sandran 2022). Prior negative experiences with the regime's authoritarian methods were a further factor.[33] Nonetheless, a sub-population of protesters associated the causes of the 2022 crisis with long-term political mismanagement and authoritarianism under the Rajapaksa regime, which also underpinned many of the civil war grievances. These interactions led to broad-based reformative apomediation dynamics.

Moreover, the many vivid exchanges on social media also provided an opportunity for integrative and programmatic apomediation. The protests were accompanied by countless everyday online interactions that contributed to challenging how ordinary Sri Lankans thought and felt about inter-group relations and the long-standing causes of violence between the Sinhala majority and the country's religious and ethnic minorities. A sizeable number of minority activists joined the protest movement, which they perceived as an opportunity to raise topics such as religion, ethnicity, and inter-group relations, and their role in Sri Lanka's authoritarian politics.[34] Through social media posts and in chat groups, the protests were turned into an opportunity to demonstrate, discuss, and experience unity across Sri Lanka's diverse population groups through integrative interactions that linked grievance narratives across diverse religious and ethnic groups. Many small acts of creating, sharing, and posting content online, some by established CSOs, but many by ordinary citizens, contributed to establishing understandings of conflict that framed the regime's oppressive and divisionist policies as a key factor for the crisis, while drawing parallels to Sri Lanka's violent past. Against this background, inter-ethnic collaboration and unity could be presented as a possible way out of the crisis, and as a pivotal step towards a more peaceful political order.

Many social media posts drew parallels between the experiences of Sri Lanka's various groups with the regime's violent response to dissent, by demonstrating that the authorities employed the same repressive methods to quell the Aragalaya as they had

[32] Interview 83, Director, Civil Society Organisation, Colombo, Sri Lanka, 9 February 2023.
[33] Interview 68.
[34] Focus Group Discussion, Kandy, Sri Lanka, conducted by Hasini Haputhanthri, 29 May 2023.

Figure 6.3 Social media meme in response to regime violence against Aragalaya protesters, posted by Sayanthan Kathir, 4 April 2022, reproduced with the permission of the author

applied to minority groups in the years before. For instance, many protestors were arrested under the PTA legislation, which was previously only applied to arrest Tamil and Muslim regime critics and charge them with terrorism.[35] Moreover, the regime aimed to intimidate protesters through violent crackdowns, using tactics that echoed the oppressive playbook used against Tamils and Muslims. Amongst others, these parallels were amplified by social media memes that employed irony and, at times, sarcasm to confront the Sinhala majority with their silent complicity. I have discussed how memes contribute to everyday online reconciliation efforts in greater detail elsewhere (Hirblinger et al. 2024). The following example from this research reproduces the power of integrative social media practices particularly well. The meme in Figure 6.3 was posted on a private social media account together with a comment referencing a news article about Aragalaya protesters being attacked by camouflaged military on motorbikes and stating, 'Meanwhile minorities in Sri Lanka'.

The meme was widely shared in the user's network, linking the experience of the protesters to similar attacks against Tamil communities in the decades before. The question 'first time' provides a sarcastic undertone and invokes differences in the communities' experiences by alluding to a repetitive (and potentially longer) exposure to political oppression. Several comments added under the meme suggested that the meme could generate mixed reactions. For instance, one user apologized for the violence against the Tamil people, while another admitted that it had 'hit hard'. A further user commented, 'when the North screamed out about these things, the South laughed. Now the South is seeing this with their own eyes. Is the South laughing anymore?'.

[35] Interview 67.

The creator of the meme, who identifies with the Tamil-speaking diaspora, revealed a similar consideration in a written exchange. '[Acts of state-sponsored terrorism in the North] took place on the island, in the same country where Sinhalese people also have their homes. However, the large bulk of them have never cared and have never reacted'. After he had created and shared the meme, however, he felt a 'spark of hope' because the meme was shared widely among Sinhalese youth. Responding to the original post, another user clarified that similar incidents had indeed happened in the Southern part of the country during a youth insurrection in 1988–89, thus recalling and invoking a longer joint history of oppression. Several civil society activists with whom I discussed this meme echoed this sentiment. One of them described it as capturing an 'eye-opening' moment because people (the Sinhala majority) hadn't been aware of the impact of militarization of the state on the communities in the North and East during and after the civil war.[36] According to another activist, the meme displayed similarities between 'what they go through in the North and East and what others go through in the rest of the country'.[37]

The protests were also marked by many explicit, and sometimes planned and staged, demonstrations of ethno-religious unity (Fernando 2022; Modin-Lundin 2022). Many of these were livestreamed or documented online and via social media. These acts furthered integrative apomediation between protestors, activists, and regime critics identifying with a variety of ethnic and religious groups, by demonstrating how they were similarly affected by the regime's oppressive tactics and violence. This often also entailed efforts to document, share, and enhance offline events and experiences on social media. For instance, some peacebuilding organizations formally joined the protest online and at the physical protest sites, and many of their employees contributed to the protests in an informal and private capacity. Coordinating protest activities on-site and online, and reporting from the protests, these individuals aimed to use the grievances expressed by the Sinhala-dominated middle class to raise grievances that ethnic minorities and socially marginalized parts of the population held. By connecting the two grievance narratives, these networks of activists produced information and content that emphasized a common cause of conflict: the Rajapaksa regime and its Sinhala-nationalist politics.

With all this came the realization that the regime's authoritarian politics affected all parts of the population. Through street and online action, the protest movement drew many more parallels between the 2022 crisis and the civil conflict through the symbolic representation of minority concerns at the protests, as well as the memorialization of past abuses against the Tamil minority. For instance, protesters commemorated the regime's bombing campaign that had brought the armed conflict to a violent end in May 2008. They also established a small library in the main protest camp and livestreamed events from there to collectively remember the burning of the

[36] Participant, Focus Group Discussion, Colombo, Sri Lanka, conducted with Hasini Haputhanthri, 8 February 2023.

[37] Participant, Focus Group Discussion, Colombo, Sri Lanka, conducted with Hasini Haputhanthri, 8 February 2023.

Jaffna library during the civil war. Such symbolic gestures helped capture a deeper understanding of the conflict, and they could provide the ground for overcoming past divisions by speaking to the concerns of the Tamil and Muslim minorities. One Sinhala activist recalled the words of a Tamil colleague about the impact of these social media interactions as follows:

> Now, Colombo knows all our problems. Now you are sensitized about the issues we are facing. Now we are on 'common ground'—somebody put it like that. Now, we are on equal footing. And now you know that the political elite is not only against the people of the North and East, but also against you![38]

Some of the organizations involved in the Aragalaya also activated inter-ethnic and inter-religious social media networks built during earlier offline and in-person peace-building activities, which further enabled such integrative interactions. For instance, the community-based organization Sri Lanka Unites used its vast network of former volunteers and training participants to coordinate protest action, plan, implement, and stage integrative events, and create a multi-ethnic protest platform. It also represented the integrative nature of the popular protests through social media content (see Figure 6.4). Many users joined these groups, and the content they shared contributed to integrative apomediation in an organic manner.

Furthermore, the protests also provided an opportunity for programmatic apomediation efforts to extend their reach. Peace activists used social media to draw attention to more complex issues that stood in the way of reconciliation. For instance, the main protest camp in Colombo featured a tent hosting a Tamil victims' association, the Families of the Disappeared (sometimes referred to as 'Mothers of the Disappeared), representing parents whose children had been abducted and possibly killed during the civil war. Images of the tent were widely shared online, and events and speeches at the tent were reported on social media and online news pages, which drew parallels between different historical moments of struggle against state oppression. An article on the well-known online news page Groundviews described 'the families of the disappeared' as 'reminders, reliving the pain of a collective history that can't be closed like a chapter in a book' (Wirasinha 2022). Furthermore, Tamil activists used social media to revive concerns about past human rights abuses, the need for transitional justice, and the implementation of political decentralization (Fedricks et al. 2022). This meant, as one peace activist put it, that online conversations that had previously been directed to 'Northern audience' would now involve the majority population, including Sinhala and Buddhist people based in Colombo 'listening to their conversations and activism'.[39] One of the consequences of this development was that political demands previously uttered only by minority activists were now also discussed by Sinhala peace activists, such as regarding the implementation of the 13th Amendment.

[38] Interview 83.
[39] Interview 83.

Figure 6.4 Social media meme calling for unity among Sri Lanka's religious groups, posted by Prashan de Visser, 14 July 2022, reproduced with permission of the author

However, some observers remained sceptical about how much these concerns mattered overall. In the view of another prominent civil rights activist, minority perspectives and demands, such as regarding the devolution of power, 'may have been brought in, but they first slid into a "secondary position" and then became practically non-existent, because (the protest movement put), very much the emphasis on system change'.[40] Thus, in his view, minority perspectives were subsumed under the larger objective of systemic political change. Other observers felt that most 'majority protesters' maintained a sentiment that they had 'won the war', which meant that there was no need to consider minority grievances and demands for political change.[41] This demonstrates how apomediated understandings of conflict and peace, and their impact on the political crisis and transition, remained highly subjective.

Overall, these apomediation dynamics had a much more profound impact on the long-term prospects of conflict transformation than those I observed in Northern Ireland and South Sudan. I suggest describing the apomediation dynamics around the 2022 Aragayala protests as *progressive* because they managed to create forward momentum for political change by linking reformative, integrative, and

[40] Interview 85, Executive Director, National Civil Society Organization, Colombo, Sri Lanka, 9 February 2022.
[41] Interview 94.

programmatic apomediaries, which created an understanding of conflict and crisis that blamed the authoritarian and divisive politics of the incumbent regime. While 'peace' and 'peacebuilding' may not have prominently featured in verbatim, these progressive dynamics nonetheless encouraged a sense that the crisis of 2022 would be overcome if the drivers underpinning the decade-long conflict would be addressed—chiefly, the exploitation of ethnic difference through DD&R. Since the ousting of the Rajapaksa regime in July 2022, Sri Lanka's new head of state, Ranil Wickremesinghe, has made efforts to further slant the playing field by advancing the introduction of new and more restrictive social media legislation (Ada Derana 2023). However, he has also re-initiated discussions around implementing the 13th Constitutional Amendment, promising to further devolve power within the state and to Sri Lanka's minority groups, thus somewhat responding to the protesters' demands.

As I conclude the work on this chapter, there are further signs that the progressive apomediation dynamics that unfolded online during the Aragalaya could indeed translate into tangible political change. Sri Lanka's 2024 parliamentary elections resulted in an overwhelming victory for a new political coalition, the *Jathika Jana Balawegaya*, or National People's Power (NPP) alliance. The presidential elections were also won by the NPP's leader, Anura Kumara Dissanayake. This multi-ethnic coalition of more than twenty political parties promised to implement far-reaching political reforms (Ethirajan et al. 2024). The NPP attracted support from across the country, which many observers interpreted as a first victory along the strenuous path towards overcoming the country's ethnic and religious divides (Mallempati 2024; Croos 2024; Janardhanan 2024). The NPP has also committed to building a unified Sri Lankan nation, promoting reconciliation and truth-seeking, and adopting governance initiatives that maintain harmony and co-existence among the country's ethnic groups (Jathika Jana Balawegaya 2024, 124). NPP representatives also pledged to launch a constitutional process to tackle the thorny issue of political devolution (Samaraweera 2024). How much precisely the progressive apomediation dynamics mattered for achieving this outcome is difficult to ascertain—as are its effects on long-term political change. However, it has become clear that the potential of online interactions to contribute to conflict transformation grows considerably when organized digital peacebuilding efforts are catalysed by the momentum of a critical digital mass nourished by popular grievances and the will of ordinary citizens to seek change.

Conclusion

This final chapter demonstrated the need to rethink what conflict transformation should mean in conflict contexts shaped by digitalization and what forms of digital activism are best suited to contribute to it. To this end, it shed light on how the apomediation of conflict and peace in online publics and on social media helps or hinders efforts to transform armed conflicts. Against the backdrop of rising digital authoritarianism and DD&R tactics in contemporary armed conflicts, I have described three ideal-typical apomediaries that differ in terms of the relations they form and the understandings of conflict and peace they produce and convey. First, programmatic apomediaries, shaped by peacebuilding professionals and policymakers, centre on specific peacebuilding topics and construct technical understandings of the causes of conflict and their resolution. Second, reformative apomediaries, composed of civil society activists, the political opposition, and diaspora members, develop vertical relations by encouraging large numbers of ordinary social media users to voice their grievances. They associate the causes of conflict directly with the incumbent regime and seek political change as a way of solving the conflict. Third, integrative apomediaries, composed of professional peacebuilding actors, their volunteer networks, and other inclusive societal groups, develop horizontal relations, thus associating conflict with societal cleavages and aiming to connect citizens across societal divisions to build peace.

Everyday online interactions of information exchange matter in public efforts to transform conflict because they shape how conflict parties, stakeholders, and the general population make sense of conflict causes and options for conflict resolution. However, comparing apomediation dynamics in Northern Ireland, South Sudan, and Sri Lanka, I also demonstrated that variations in the digi-political playing field have a crucial impact on these dynamics. In contexts with low digitalization and efforts to slant the playing field, programmatic apomediation tends to have a weak rooting in politics and society and remains toothless in the face of the DD&R tactics of authoritarian governments. In the organized digital peacebuilding efforts of professional organizations aimed at a change in inter-group relations, what is needed for fundamental political change thus remains missing-from-hand. Much of this may have to do with the fact that progressive apomediation dynamics cannot be simply planned, designed, and algorithmically implemented. As the chapter has demonstrated, they commonly require an intricate interplay across the socio-technical canvas of conflict-affected societies. However, everyday conflict transformation efforts can gain traction if programmatic efforts are joined by integrative and reformative apomediaries. However, those moments are scarce, particularly in places where the digi-political playing field is strongly slanted and little adaptive.

The apomediation dynamics that unfolded during Sri Lanka's 2022 political and economic crisis are particularly illustrative for exploring when and how apomediation may help build peace by countering DD&R. Compared to South Sudan, Northern Ireland, and pre-crisis Sri Lanka, these dynamics were characterized by the presence of a strong reformative apomediary that contributed to an intermediate transformative outcome. Sri Lanka's context is not unique. Many conflict-affected contexts with a slanted playing field will, at some point, see a continued growth in digital adaptability, which powers reformative apomediation by enabling a broad-based engagement with the complexity of conflict causes while providing some degree of protection against digital authoritarianism. This makes further research on this topic very important. I demonstrated that the interplay between programmatic, reformative, and integrative apomediaries advanced understandings of conflict and peace that diverged from the regime's narrative, making suggestions for how to transform community relations and the political institutions through which they are governed. Such progressive apomediation was enabled by the relatively high digital adaptability of the playing field, which allowed for sustaining social media interactions that the government's DD&R tactics would otherwise have curtailed. However, it was also driven by concrete socio-economic pressures and a mass mobilization that changed the cost-benefit calculations of individual social media users. The chapter also demonstrated that the presence of social institutions, including organized civil society, and tangible socio-material factors, such as financial inflation or fuel shortages, matter greatly in shaping apomediation dynamics. This suggests that efforts to transform conflicts and promote reconciliation that exclusively target the digital realm will have little sway. They need to connect with the everyday, tangible, and lived experiences of ordinary citizens as they struggle for a brighter and less divisive political future.

Conclusion

There has been an immense investment in digital peacebuilding approaches, and with it, a myriad of efforts to design, build, and employ digital technologies to support efforts to prevent and end armed conflict. This book has covered only a small fraction of them, but with the intention of bringing out theoretical perspectives and empirical learning examples that are relevant for studying the whole spectrum of digital peacebuilding, including future applications and dynamics. The book made clear that the application of digital technologies for peacebuilding tends to be grounded in various understandings of, and approaches to, peacebuilding, and sometimes in a lack thereof. It also demonstrated that technologies often inspire tech-solutionist thinking, which affects how tech designers and tech users think about peace and peacebuilding.

The book's core task was to explore how, in digital peacebuilding, decentred networks of human and machine agents shape the understandings of conflict and peace that condition how peacebuilding is done. The understandings of conflict and peace that emerge and circulate, for instance, on social media, in conflict early warning networks or during digital dialogues, correspond with specific digital peacebuilding approaches: deleting harmful speech to prevent violent riots against ethnic minorities, collecting timely information from field-based conflict monitors, or algorithmically establishing what peace agreement provisions could have broad-based support, to mention only a few examples covered in this book. Which understandings of peace and peacebuilding emerge in these networks and which are implemented is also conditioned by the digi-political playing field on which apomediation unfolds: the degree to which governments and political or military actors constrain interactions and impede their translation into offline action, and the ability of ordinary citizens to cope with these constraints and counter them.

Some of the digital peacebuilding practices that this book has brought to light have transformative ambitions: speculating, claiming, or describing how digital technologies could do more than prevent or terminate war and violence by contributing to a change in relationships between the parties and people in conflict—and also between people and technology. Sometimes, the ambition is to directly influence how those in conflict perceive or portray each other, for instance, through social media monitoring and counter-speech campaigns. Other times, peacebuilding actors aim to increase their capacity to find non-violent means to respond to conflict, for example, through training or the roll-out of mobile applications. In all such cases, a specific role is dedicated to digital technology, for instance, in enabling digital connectivity and data

Digital Peacebuilding?. Andreas T. Hirblinger, Oxford University Press. © Andreas T. Hirblinger (2026).
DOI: 10.1093/9780198986553.003.0008

processing capabilities. The digitalization of peacebuilding has thus rendered conflict transformation a task of enabling socio-technical relations that make it less likely that conflict parties resort to violence that denies the opponent's rights and needs, or worse, their right to existence. This requires, in one way or another, that the opponents recognize each other as legitimate adversaries so that conflicting needs and interests can be reconciled. While this ideal of agonistic peacebuilding is often difficult to achieve, conflict transformation could entail the no less important goal of strengthening the resilience and capacity for survival of vulnerable groups, so that violent antagonisms can't run their course.

However, the book's empirical chapters have pointed to the manifold reasons for which something in digital peacebuilding has become, or remains, missing-from-hand for such transformative approaches, thus leaving a sense of void among those who aimed to employ digital technologies to work towards a lasting peace that addresses the underlying causes of antagonistic relationships.

In Chapter 3, I discussed civil society initiatives aimed at fighting harmful speech on social media, which represent a relatively new approach to preventing conflict. I discussed how these initiatives are primarily occupied with monitoring and documenting harmful content and shed light on how efforts to counter such content can be challenging, especially in contexts with a digi-political playing field that is slanted and little adaptive. The chapter demonstrated how social media activism tackling harmful speech often struggles to address the deeper causes of hate. In both South Sudan and Sri Lanka, political elites benefit from, and sometimes further, harmful speech because they maintain ethnic divisions that help them to maintain a loyal following. Partnerships with social media platforms and sophisticated social media data collection and analysis methods may help detect and filter harmful speech at scale, yet they provide few resources for engaging with what powers every single hateful post or picture. The apomediation dynamics they enable thus generate peacebuilding practices that are primarily concerned with digitally produced and shared speech acts. The datafication of hate somewhat severs such harmful speech from the conditions of its emergence, and the technical solutions to dealing with such content likewise only scratch the surface. This means that the historically grown and culturally, as well as materially, sustained relations that fuel the violent antagonisms will remain in place. The work of preventing conflict on social media is thus intricately shaped by this missing-from-handedness of socio-technical relations that could do the deeper work necessary for transforming conflict.

In Chapter 4, I shed light on several conflict early warning initiatives that employ mobile phones to collect signs of impending conflict via text messages or mobile applications to enable mitigative efforts. Such systems are widely acknowledged as essential for early response, which may help de-escalate the conflict and, ideally, reduce and resolve tensions and could contribute to an agonistic peace by stopping opponents from fighting out their differences by using force. However, responding to emerging conflict in ways that help overcome antagonisms in lasting ways requires more than obtaining instant early warning signals. The chapter highlighted the need for apomediation arrangements that reach deep into conflict-affected contexts and provide contextually rich information to localized response networks. This requirement was already somewhat reflected in the worries of some of the early pioneers of computer-assisted conflict early warning systems, namely that their functions and data outputs would not be in the hands of the most vulnerable, that is, those affected by conflict. However, the ambition—of socio-technical relations that would radically challenge power relations and empower ordinary people to carry out mitigative action—has remained largely missing-from-hand in contemporary early warning. Retracing the evolution of early warning across various systems, the chapter demonstrated how concerns with power seemed particularly apt against the backdrop of the so-called early warning–early response gap, which was attributed mainly to a lack of political will of high-level decision-makers to act on locally collected early warning information. Visions of community-centred early warning can also be found in idealistic notions of crowdsourcing, and they may temporarily manifest in the spontaneous and organically unfolding community-based initiatives that integrate mobile technologies to enhance existing social relationships rather than building entirely new ones. However, the chapter also demonstrated that efforts to make conflict early warning systems more sophisticated, for instance, by introducing structured and fine-grained data collection methods, may well have the opposite effect, namely, stifling spontaneous and immediate action. The early warning application rolled out in South Sudan, for instance, tended to produce data that lacked the contextual richness necessary to inform transformative work. At the same time, the system became closely embedded within various levels of government and the security apparatus. This increased the risk that its data could be employed for a repressive response. The system was thus ill-suited to strengthening de-escalatory, mitigative measures that could pave the way for a lasting, transformative change.

Chapter 5 shed light on the increasing use of digital technologies in support of peace mediation and dialogue. It described how this area of engagement has been driven by the tech-solutionist promise that increases in digital connectivity and the data processing power of artificial intelligence (AI) could enable peacemaking efforts that are not only more inclusive but also faster and more effective. Moreover, I discussed how discussions of (digital) inclusion were also partly driven by the hope that broad-based peace processes could empower marginalized stakeholders and strengthen 'bottom-up' efforts. In this context, I argued that data must contain 'voice' to ensure that people can intentionally contribute to peacemaking and change the status quo they object to. However, in the empirical cases discussed in the chapter, the necessary socio-technical constellations remained missing-from-hand for such transformative ambitions. This is due to several reasons, including the fact that innovative approaches such as digital consultations are often used to serve pre-defined goals and to fit established organizational mandates and ways of operation that privilege high-level peacemaking efforts. Furthermore, the chapter focused on how algorithmic models embedded in digital peacebuilding approaches can influence the ways data are collected. This, in turn, affects how the dialogue participants of these processes can interact with one another, reducing the opportunities for dialogical encounters in which the participants could grapple with the complexity of their various needs, interests, and identities. Yet, such work is necessary to overcome hardened worldviews and narratives that underpin the antagonistic relations between parties and people in conflict. The chapter closed with a discussion of how 'subjunctive' uses of technology and data could provide a viable alternative, illustrated by a case study of an intra-Palestinian dialogue process. It demonstrated the need to consider the role of society and culture when designing transformative dialogues, and showed that digital approaches will have a limited impact if the wider context remains unconducive to peace.

Finally, Chapter 6 described efforts to support the long-term transformation of group relations through everyday online and social media practices. More specifically, it asked how apomediation could help counter the efforts of governing regimes to reproduce, manipulate, and exploit group differences to maintain their grip on power, which I described as a policy of 'digital divide and rule' (DD&R). The chapter showed how organized digital peacebuilding efforts may contribute to programmatic and integrative apomediation dynamics, the former enabling subject-centred information exchanges among small groups of experts and the latter enabling horizontal exchanges between users identifying with different groups. Despite such efforts, socio-technical constellations that could challenge DD&R often remain missing-from-hand. Comparing social media peace activism in South Sudan and Sri Lanka, I pointed to the need for 'reformative' apomediaries that create a vertical thrust by linking popular grievances to the politics of governing elites. More specifically, I showed how a critical mass of such connections might have the power to counter the incumbent regimes' efforts to keep the playing field slanted. This can enable what I described as transformative apomediation dynamics that draw on the strengths of programmatic, integrative, and reformative apomediaries. In such circumstances, social media and other online spaces are used to advance understandings of conflict across a diversity of users who identify with the antagonistic groups. Such constellations may generate enough momentum to counter authoritarian divide-and-rule politics more effectively because they identify the elite's divisionist techniques as a cause of conflict and formulate visions of unity and peaceful co-existence as a peacebuilding strategy. In Sri Lanka, this constellation was temporarily achieved during the Aragalaya. The discussion of this case also demonstrated that lasting change must reach deeper into politics and society, including through implementing political reforms and acts of justice and reconciliation that go beyond the digital sphere.

My research suggests that the diverse socio-technical constellations that shape the apomediation of peacebuilding commonly contribute very little to help those in, and affected by, armed conflict in their effort to move from relations of enmity to relations of adversity. This is, to no small degree, because those who design and use technologies for peacebuilding do so based on problem understandings and theories of change that often simply care very little about the particular approach of conflict transformation I introduced at the beginning of this book. However, *some* visions and hopes about the power of PeaceTech and digital peacebuilding speak to such transformative ideals, yet they often remain unfulfilled. This makes a critical reflexive perspective on digital peacebuilding especially worthwhile, because it helps to shed light on how the design and use of technologies are conditioned by specific claims about technology and society – and many of these do not sit well with theories of conflict transformation. The four empirical chapters have variously pointed to the power of these claims, such as when portraying hate as something that is primarily contained in social media posts, when arguing that the lack of effectiveness of early warning is due to the lack of context-specific and grounded information, or when claiming that AI can help reach peace agreements faster because it can facilitate broad-based negotiations in record time.

I have also shown that such claims are not universally powerful. They may help create or justify new socio-technical relations, as was the case with the invention of local early warning systems or digital dialogues, and they can also alter existing ones, such as when forging alliances between conventional peacebuilding donors, new civil society initiatives, and social media platforms. However, the impact of these claims is likewise conditioned by the socio-technical context in which they operate: the data of local early warning systems are skimmed by security organs and intelligence services that effectively perpetuate conflict and violence, the results of digital dialogues are re-inserted into stalling or ineffective high-level peacemaking processes, and efforts to identify and counter harmful speech are curtailed by authoritarian regimes or worn down by incompatible narratives and myths that have existed for decades and sometimes centuries.

Transformative socio-technical relations are also often made more difficult by sluggish and ill-fitting technology, including pre-existing applications and off-the-shelf algorithms that are cheap, convenient to use, and ready-to-hand. The utility of technology enframes digital peacebuilding in ways that make its transformative potential more obscure and often simply lacking. In consequence, peacebuilding technology becomes utilized for one purpose, such as establishing which statements are representative of a stakeholder group or getting real-time information about armed skirmishes in remote areas, but not necessarily for the purpose of transforming conflict. This would require digital dialogues that enable encounters that help conflict parties and stakeholders to work on changing their cemented and essentializing beliefs about the Other, or local response networks that can use context-specific information in more intuitive ways.

Perhaps this indicates the need to move beyond a perspective that is critical and reflexive and towards a more activist stance that not only studies how claims about technology and society condition conflict transformation but also commits to the work of 'claiming' transformative constellations—in the sense of *demanding* them. What would that entail? I have described the state of missing-from-hand as one of a state of obscurity: we don't know, we can't entirely see, we can't wholly feel what is required to enable a socio-technical constellation in which digital peacebuilding can indeed power a transformative effort. However, we also don't want tech-solutionist claims about how certain 'tools' can transform conflict because we have learned about the social embeddedness of technology and that it shapes how technology is ultimately employed and with what effects. On the other hand, there's no return to a pre-digital age either. Therefore, a pragmatic *and* critical engagement with the digitalization of peacebuilding might require that we accept that some tech-solutionist claims are better than others while pointing out where the purposes of technology start to deviate as claims are translated into concrete peacebuilding approaches. This also means remaining sceptical about the idea that a single technology could indeed make a (or the) difference. The transformation of conflict will always unfold in socio-technical constellations.

The Death of Socio-Technicality?

This book studied digital peacebuilding through the lens of apomediation, as conditioned by the production, sharing, and consumption of information about conflict and peace in peer-to-peer networks that are as much social as they are technical because they involve both humans and machines. One of the fundamental assumptions behind apomediation is that digitalization has given rise to decentred communication infrastructures, chiefly the internet and social media, that enable information exchanges which circumvent conventional peacebuilding actors. This is what the preposition 'apo' is meant to capture. This assumption has been partly confirmed across the book's empirical explorations, but not always and not without important caveats. Digital peacebuilding initiatives often brought new actors and alliances to the fore, shifting resources and agency to civil society organizations and the private sector. Yet, long-established actors have increasingly adapted and now commonly seek to leverage digital technologies to meet their mandates and objectives. Powerful data collection and processing tools can change who has the authority to make claims about conflict and peace and who can shape peacebuilding responses, as is, for instance, the case with the initiatives aimed at countering harmful speech on social media, discussed in Chapter 3. However, they can likewise reproduce or even strengthen the authority of conventional peacebuilding organizations and established approaches, as is the case with the high-level mediation initiatives discussed in Chapter 4.

Even more crucial is that apomediation awards greater power to technical agents by directing information away from human agents. This is especially the case where technology is used for more than merely creating communication channels between humans, namely for gathering and processing information in ways that shape understandings of conflict and peace. Some of the examples discussed in the book include tools for social media analysis, data dashboards that display information produced by text-mining algorithms, or mobile applications that structure data inputs from field monitors. All these technical elements alter and, at times, create information and thus have the agency to influence the understandings of conflict and peace conveyed through them. When recounting how some PeaceTech pioneers pursued a tool-centred look that portrayed digital technologies as seemingly neutral and detached from society, I argued for the need to grapple with the social embeddedness of technologies as a way of making visible how digital peacebuilding dynamics and outcomes are never conditioned by a tool alone. However, this doesn't mean that tools have no agency—indeed, the opposite is the case. The Heideggerian idea that technology 'enframes' the world and 'destines' being *because* they have utility captures this point well—and so does the negative image of socio-technical constellations that remain missing-from-hand for conflict transformation.

My idea, when exploring how digital peacebuilding leaves a sense of 'void' was to bring attention to the lack of socio-technical constellations that are necessary for conflict transformation. To be sure, there is no perfect tool that can simply 'generate' an agonistic peace, but many tools may stand in the way of this ambition and with it, social practices of conceiving, designing, and applying these tools that constitute digital peacebuilding in a way that is ultimately not conflict transformative. This emphasis on this social embeddedness of technology enables us to consider how digital peacebuilding is conditioned by a context that is as much technical as it is social. I have attempted to describe this via the conceptual heuristic of the digi-political playing field, but there are other aspects that have come to light throughout the book, such as the relevance of global politics structured by the continued dominance of the Global North.

It is important to stress that many of the designers and users of technology I encountered during my research were, to some extent, aware of how the digi-political playing field shaped their efforts to employ digital technologies for peacebuilding. Sometimes, they accepted these context conditions as given, and sometimes they tried to challenge them. For instance, communication professionals and spokespersons of major international actors in South Sudan simply didn't attempt to use social media as a forum to facilitate dialogues with ordinary South Sudanese because they perceived the online conversations as too polarized and too compromised by the repressive political climate. Some involved in the design of early warning mobile application believed that the data collection and response mechanism should be closely embedded within existing administrative structures. In both cases, the acquiescence with the digi-political status quo led to dynamics of apomediated peacebuilding that proved poorly suited to support the deeper work on the relations underpinning the conflict in South Sudan. However, many peacebuilding actors sought to challenge the context conditions of their digital approaches. Some did so by influencing the digital adaptability of the playing field, including by enhancing digital connectivity through internet hubs and data vouchers so that a larger number of people could contribute to activities such as early warning, consultation, or dialogues. They also launched capacity building measures aimed at increasing digital skills and other relevant knowledge necessary to counter online polarization and hatred that was often promoted, and at least played into the hands of, powerful political leaders and incumbent governments. Such efforts had the potential to influence the dynamics of relation-making and meaning-making that shaped the apomediation of peacebuilding. They also reflected the insight that digital peacebuilding was not something that could simply be done by employing tools but that it required working on the context conditions that determined how these tools could be used and with what effects.

Many staff, volunteers, and participants in digital peacebuilding activities were not acquiescent when it came to the design of the specific technologies they employed. Instead, they were often concerned with how the digital methods and applications they employed would change their approach and operations. For instance, initiatives that curated online museums, archives, or memorial sites to document Sri Lanka's violent past were aware that the particular media they chose, the aesthetics of visualizations and photos, or long-form text would mean that these efforts would resonate broadly with the country's urban, well-educated elites. Yet, they made intentional design choices, sustained by the idea that the content they wanted to convey required a specific digital environment. Other initiatives were discontinued or aimed to improve their technical solutions once they perceived that they didn't deliver as intended. Others organized open competitions and iterative design processes to ensure technical solutions would fit the purpose, which meant that the social embeddedness of technology didn't only inform the selection and use of tools but their very design. Yet, in many other cases, technologies proved difficult to modify because they were acquired off-the-shelf, proprietary, or both. For instance, many of the social media analysts I spoke to were aware of the limitations of the specific monitoring software and the data collection routine they employed, as well as the technical functionalities and gateways created by the social media platforms for reporting harmful content. Some also supported the idea that countering harmful speech required more than merely clicking it away.

However, tool-centrism may be poised to return to us in entirely different ways. Indeed, the vision of digital peacebuilding as something that is done with digital tools and without much of a role left for society and politics might increasingly be implemented by design. Make no mistake: we can't think of digital peacebuilding as only involving tools because these tools need to be used by somebody and for some purpose, which gives the social embeddedness of PeaceTech an almost axiomatic status. However, there is a risk that future peacebuilding will increasingly rely on digital methods that operate with a probabilistic model of society inscribed into AI models, which cannot be easily challenged by ordinary technology users. This would mean that the 'social' in socio-technicality, at least the social bits inscribed into AI methods that power data processing, will become increasingly lifeless. The strand of technologies that could make such a development more likely is commonly subsumed under the term generative AI, powered by foundational models built through unsupervised learning techniques. Clearly, these models are socially embedded, since these models have been built by humans, and have learned from human-generated data. However, once a model is publicly released, it no longer has the ability to learn from the social context that it is meant to serve. Building foundational models is an extremely resource-intensive task that requires highly skilled labour. This means that most advanced models operate in ways that are not only intransparent to most of their users but also unchallengeable. While there are options to fine-tune and customize generative AI, the foundational model underpinning them will process information according to algorithms that peacebuilding actors usually won't be able to modify. By extension, this also limits the scope of impact for critical and reflexive engagements with technology.

While generative AI runs on the fuel of social, digital artefacts, the social component in this socio-technical relation becomes hardened and lifeless once such models are brought to market. Chatbots powered by large language models (LLMs) may seem 'social' because of their ability to interact with us through natural language. However, are they really social if we can't alter the foundation of their inference processes because their underlying data infrastructure remains rigid and opaque? Importantly, generative AI will power an increasing number of digital peacebuilding applications that support a large variety of tasks, ranging from strategic communication to providing resources and knowledge to facilitating dialogues and negotiations. This development is particularly worrying against the backdrop of the global race for AI technologies, which seems to unfold largely among only a handful of major economies, especially the United States and China. It means that digital peacebuilding initiatives emerging elsewhere, especially those in contexts with resource constraints, will have to accept ready-made tools that run on data not generated in this context. The problem is exacerbated by the fact that many peacebuilding contexts have low levels of digitalization and thus generate little data that could be used for AI development. It is therefore likely that globally dominant models employed in digital peacebuilding will not be, or only insufficiently, trained on context-specific information because the data necessary to fine-tune them are simply not available. This will affect the ability of generative AI-powered technologies to operate in ways that fit their context of operation, for instance, to receive inputs and generate outputs that reflect the language and ways of reasoning of communities in digitally poorly connected areas. It will also mean that the inference processes that are employed, such as those for the analysis of natural language inputs of dialogue participants discussed in Chapter 5 of this book, may struggle to align with the ways in which the people in conflict based outside of global centres of innovation conceive of the work of overcoming armed conflict and the antagonisms that sustain it.

'Hyperreal' Peacebuilding?

In a world increasingly dominated by digital communication, it is through the very acts of producing and sharing data that humans can have a voice, which I described as the effort of giving an account of oneself in the face of adversity. This ability is essential to conflict transformation because it furthers agonistic peace, enabling modes of conflict in which conflict parties will view the other no longer as an enemy that needs to be annihilated but as an adversary that can be 'fought' by uploading data via mobile applications, during online conferences, or through social media memes. While many people I spoke to wouldn't describe what they do as 'apomediation', they are aware of how the mediating powers of digital technologies shape what they can say and with what effects.

Apomediated peacebuilding may help with conflict transformation by enabling those affected by conflict and violence to produce and share their voice and make sense of that of their opponents. When opponents express themselves through digital means, they can make their subjective experiences of the tangible and material aspects of the conflict perceivable to their opponents. This may be the starting point of transformative efforts that go beyond what is so often conceived as the 'online' or 'virtual' sphere, potentially supporting comprehensive political change. The concept of apomediation also carries the hope that ordinary people can have *more* voice—and influence—because apomediation broadens and decentres communication, which can undermine the power of those who conventionally dominate understandings of conflict and peace. However, this book made clear that this promise often doesn't materialize. Many powerful actors can continue to undermine conflict transformation, be it well-intended peacebuilding organizations taking a technocratic and depoliticizing approach or authoritarian leaders who encourage harmful online speech to rule through divisionist politics or use digital methods to suppress dissent.

Future innovations will likely increase the likelihood that digital peacebuilding approaches stifle the voice of those affected by conflict and violence, especially those leveraging generative AI. However, the risk with digital peacebuilding is *not* that it could lead to a 'virtual' peace that is solely built 'online' realm, thus seemingly detached from the 'offline' lives of those in conflict. This book has made abundantly clear that such a distinction between the 'analogue' and 'digital' is analytically of little help because both spheres shape peacebuilding dynamics and outcomes in intricate ways. Rather, the risk is that the current trajectory of innovation in AI will enable peacebuilding approaches that are increasingly disconnected from what ordinary people say they want or need – yet, influential in shaping their lives and prospects of peace. This is because generative AI threatens to enable what I suggest calling a 'hyperreal' peace, in which what is regarded as real and what is regarded as fictional, generated, or fabricated, become seamlessly entwined (Ferrari and McKelvey 2023; Baudrillard 1994). Baudrillard described 'hyperreality' as 'the generation by models of a real without origin or reality'—in other words, the representation of the world without original referent. Discussing the impact of what is now often referred to as 'legacy media', such as television, on people's lives, he suggested that such innovations will enable the making of a world composed of simulated 'stimuli' and nothing else.

The central danger that comes with the hyperreal is not a mismatch between representations of the world and how it 'really' is, or what is 'true', but that the world behind the representation begins to vanish. Today, we may be less worried about how TV shows can lead to lives in hyperreality, largely because we have integrated hero stories and soap opera romances with our sense of reality. At the same time, disinformation on social media and the axing of fact-checking efforts on major platforms such as X and Facebook have made it increasingly difficult to assert facts and discern what's true. To be clear, conflict transformation doesn't require that conflict parties establish a sense of objectivity that would determine what is 'true'—in fact, such a sincere endeavour will be bound to fail most of the time. However, what conflict transformation requires is that those affected by conflict and violence be able to voice what they think is true and how this relates to the grievances they hold.

However, as generative AI plays a growing role in digital peacebuilding, the meaning generated in apomediation will be increasingly shaped by a data layer that renders peace an artefact of probabilistic correlations. The many ways in which AI-generated information will affect apomediation are likely manifold. We can see already today that it can be employed to rewrite, overwrite, or replace the voice of those affected by conflict and violence. Chapter 5 highlighted how advanced algorithms can be used to fill gaps in the data that describe a conflict party's or stakeholder's identity, preferences, and needs, using the model's predictions to generate synthetic data instead of empirical data. Yet, there are many other examples of technology that add a hyperreal data layer into apomediated understandings of conflict and peace. For instance, LLMs can be used to standardize natural language inputs collected through online consultation or in live dialogues so that they can be further algorithmically processed. This process erases the nuance of the human input, removing anything from typos to rhetorical devices to inconsistencies or mistakes that are telling of the human behind the input, their identity, and their needs. In the not-so-distant future, we will also see more experimentation and utilization of 'digital twins', which have already become increasingly common in public policy and planning (Kopponen et al. 2022; Shepherd 2023). In this case, existing data sets, such as on past conflicts and peace processes, will be utilized in machine learning to create virtual representations of conflict contexts, peace processes, or aspects of them to simulate scenarios, test the effectiveness of measures, and predict outcomes, to determine what is necessary to terminate a conflict. At the same time, the need to engage with the reality of conflict, and how conflict parties and stakeholders make sense of it, will seemingly decrease.

It is not inconceivable that such approaches could inform the drafting of a peace agreement or the design of a transitional process, for instance. Digital twins can utilize new or even real-time data, such as from sensors or social media APIs. This kind of data tends to be the kind of 'large' or 'big' data needed to fuel machine learning algorithms that can identify patterns and make predictions. Think, for instance, about satellite data to analyse environmental factors such as the climate, troop movements, or social media data to analyse socio-economic grievances. Thus, the hyperreality that such approaches would bring to peacemaking would not lead to a complete disconnect between representation and its origins. However, it would be one that is less and less dependent on the subjective truths that determine how conflict parties and stakeholders perceive the conflict they find themselves in. It would be 'skimmed' in the same way as the data employed in the early warning systems discussed in Chapter 4. As with the hyperreality that is generated when we stream a soap opera that seemingly tells us about the true nature of romantic relationships or what it means to be a hero, it is not at all the case that conflict parties may necessarily object to such representations. Nonetheless, this hyperreality would stand in the way of any transformative effort that seeks to facilitate a change in the relationships between the conflict parties in ways that take their voice into account. It would make it seem unnecessary to conduct peacebuilding efforts that are grounded in the contemporary experiences of those in conflict, and their objections to the status quo that brought about not only conflict but enmity and violence.

Conflict transformation is an inherently political act through which established political institutions are contested. Yet, in a future increasingly shaped by AI enhancements, experiences of marginalization, deprivation, and insecurity that could inform such transformative efforts risk being reduced to mere input data for the probabilistic fabrication that informs foundational models. Therefore, the human voice needed to transform conflict could literally get lost in the 'black box'. This makes clear that the prospects of transforming conflict, now and in the future, will depend on our ability to shape a socio-technicality of digital peacebuilding in ways that ensure that digital technologies remain receptive to, and pliable by, the voice of those affected by conflict and violence. As I finalize this book in August 2025, further research on the apomediation of peacebuilding thus seems very much needed. However, future studies of apomediated peacebuilding may be less occupied with how decentralized networks of humans and machines could challenge the authority of conventional peacebuilding actors. Instead, apomediated peacebuilding may come to mean that humans remain 'in the loop' of increasingly automated peacebuilding efforts—especially those humans affected by conflict and violence. Humans who remain in the loop of digital peacebuilding will be able to challenge machine-mediated and -generated understandings of conflict—and approaches to building peace—that stand in the way of conflict transformation. This promise may not only be worthwhile to study. It might increasingly become necessary to defend it.

References

Abubakar, Naima Hafiz, and Salihu Ibrahim Dasuki. 2018. "Empowerment in Their Hands: Use of WhatsApp by Women in Nigeria." *Gender, Technology and Development* 22 (2): 164–183. https://doi.org/10.1080/09718524.2018.1509490.

Accord. 2023. "CEWARN's Strategy Framework and Civil Society Partnership: Lessons for Early-Warning and Response Systems." *ACCORD*. https://www.accord.org.za/conflict-trends/cewarns-strategy-framework-and-civil-society-partnership-lessons-for-early-warning-and-response-systems/.

Adichie, Chimamanda Ngozi. 2009. "The Danger of a Single Story | TED Talk." https://www.ted.com/talks/chimamanda_ngozi_adichie_the_danger_of_a_single_story.

Adler, Emanuel, and Vincent Pouliot. 2011. "International Practices." *International Theory* 3 (1): 1–36. https://doi.org/10.1017/S175297191000031X.

Adler-Nissen, Rebecca, and Kristin Anabel Eggeling. 2022. "Blended Diplomacy: The Entanglement and Contestation of Digital Technologies in Everyday Diplomatic Practice." *European Journal of International Relations* 28 (3): 640–666. London: Sage UK. https://doi.org/10.1177/13540661221107837.

African Union Commission. 2024. "Regional Economic Communities." https://au.int/en/recs.

Aggestam, Karin, and Elsa Hedling. 2023. "Digital Disruption and Rethinking the 'Script' of Peace Mediation." *The Hague Journal of Diplomacy* 1 (aop): 1–31. https://doi.org/10.1163/1871191x-bja10158.

Åkebo, Malin, and Sunil Bastian. 2020. "Beyond Liberal Peace in Sri Lanka: Victory, Politics, and State Formation." *Journal of Peacebuilding & Development*, December 6, 1542316620976121. https://doi.org/10.1177/1542316620976121.

Alavi, Masood, Daanish, Martin Wählisch, Colin Irwin, and Andrew Konya. 2022. "Using Artificial Intelligence for Peacebuilding." *Journal of Peacebuilding & Development*, May 23, 15423166221102757. https://doi.org/10.1177/15423166221102757.

Alexander, David E. 2014. "Social Media in Disaster Risk Reduction and Crisis Management." *Science and Engineering Ethics* 20 (3): 717–733. https://doi.org/10.1007/s11948-013-9502-z.

Alliance for Peacebuilding. 2021. "Digital Peacebuilding Community of Practice." 2021. https://www.allianceforpeacebuilding.org/digital-peacebuilding-cop.

Al-Zaman, Md. Sayeed. 2019. *Digital Disinformation and Communalism in Bangladesh*. Preprint. SocArXiv. https://doi.org/10.31235/osf.io/8s6jd.

American University of Beirut. 2023. "Making Peace More Inclusive: New Research Helps AI-Enabled Computers Understand Arabic Dialects." American University of Beirut. https://aub.edu.lb/articles/Pages/new_research_helps_AI_enabled_computers_understand_Arabic_dialects.aspx.

Amnesty International. 2021. *South Sudan: "These Walls Have Ears": The Chilling Effect of Surveillance in South Sudan*. Amnesty International. https://www.amnesty.org/en/documents/afr65/3577/2021/en/.

Amnesty International. 2021a. "South Sudan: Rampant Abusive Surveillance by NSS Instils Climate of Fear." Amnesty International, February 2. https://www.amnesty.org/en/latest/press-release/2021/02/south-sudan-abusive-surveillance-by-national-security-service-climate-of-fear/.

Amnesty International. 2021b. *South Sudan: Unfreeze Civil Society and Political Activist's Bank Accounts*. Amnesty International. https://www.amnesty.org/en/documents/afr65/5017/2021/en/.

Amnesty International. 2022. *Myanmar: The Social Atrocity: Meta and the Right to Remedy for the Rohingya*. Amnesty International. https://www.amnesty.org/en/documents/asa16/5933/2022/en/.

Amnesty International. 2024. *Sri Lanka: Online Safety Act Major Blow to Freedom of Expression*. https://www.amnesty.org/en/latest/news/2024/01/sri-lanka-online-safety-act-major-blow-to-freedom-of-expression/.

Amoore, Louise, Alexander Campolo, Benjamin Jacobsen, and Ludovico Rella. 2024. "A World Model: On the Political Logics of Generative AI." *Political Geography* 113 (August): 103134. https://doi.org/10.1016/j.polgeo.2024.103134.

Arab League, African Union, and European Union. 2020. "The Berlin Conference on Libya—Conference Conclusions (19 January 2020)—Libya | ReliefWeb." https://reliefweb.int/report/libya/berlin-conference-libya-conference-conclusions-19-january-2020.

Arana-Catania, M., F. A. Van Lier, and Rob Procter. 2021. "Machine Learning for Mediation in Armed Conflicts." arXiv:2108.11942 [Cs], August 26. http://arxiv.org/abs/2108.11942.

Article One Advisors. 2018. *Assessing the Human Rights Impact of the Facebook Platform in Sri Lanka*. Article One Advisors. https://about.fb.com/wp-content/uploads/2020/05/Sri-Lanka-HRIA-Executive-Summary-v82.pdf.

Ashcroft, Bill. 2001. *On Post-Colonial Futures: Transformations of Colonial Culture*. Writing Past Colonialism. London: Continuum.

Asseburg, Muriel, Wolfram Lacher, and Mareike Transfeld. 2018. *Mission Impossible? UN Mediation in Libya, Syria and Yemen*. SWP Research Paper 8. German Institute for International and Security Affairs.

Ataci, Tugce. 2022. "Memetizing Genocides and Post-Genocide Peacebuilding: Ambivalent Implications of Memes for Youth Participation and Imaginaries in Rwanda." *Information, Communication & Society* 26(16), 3157–3178. https://doi.org/10.1080/1369118X.2022.2146984.

Austin, Alexander. 2004. "Early Warning and The Field: A Cargo Cult Science?" In *Transforming Ethnopolitical Conflict*, edited by Alex Austin, Martina Fischer, and Norbert Ropers. VS Verlag für Sozialwissenschaften. https://doi.org/10.1007/978-3-663-05642-3_7.

Avgerou, Chrisanthi. 2008. "Information Systems in Developing Countries: A Critical Research Review." *Journal of Information Technology* 23 (3): 133–146. https://doi.org/10.1057/palgrave.jit.2000136.

Babaud, Sébastien, and Natalia Mirimanova. 2011. *The European Commission Early-Warning Architecture and Crisis-Response Capacity*. Initiative for Peacebuilding c/o International Alert. Brussels, Belgium.

Badawy, A., E. Ferrara, and K. Lerman. 2018. "Analyzing the Digital Traces of Political Manipulation: The 2016 Russian Interference Twitter Campaign." In *2018 IEEE/ACM International Conference on Advances in Social Networks Analysis and Mining (ASONAM)*, 258–265. https://doi.org/10.1109/ASONAM.2018.8508646.

Baechler, Günther, Andreas Kohlschütter, Volker Böge, and Wilhelm Nolte. 1998. *FAST: Frühanalyse von Spannungen und Tatsachenermittlung: Abschlussbericht zum Pilotprojekt FAST*. With Schweizerische Friedensstiftung. Working paper/Schweizerische Friedensstiftung 26. Schweizerische Friedensstiftung.

Baele, Stephane J., and Gregorio Bettiza. 2021. "'Turning' Everywhere in IR: On the Sociological Underpinnings of the Field's Proliferating Turns." *International Theory* 13 (2): 314–340. https://doi.org/10.1017/S1752971920000172.

Jathika Jana Balawegaya. 2024. *A Thriving Nation. A Beautiful Life*. Pelawatte, Battaramulla, Sri Lanka. https://www.npp.lk/up/policies/en/npppolicystatement.pdf.

Bargués, Pol. 2020. "Peacebuilding without Peace? On How Pragmatism Complicates the Practice of International Intervention." *Review of International Studies* 46 (2): 237–255. https://doi.org/10.1017/S0260210520000042.

Barkawi, Tarak, and Mark Laffey. 1999. "The Imperial Peace: Democracy, Force and Globalization." *European Journal of International Relations* 5 (4): 403–434. https://doi.org/10.1177/1354066199005004001.

Barnett, Michael, and Raymond Duvall, eds. 2004. *Power in Global Governance.* Cambridge Studies in International Relations. Cambridge University Press. https://doi.org/10.1017/CBO9780511491207.

Barnett, Michael, Songying Fang, and Christoph Zürcher. 2014. "Compromised Peacebuilding." *International Studies Quarterly* 58 (3): 608–620. https://doi.org/10.1111/isqu.12137.

Bastos, Marco T., Dan Mercea, and Arthur Charpentier. 2015. "Tents, Tweets, and Events: The Interplay Between Ongoing Protests and Social Media." *Journal of Communication* 65 (2): 320–350. https://doi.org/10.1111/jcom.12145.

Batrinca, Bogdan, and Philip C. Treleaven. 2015. "Social Media Analytics: A Survey of Techniques, Tools and Platforms." *AI & SOCIETY* 30 (1): 89–116. https://doi.org/10.1007/s00146-014-0549-4.

Baudrillard, Jean. 1994. *Simulacra and Simulation.* Translated by Sheila Faria Glaser. Later prt. edition. Ann Arbor, Mich: University of Michigan Press.

Bell, Christine. 2024. *PeaceTech: Digital Transformation to End Wars.* Cham: Springer International Publishing. https://doi.org/10.1007/978-3-031-38894-1.

Bennett, W. Lance, and Alexandra Segerberg. 2012. "The Logic of Connective Action." *Information, Communication & Society* 15 (5): 739–768. https://doi.org/10.1080/1369118X.2012.670661.

Bermeo, Nancy. 2016. "On Democratic Backsliding." *Journal of Democracy* 27 (1): 5–19. https://doi.org/10.1353/jod.2016.0012.

Bigo, Didier. 2011. "Pierre Bourdieu and International Relations: Power of Practices, Practices of Power." *International Political Sociology* 5 (3): 225–258. https://doi.org/10.1111/j.1749-5687.2011.00132.x.

Bilich, J., Varga, M., Masood, D., and Konya, A. 2019. Faster Peace via Inclusivity: An Efficient Paradigm to Understand Populations in Conflict Zones. AI for Social Good workshop at NeurIPS, Vancouver, Canada.

Bjola, Corneliu, and Markus Kornprobst, eds. 2023. *Digital International Relations: Technology, Agency and Order.* London: Routledge. https://doi.org/10.4324/9781003437963.

Björkdahl, Annika, and Kristine Höglund. 2013. "Precarious Peacebuilding: Friction in Global-Local Encounters." *Peacebuilding* 1 (3): 289–299.

Blaney, David L., and Arlene B. Tickner. 2017. "Worlding, Ontological Politics and the Possibility of a Decolonial IR." *Millennium* 45 (3): 293–311. https://doi.org/10.1177/0305829817702446.

Bodie, Graham D., Debra L. Worthington, and Zenebe Beyene, eds. 2023. *Listening, Community Engagement, and Peacebuilding: International Perspectives.* New York: Routledge. https://doi.org/10.4324/9781003214465.

Special EU Programmes Body. 2019. *Your EU! An Update on the European Union's PEACE and INTERREG Programmes.* https://niopa.qub.ac.uk/bitstream/NIOPA/17848/10/Your_EU_Spring_2019.pdf.

Bond, Doug, Joe Bond, Churl Oh, J. Craig Jenkins, and Charles Lewis Taylor. 2003. "Integrated Data for Events Analysis (IDEA): An Event Typology for Automated Events Data Development." *Journal of Peace Research* 40 (6): 733–745.

Boutros-Ghali, B. 1992. "An Agenda for Peace: Preventive Diplomacy, Peacemaking and Peace-keeping." *International Relations*, 11(3), 201–218. https://doi.org/10.1177/004711789201100302

Bowker, Geoffrey C., and Susan Leigh Star. 2008. *Sorting Things out: Classification and Its Consequences*. 1. paperback ed., 8. print. Inside Technology. Cambridge, Mass.: MIT Press.

Bramsen, Isabel, and Anine Hagemann. 2021. "The Missing Sense of Peace: Diplomatic Approachment and Virtualization during the COVID-19 Lockdown." *International Affairs* 97 (2): 539–560. https://doi.org/10.1093/ia/iiaa229.

The Bytes Project. 2022. *Programmes – Bytes*. February 8. https://www.bytes.org/programmes/.

Cadwalladr, Carole, and Emma Graham-Harrison. 2018. "Revealed: 50 Million Facebook Profiles Harvested for Cambridge Analytica in Major Data Breach." News. *The Guardian*, March 17. https://www.theguardian.com/news/2018/mar/17/cambridge-analytica-facebook-influence-us-election.

CAIN. 2024. "Introduction to the CAIN Web Site." https://cain.ulster.ac.uk/cainbgn/index.html.

Caplan, Richard. 2005. *International Governance of War-Torn Territories: Rule and Reconstruction*. Oxford: Oxford University Press.

Carey, Henry F. 2020. *Peacebuilding Paradigms: The Impact of Theoretical Diversity on Implementing Sustainable Peace*. Cambridge: Cambridge University Press.

Kessel, Stijn van, and Remco Castelein. 2016. "Shifting the Blame. Populist Politicians' Use of Twitter as a Tool of Opposition." *Journal of Contemporary European Research* 12 (2): 595–614. https://doi.org/10.30950/jcer.v12i2.709.

Castells, Manuel. 2013. *Communication Power*. Oxford: Oxford University Press.

Castells, Manuel. 2015. *Networks of Outrage and Hope: Social Movements in the Internet Age*. 2nd., [Rev.] ed. Cambridge: Polity.

Cederman, Lars-Erik, Kristian Skrede Gleditsch, and Halvard Buhaug. 2013. *Inequality, Grievances, and Civil War*. Cambridge Studies in Contentious Politics. Cambridge: Cambridge University Press.

Cederman, Lars-Erik, and Nils B. Weidmann. 2017. "Predicting Armed Conflict: Time to Adjust Our Expectations?" *Science* 355 (6324): 474–476. https://doi.org/10.1126/science.aal4483.

Centre for Policy Alternatives. 2023. *A Brief Analysis of the Aragalaya*. Centre for Policy Alternatives. https://www.cpalanka.org/wp-content/uploads/2023/05/A-Brief-Analysis-of-the-Aragalaya_Final-Report.pdf.

UNDPPA, and HD Centre. 2019. *Digital Technologies and Mediation in Armed Conflict*. United Nations Department of Political and Peacebuilding Affairs and Centre for Humanitarian Dialogue.

Centre on Constitutional Change. 2024. "The Same, but Different? A Return to Power-Sharing in Northern Ireland | Centre on Constitutional Change." https://www.centreonconstitutionalchange.ac.uk/news-and-opinion/same-different-return-power-sharing-northern-ireland.

Chadwick, Wren, Tobias Debiel, Frank Gadinger, and David Chandler, eds. 2013. "Relational Sensibilities: The End of the Road for 'Liberal Peace.'" In *Relational Sensibility and the 'Turn to the Local': Prospects for the Future of Peacebuilding*, 19–26. Global Dialogues 2. Duisburg: Centre or Global Cooperation Research.

Chandler, David. 2013. "Peacebuilding and the Politics of Non-Linearity: Rethinking 'Hidden' Agency and 'Resistance.'" *Peacebuilding* 1 (1): 17–32. https://doi.org/10.1080/21647259.2013.756256.

Chandler, David. 2017a. *Peacebuilding: The Twenty Years' Crisis, 1997–2017*. Rethinking Peace and Conflict Studies. Cham: Springer International Publishing. https://doi.org/10.1007/978-3-319-50322-6.

Chandler, David. 2017b. "Securing the Anthropocene? International Policy Experiments in Digital Hacktivism: A Case Study of Jakarta." *Security Dialogue* 48 (2): 113–130. https://doi.org/10.1177/0967010616677714.

De Choudhury, Munmun, Shagun Jhaver, Benjamin Sugar, and Ingmar Weber. 2016. "Social Media Participation in an Activist Movement for Racial Equality." *Proceedings of the International AAAI Conference on Weblogs and Social Media. International AAAI Conference on Weblogs and Social Media* 2016 (May):92–101.

Christensen, Britt. 2019. "Cyber State Capacity: A Model of Authoritarian Durability, ICTs, and Emerging Media." *Government Information Quarterly* 36 (3): 460–468. https://doi.org/10.1016/j.giq.2019.04.004.

Cilliers, Jakkie. 2005. *Towards a Continental Early Warning System for Africa.*

The Civil Action SSD [@CivilActionSSD]. 2021a. "3/3 https://t.co/PHyDTFvGtp." Tweet. Twitter, August 26. https://x.com/CivilActionSSD/status/1430876927318020099/video/1

The Civil Action SSD [@CivilActionSSD]. 2021b. "https://t.co/H1yM0UDV6K." Tweet. Twitter, March 8. https://nupino-my.sharepoint.com/:w:/g/personal/andreas_hirblinger_nupi_no/IQCMWyfJTmHpTY4hwME19ZH1AWhJ8oRznLwYxnD1rr65sD4?e=BtWRf4&nav=eyJjIjo2NjQ5Mjg5NzR9.

Clark, Phil. 2014. "Negotiating Reconciliation in Rwanda: Popular Challenges to the Official Discourse of Post-Genocide National Unity." *Journal of Intervention and Statebuilding* 8 (4): 303–320. https://doi.org/10.1080/17502977.2014.958309.

Clark, Serena, and Claudio Alberti. 2021. "COVID-19 and Adapting to the New Normal: Lessons Learned for Peacebuilding." *Journal of Peacebuilding & Development*, November, 15423166211052832. https://doi.org/10.1177/15423166211052832.

Clayton, Russell B., Glenn Leshner, and Anthony Almond. 2015. "The Extended iSelf: The Impact of iPhone Separation on Cognition, Emotion, and Physiology." *Journal of Computer-Mediated Communication* 20 (2): 119–135. https://doi.org/10.1111/jcc4.12109.

Close, Sophia, Hesta Groenewald, and Diana Trimiño Mora. 2020. *Facilitation Guide Gender-Sensitive Conflict Analysis.* Conciliation Resources and Saferworld.

CMI. 2024. "Future(s) of Cyberpeace: Foresight and Digital Technology in Peace Mediation." *CMI* (blog). February 8, 2024. https://cmi.fi/2024/02/08/futures-of-cyberpeace-foresight-and-digital-technology-in-peace-mediation/.

Coning, Cedric de. 2018. "Adaptive Peacebuilding." *International Affairs* 94 (2): 301–317. https://doi.org/10.1093/ia/iix251.

Coning, Cedric de, Rui Saraiva, and Ako Muto. 2023. *Adaptive Peacebuilding: A New Approach to Sustaining Peace in the 21st Century.* Cham: Springer Nature.

Coppedge, Michael, John Gerring, Carl Henrik Knutsen, Staffan I. Lindberg, and Jan Teorell. 2021. "V-Dem Methodology V11.1". Varieties of Democracy (V-Dem) Project. https://v-dem.net/static/website/img/refs/methodologyv111.pdf.

Carnegie Corporation. 1997. *Preventing Deadly Conflict: Final Report.* https://media.carnegie.org/filer_public/b2/0e/b20e1080-7830-4f2b-9410-51c14171809b/ccny_report_1997_ccpdc_final.pdf.

Cox, Robert W. 1981. "Social Forces, States and World Orders: Beyond International Relations Theory." *Millennium—Journal of International Studies* 10 (2): 126–155. https://doi.org/10.1177/03058298810100020501.

Cramer, Florian. 2015. "What Is 'Post-Digital'?" In *Postdigital Aesthetics: Art, Computation and Design*, edited by David M. Berry and Michael Dieter. London: Palgrave Macmillan UK. https://doi.org/10.1057/9781137437204_2.

Croos, Sharon. 2024. "NPP's Historic Victory Sparks Hope and Unity." *Groundviews*, November 16. https://groundviews.org/2024/11/16/npps-historic-victory-sparks-hope-and-unity/.

CrowdTangle. 2021. "CrowdTangle | Content Discovery and Social Monitoring Made Easy." CrowdTangle. https://www.crowdtangle.com//.

Cuhadar-Gürkaynak, E. 2009. "Assessing Transfer from Track Two Diplomacy: The Cases of Water and Jerusalem." *Journal of Peace Research* 46 (5). 641–658.

Dajer, Diana. 2018. "Cracking the Code of Tech for Peace: International Perspectives of Peacetech Research and Practice." In *Reflections on Building Inclusive and Sustainable Peace*, edited by Christine Wilson 70–79. London: British Council. https://www.britishcouncil.org/sites/default/files/j063_peace_and_beyond_essays_final_web_new_0.pdf.

Dangerous Speech Project. 2021. *Dangerous Speech: A Practical Guide.* https://dangerousspeech.org/about-dangerous-speech/.

Daniels, Jessie. 2018. "The Algorithmic Rise of the 'Alt-Right.'" *Contexts* 17 (1): 60–65. https://doi.org/10.1177/1536504218766547.

Darby, John, and Roger Mac Ginty. 2008. *Contemporary Peacemaking: Conflict, Peace Processes and Post-War Reconstruction.* 2nd ed. New York: Palgrave Macmillan. http://catdir.loc.gov/catdir/toc/ecip0815/2008015879.html.

Davey, Jacob, and Julia Ebner. 2019. *The 'Great Replacement': The Violent Consequences of Mainstreamed Extremism.* London: Institute for Strategic Dialogue.

David, Marian. 2022. "The Correspondence Theory of Truth." In *The Stanford Encyclopedia of Philosophy*, Summer 2022, edited by Edward N. Zalta. Metaphysics Research Lab, Stanford University. https://plato.stanford.edu/entries/truth-correspondence/, Accessed May 2, 2016.

Dawkins, Sophia, and Liz Gaere. 2014. "Building the House of Governance: Towards Sustainable Peace in Jonglei State." *Conflict Dynamics International.* https://www.cmi.no/file/2951-BHoG-Jonglei---Briefing-Paper-II---FINAL.pdf.

Debiel, Tobias, Thomas Held, and Ulrich Schneckener, eds. 2016. "The Future of Peacebuilding." In *Peacebuilding in Crisis: Rethinking Paradigms and Practices of Transnational Cooperation*, by David Chandler. London, New York: Routledge.

Deiana, Maria-Adriana, and Heidi Riley. 2024. *Toolkit for Transformative Mediation in Peacebuilding.* Talk4Peace.

Ada Derana. 2023. "Sri Lanka to Introduce New Laws to Regulate Social Media." Adaderana.Lk. https://www.adaderana.lk/news.php?nid=87400.

Deutsche Welle, dir. 2018. *24 Reactions · 5 Comments | "My Old Friend Wani Michael Have Been Meeting Every Month with His Friends to Clean #Juba under the Banner #NadafaLeBeledna Which Means "Cleaning Our . . . | By Juba International Airport | Facebook.* https://www.facebook.com/jubaairport/videos/my-old-friend-wani-michael-have-been-meeting-every-month-with-his-friends-to-cle/2817784598247787/.

DeVotta, Neil. 2021. "Sri Lanka: The Return to Ethnocracy." *Journal of Democracy* 32 (1): 96–110.

Dingli, Sophia. 2015. "We Need to Talk about Silence: Re-Examining Silence in International Relations Theory." *European Journal of International Relations* 21 (4): 721–742. https://doi.org/10.1177/1354066114568033.

Diplo. 2021. "The Telegraph: How It Changed Diplomacy." September 13. https://www.diplomacy.edu/histories/the-telegraph-how-it-changed-diplomacy/.

Doyle, J., and Connolly, E. 2019. The Effects of Brexit on the Good Friday Agreement and the Northern Ireland Peace Process. In C.-A. Baciu & J. Doyle (Eds.), *Peace, Security and Defence Cooperation in Post-Brexit Europe: Risks and Opportunities* (pp. 79–95). Springer International Publishing. https://doi.org/10.1007/978-3-030-12418-2_4

DPO, Dos, and DMSPC. 2021. "Strategy for the Digital Transformation of UN Peacekeeping." New York, NY: United Nations. https://peacekeeping.un.org/en/strategy-digital-transformation-of-un-peacekeeping.

Dudouet, Véronique, ed. 2015. *Civil Resistance and Conflict Transformation.* Oxon and New York: Routledge. https://doi.org/10.4324/9781315778877.

Duffield, Mark. 2016. "The Resilience of the Ruins: Towards a Critique of Digital Humanitarianism." *Resilience* 4 (3): 147–165. https://doi.org/10.1080/21693293.2016.1153772.

Dulanjana, W. Sachinda. 2024. "Finding a Way Forward amidst the Contemporary Challenges to Sri Lankan Student Politics." In *The Bloomsbury Handbook of Student Politics and Representation in Higher Education*, 1st ed., edited by Manja Klemenčič. London: Bloomsbury Publishing Plc. https://doi.org/10.5040/9781350376007.

Dutton, William H., and Bianca C. Reisdorf. 2019. "Cultural Divides and Digital Inequalities: Attitudes Shaping Internet and Social Media Divides." *Information, Communication & Society* 22 (1): 18–38. https://doi.org/10.1080/1369118X.2017.1353640.

Duursma, Allard, and John Karlsrud. 2019. "Predictive Peacekeeping: Strengthening Predictive Analysis in UN Peace Operations." *Stability: International Journal of Security and Development* 8 (1): 1. https://doi.org/10.5334/sta.663.

Duursma, Allard, and John Karlsrud. 2021. "Technologies of Peace." *The Oxford Handbook of Peacebuilding, Statebuilding, and Peace Formation*, July, 414–427. https://doi.org/10.1093/oxfordhb/9780190904418.013.28.

The Economist. 2023. "Sri Lankans Are Squabbling over Monuments." November 30. https://www.economist.com/asia/2023/11/30/sri-lankans-are-squabbling-over-monuments.

Eggeling, Kristin Anabel, and Larissa Versloot. 2023. "Taking Trust Online: Digitalisation and the Practice of Information Sharing in Diplomatic Negotiations." *Review of International Studies* 49 (4): 637–656. https://doi.org/10.1017/S0260210522000559.

Ekman, Joakim. 2009. "Political Participation and Regime Stability: A Framework for Analyzing Hybrid Regimes." *International Political Science Review/Revue Internationale de Science Politique* 30 (1): 7–31.

Ellis-Petersen, Hannah, and Rubatheesan Sandran. 2022. "'We Want Justice, Not Fuel': Sri Lanka's Tamils on North–South Divide." World News. *The Guardian*, June 22. https://www.theguardian.com/world/2022/jun/22/sri-lanka-tamils-protests-economic-crisis.

Elman, Colin. 2005. "Explanatory Typologies in Qualitative Studies of International Politics." *International Organization* 59 (2): 293–326. https://doi.org/10.1017/S0020818305050101.

Elswah, Mona. 2024. "Investigating Content Moderation Systems in the Global South." *Center for Democracy and Technology*, January 30. https://cdt.org/insights/investigating-content-moderation-systems-in-the-global-south/.

Engel, Ulf. 2018. "Knowledge Production on Conflict Early Warning at the African Union." *South African Journal of International Affairs* 25 (1): 117–132. https://doi.org/10.1080/10220461.2018.1417902.

Engel, Ulf. 2019. "Early Warning and Conflict Prevention." In *The Routledge Handbook of Transregional Studies*, edited by Matthias Middell. London: Routledge.

Engesser, Sven, Nicole Ernst, Frank Esser, and Florin Büchel. 2017. "Populism and Social Media: How Politicians Spread a Fragmented Ideology." *Information, Communication & Society* 20 (8): 1109–1126. https://doi.org/10.1080/1369118X.2016.1207697.

Ethirajan, Anbarasan, Swaminathan Natarajan, and Ishara Danasekara. 2024. "Sri Lanka: Left-Leaning Leader's Coalition Secures Landslide Victory." *BBC News*, November 15. https://www.bbc.com/news/articles/crr9n2w0lyzo.

EU Commission. 2023. *No Place for Hate: A Europe United against Hatred*. https://doi.org/10.1093/law-oeeul/e66013.66.

European Council of the European Union. 2024. "Northern Ireland Protocol and Windsor Framework Explained." Consilium. https://www.consilium.europa.eu/en/policies/the-protocol-on-ireland-and-northern-ireland-explained/.

European Union Institute for Security Studies. 2014. *Crisis Rooms :Towards a Global Network*. Publications Office. https://data.europa.eu/doi/10.2815/37165.

Evans, Sam Weiss, Matthias Leese, and Dagmar Rychnovská. 2021. "Science, Technology, Security: Towards Critical Collaboration." *Social Studies of Science* 51 (2): 189–213. https://doi.org/10.1177/0306312720953515.

Eysenbach, Gunther. 2007. "From Intermediation to Disintermediation and Apomediation: New Models for Consumers to Access and Assess the Credibility of Health Information in the Age of Web2.0." *MEDINFO*, 162–166.

Eysenbach, Gunther. 2008. "Medicine 2.0: Social Networking, Collaboration, Participation, Apomediation, and Openness." *Journal of Medical Internet Research* 10 (3): 59. https://doi.org/10.2196/jmir.1030.

Ezeibe, Christian. 2021. "Hate Speech and Election Violence in Nigeria." *Journal of Asian and African Studies* 56 (4): 919–935. https://doi.org/10.1177/0021909620951208.

FactCheckNI. 2023. "About." FactCheckNI. https://factcheckni.org/about/.

Palmiano Federer, Julia, Julia Pickhardt, Philipp Lustenberger, Christian Altpeter, and Katrina Abatis. 2019. *Beyond the Tracks? Reflections on Multitrack Approaches to Peace Processes.* Report. ETH Zurich. https://doi.org/10.3929/ethz-b-000389119.

Fedricks, Krishantha, Farzana Haniffa, Anushka Kahandagamage, Chulani Kodikara, Kaushalya Kumarasinghe, and Jonathan Spencer. 2022. "Snapshots from the Struggle, Sri Lanka April–May 2022." *Anthropology Now* 14 (1–2): 21–38. https://doi.org/10.1080/19428200.2022.2123194.

Feltwell, Thomas Edgar Ray. 2020. "Agonistic Interfaces: Prompting Debate And Reflection On Socio-Political Issues Using Digital Technology." Doctoral thesis, Northumbria University.

Fernando, Dushanti. 2020. *Support for Sri Lanka to Design Transitional Justice Mechanisms: Phase 1.* Center for Poverty Analysis. https://www.un.org/peacebuilding/sites/www.un.org.peacebuilding/files/documents/pbf_lka_a-1_final_report_mar18_1.pdf.

Fernando, Jude. 2022. "The Citizen's Uprising in Sri Lanka: A Watershed Moment against Racial Capitalism." *Human Geography*, September 1, 19427786221123004. https://doi.org/10.1177/19427786221123004.

Ferrari, Fabian, and Fenwick McKelvey. 2023. "Hyperproduction: A Social Theory of Deep Generative Models." *Distinktion: Journal of Social Theory* 24 (2): 338–360. https://doi.org/10.1080/1600910X.2022.2137546.

Ferroggiaro, W. 2021. *Social Media, Conflict, and Peacebuilding: Issues and Challenges.* Discussion Paper. Mercy Corps. https://europe.mercycorps.org/sites/default/files/2022-01/Social-Media-Discussion-Paper-9-Dec-1.pdf.

Fewer Secretariat. 1997. "Forum on Early Warning and Early Response:A New Collaborative Effort." *Refuge: Canada's Journal on Refugees*, May 1, 24–28. https://doi.org/10.25071/1920-7336.21910.

Fink, Christina. 2018. "Dangerous Speech, Anti-Muslim Violence, and Facebook in Myanmar." *Journal of International Affairs* 71 (1.5): 43–52.

Firchow, Pamina, Charles Martin-Shields, Atalia Omer, and Roger Mac Ginty. 2016. "PeaceTech: The Liminal Spaces of Digital Technology in Peacebuilding." *International Studies Perspectives*, December, ekw007. https://doi.org/10.1093/isp/ekw007.

Firchow, Pamina, Charles Martin-Shields, Atalia Omer, and Roger Mac Ginty. 2017. "PeaceTech: The Liminal Spaces of Digital Technology in Peacebuilding." *International Studies Perspectives* 18 (1): 4–42. https://doi.org/10.1093/isp/ekw007.

Flyvbjerg, Bent. 2001. *Making Social Science Matter. Why Social Inquiry Fails and How It Can Succeed Again.* Oxford, New York: Cambridge University Press. http://site.ebrary.com/id/10069912.

Flyvbjerg, Bent. 2006. "Five Misunderstandings About Case-Study Research." *Qualitative Inquiry* 12 (2): 219–245. https://doi.org/10.1177/1077800405284363.

Folger, Joseph P., and Robert A. Baruch Bush. 2007. "Transformative Mediation and Third-party Intervention: Ten Hallmarks of a Transformative Approach to Practice." *Mediation Quarterly* 13 (4): 263–278. https://doi.org/10.1002/crq.3900130403.

Foucault, Michel1994. *The Birth of the Clinic: An Archaeology of Medical Perception.* New York: Vintage Books.

Frankenberger, William. 2024. "Relational Conflicts Experienced through Digital Platforms among Generation Z Sorority Women." Kent State University, https://coe.k-state.edu/research/documents/Frankenberger%20William.pdf, Accessed May 2, 2026.

Freelon, Deen, Charlton McIlwain, and Meredith Clark. 2018. "Quantifying the Power and Consequences of Social Media Protest." *New Media & Society* 20 (3): 990–1011. https://doi.org/10.1177/1461444816676646.
Gallagher, Michael, and Jeremy Knox. 2019. "Global Technologies, Local Practices." *Learning, Media and Technology* 44 (3): 225–234. https://doi.org/10.1080/17439884.2019.1640741.
Gamage, Gayathri. 2023. "Aragalaya: A New Page of People's Resistance in Sri Lanka." *Protest* 2 (2): 269–277. https://doi.org/10.1163/2667372X-02020006.
Garbe, Lisa, Lisa-Marie Selvik, and Pauline Lemaire. 2023. "How African Countries Respond to Fake News and Hate Speech." *Information, Communication & Society* 26 (1): 86–103. https://doi.org/10.1080/1369118X.2021.1994623.
García-Gómez, Antonio. 2020. "Intercultural and Interpersonal Communication Failures: Analyzing Hostile Interactions among British and Spanish University Students on WhatsApp." *Intercultural Pragmatics* 17 (1): 27–51. https://doi.org/10.1515/ip-2020-0002.
General Assembly Security Council. 2018. *Peacebuilding and Sustaining Peace: Report of the Secretary-General.* New York.
Hashtag Generation. 2023. "ඕනෑම ආගමික බැතිමතකුට කුරුඳුමලේ ආගමික පුදබිම..." Facebook. https://www.facebook.com/hashtaggenerationsl/posts/pfbid02d9oRT4LxER57gGaChVQiT7VLMnCFuRCnip1j4ZDYwENeMg4iY7pCLnJ9JrC4cDgGl.
George, Alexander L., and Jane E. Holl. 1997. "The Warning-Response Problem and Missed Opportunities in Preventive Diplomacy. A Report to the Carnegie Commission on Preventing Deadly Conflict." Carnegie Corporation of New York.
George, Nicole. 2020. "Conflict Transition, Emplaced Identity and the Gendered Politics of Scale in Solomon Islands." *Cooperation and Conflict* 55 (4): 518–534. https://doi.org/10.1177/0010836720954476.
Gibson, James L. 2004. "Truth, Reconciliation, and the Creation of a Human Rights Culture in South Africa." *Law & Society Review* 38 (1): 5–40. https://doi.org/10.1111/j.0023-9216.2004.03801001.x.
Mac Ginty, Roger 2021. *Everyday Peace: How So-Called Ordinary People Can Disrupt Violent Conflict.* Studies in Strategic Peacebuilding. Oxford: Oxford University Press.
GIZ. n.d. "Strengthening Reconciliation Processes in Sri Lanka." Accessed April 15, 2025. https://www.giz.de/en/worldwide/18405.html.
Gladstone, Rick, and Peter Robins. 2021. "The Ghosts of Northern Ireland's Troubles Are Back. What's Going On?" World. *The New York Times*, April 12. https://www.nytimes.com/2021/04/12/world/europe/Northern-Ireland-Brexit-Covid-Troubles.html.
Global Disaster Preparedness Center. 2017. "Crisis Mapping." *PrepareCenter.* https://preparecenter.org/topic/crisis-mapping/.
Gnanguenon, Amandine. 2021. *An Analysis of Continental and Regional Early Warning Systems*, https://www.iss.europa.eu/sites/default/files/EUISSFiles/Brief_3_2021_0.pdf.
Goga, Safiyya. 2015. "Beyond Binary Thinking: Making Sense of Modernity as a Historical Phenomenon." *Psychology in Society* 48: 90–96. https://doi.org/10.17159/2309-8708/2015/n48a5.
Gohdes, Anita R. 2020. "Repression Technology: Internet Accessibility and State Violence." *American Journal of Political Science* 64 (3): 488–503. https://doi.org/10.1111/ajps.12509.
Goldsmith, Paul, ed. 2020. *Conflict Early Warning in the Horn: CEWARN's Journey.* IGAD – CEWARN.
González-Bailón, Sandra, Javier Borge-Holthoefer, Alejandro Rivero, and Yamir Moreno. 2011. "The Dynamics of Protest Recruitment through an Online Network." *Scientific Reports* 1 (1), 197. https://doi.org/10.1038/srep00197.
Goodhand, Jonathan. 2010. "Stabilising a Victor's Peace? Humanitarian Action and Reconstruction in Eastern Sri Lanka." *Disasters* 34 (s3): S342–67. https://doi.org/10.1111/j.1467-7717.2010.01212.x.

Goodlad, Lauren M. E. 2023. "Editor's Introduction: Humanities in the Loop." *Critical AI* 1 (1–2). https://doi.org/10.1215/2834703X-10734016.
Gormley-Heenan, Cathy, and Arthur Aughey. 2017. "Northern Ireland and Brexit: Three Effects on 'the Border in the Mind'." *The British Journal of Politics and International Relations* 19 (3): 497–511. https://doi.org/10.1177/1369148117711060.
GPANet. 2010. *Prevention and Mitigation of Genocide and Mass Atrocities: Focus on East and Central Africa and the Islamic World.* The Genocide Prevention Advisory Network.
De Gregorio, Giovanni, and Nicole Stremlau. 2023. "Inequalities and Content Moderation." *Global Policy* 14 (5): 870–879. https://doi.org/10.1111/1758-5899.13243.
Tamil Guardian. 2022. "Sri Lankan War Criminal Unveils Giant Lion at 'Revived' Buddhist Site." https://www.tamilguardian.com/content/sri-lankan-war-criminal-unveils-giant-lion-revived-buddhist-site.
Gunasekara, Tisaranee. 2023. "Kurundi: A Runway to '56?" *Groundviews*, June 18. https://groundviews.org/2023/06/18/kurundi-a-runway-to-56/.
Guo, Weisi, Kristian Gleditsch, and Alan Wilson. 2018. "Retool AI to Forecast and Limit Wars." *Nature* 562: 331–333.
Halverson, Jeffry R., Scott W. Ruston, and Angela Trethewey. 2013. "Mediated Martyrs of the Arab Spring: New Media, Civil Religion, and Narrative in Tunisia and Egypt." *Journal of Communication* 63 (2): 312–332. https://doi.org/10.1111/jcom.12017.
Halverson, Jeffry R., and Amy K Way. 2012. "The Curious Case of Colleen LaRose: Social Margins, New Media, and Online Radicalization." *Media, War & Conflict* 5 (2): 139–153. https://doi.org/10.1177/1750635212440917.
Hamati-Ataya, I. 2013. "Reflectivity, Reflexivity, Reflexivism: IR's 'Reflexive Turn'—and Beyond." *European Journal of International Relations* 19 (4): 669–694. https://doi.org/10.1177/1354066112437770.
Hattotuwa, Sanjana. 2005. "July 2005." *Sanjana Hattotuwa* (blog). July 28. https://sanjanah.wordpress.com/2005/07/.
Hattotuwa, Sanjana. 2006. "Comments on UN ICT for Peacebuilding Report." *Sanjana Hattotuwa* (blog). January 27. https://sanjanah.wordpress.com/2006/01/27/comments-on-un-ict-for-peacebuilding-report/.
Hattotuwa, Sanjana. 2008. "ICT, Socialism and Apartheid." *ICT for Peacebuilding (#ICT4Peace)* (blog). April 7. https://ict4peace.wordpress.com/2008/04/07/ict-socialism-and-apartheid/.
Hattotuwa, Sanjana, and Roshini Wickremesinhe. 2022. "Hate Speech and Social Media in Sri Lanka." In *Acts of Media: Law and Media in Contemporary India*, edited by Siddharth Narrain. New Delhi: Sage.
Hawke, Julie, Helena Puig Larrauri, Andrew Sutjahjo, and Benjamin Cerigo. 2024. "Understanding to Intervene: The Codesign of Text Classifiers with Peace Practitioners." *Data & Policy* 6 (January): e54. https://doi.org/10.1017/dap.2024.44.
Hawkins, Virgil, and Julia Hoffmann. 2015. *Communication and Peace: Mapping an Emerging Field.* Routledge Studies in Peace and Conflict Resolution. Abingdon: Routledge.
Hearn, Sarah, Alejandra Kubitschek Bujones, and Alischa Kugel. 2014. "The United Nations 'Peacebuilding Architecture': Past, Present and Future." New York: NYU Center on International Cooperation.
Heathershaw, John, and Catherine Owen. 2019. "Authoritarian Conflict Management in Post-Colonial Eurasia." *Conflict, Security & Development* 19 (3): 269–273. https://doi.org/10.1080/14678802.2019.1608022.
Heeks, Richard. 2018. *Information and Communication Technology for Development (ICT4D).* Routledge Perspectives on Development. London, New York: Routledge (an imprint of the Taylor & Francis Group).
Hegre, Håvard, Marie Allansson, Matthias Basedau, Michael Colaresi, Mihai Croicu, Hanne Fjelde, Frederick Hoyles, Lisa Hultman, Stina Högbladh, Remco Jansen, Naima Mouhleb,

Sayyed Auwn Muhammad, Desirée Nilsson, Håvard Mokleiv Nygård, Gudlaug Olafsdottir, Kristina Petrova, David Randahl, Espen Geelmuyden Rød, Gerald Schneider, Nina von Uexkull, Jonas Vestby. 2019. "ViEWS: A Political Violence Early-Warning System." *Journal of Peace Research* 56 (2): 155–174. https://doi.org/10.1177/0022343319823860.

Hegre, Håvard, Nils W Metternich, Håvard Mokleiv Nygård, and Julian Wucherpfennig. 2017. "Introduction: Forecasting in Peace Research." *Journal of Peace Research* 54 (2): 113–124. https://doi.org/10.1177/0022343317691330.

Heidegger, Martin. 1977. *The Question Concerning Technology, and Other Essays*. New York: Garland Publishing.

Heinze, Eric. 2016. "Concepts and Contexts." In *Hate Speech and Democratic Citizenship*, edited by Eric Heinze. Oxford: Oxford University Press. https://doi.org/10.1093/acprof:oso/9780198759027.003.0002.

Hekman, Susan J. 1983. *Weber, the Ideal Type, and Contemporary Social Theory*. Oxford: University of Notre Dame Press.

Hellmüller, Sara. 2019. "Beyond Buzzwords: Civil Society Inclusion in Mediation." In *Conflict Intervention and Transformation: Theory and Practice*, edited by Ho-Won Jeong, 47–63. Maryland: Rowman & Littlefield.

Helvetas. n.d. "Empowering Youth to Initiate, Lead and Implement Community-Level Actions to Prevent Violent Extremism." Helvetas. Accessed April 15, 2025. https://www.helvetas.org/en/sri-lanka/what-we-do/how-we-work/our-projects/Asia/Sri-Lanka/EMPOWER.

Hensman, Rohini. 2022. *Sri Lanka's Democratic Revolution: The Latest Episode in a Decades-Long Drama*. Tempest Magazine, August 3.

Hettiarachchi, Radhika. 2018. *The Community Memorialisation Project,T: Memoralisaton For Non-Recccurence of Violent Conflict*. Search for Common Ground.

Himmelfarb, Sheldon. 2015. "Peacebuiding Reboot." In *#PEACETECH: EVERYTHING YOU NEED TO KNOW, FROM SOCIAL MEDIA IN AFGHANISTAN TO HUMANITARIAN DRONES IN SYRIA*, edited by Jessica Berns, 12–13. Washington, D.C.: Alliance for Peacebuilding.

Hirblinger, Andreas T. 2022. "When Mediators Need Machines (and Vice Versa): Towards a Research Agenda on Hybrid Peacemaking Intelligence." *International Negotiation*.

Hirblinger, Andreas T. 2015. "Dealing with the Trouble at the Margins: Discourses on Conflict and the Rationality of Local Government in South Sudan." PhD Dissertation, University of Cambridge.

Hirblinger, Andreas T. 2020. *Digital Inclusion in Mediated Peace Processes: How Technology Can Enhance Participation*. Peaceworks 168. United States Institute for Peace. https://www.usip.org/publications/2020/09/digital-inclusion-mediated-peace-processes-how-technology-can-enhance.

Hirblinger, Andreas T. 2022. "When Mediators Need Machines (and Vice Versa): Towards a Research Agenda on Hybrid Peacemaking Intelligence." *International Negotiation* 27 (January):1–32. https://doi.org/10.1163/15718069-bja10050.

Hirblinger, Andreas T. 2023. "When the Digits Don't Add up: Research Strategies for Post-Digital Peacebuilding." *Cooperation and Conflict* 0 (0). https://doi.org/10.1177/00108367231184727.

Hirblinger, Andreas T. 2023. "Building a Peace We Don't Know? The Power of Subjunctive Technologies in Digital Peacebuilding." *Peacebuilding* 11 (2): 113–135. https://doi.org/10.1080/21647259.2022.2154957.

Hirblinger, Andreas T. 2024. "Fighting Hate Speech on Social Media: The Challenge of 'Deep Work' in Digital Peacebuilding." In *Social Media and Peacebuilding. How Digital Spaces Shape Conflict and Peace.*, edited by Stephan Stetter and Anna Reuss. Rethinking Peace and Conflict Studies.

Hirblinger, Andreas T., Ville Brummer, and Felix Kufus. 2024. "Leveraging Digital Methods in the Quest for Peaceful Futures: The Interplay of Sincere and Subjunctive Technology

Affordances in Peace Mediation." *Information, Communication & Society* 27 (11): 2039–2058. https://doi.org/10.1080/1369118X.2023.2247070.
Hirblinger, Andreas T., Julie Marie Hansen, Kristian Hoelscher, Åshild Kolås, Kristoffer Lidén, and Bruno Oliveira Martins. 2022. "Digital Peacebuilding: A Framework for Critical–Reflexive Engagement." *International Studies Perspectives*, November 26, ekac015. https://doi.org/10.1093/isp/ekac015.
Hirblinger, Andreas T., Sara Kallis, and Hasini A. Haputhanthri. 2024. "DIY-Online Reconciliation? The Role of Memes in Navigating Inter-Group Boundaries in the Context of Sri Lanka's 2022 Political Crisis." *Social Media + Society* 10 (4): 20563051241303362. https://doi.org/10.1177/20563051241303362.
Hirblinger, Andreas T, and Dana M Landau. 2020. "Daring to Differ? Strategies of Inclusion in Peacemaking." *Security Dialogue* 51 (4): 305–322. https://doi.org/10.1177/0967010619893227.
Hirblinger, Andreas T, Martin Wählisch, Kate Keator, Chris McNaboe, Allard Duursma, John Karlsrud, Valerie Sticher, Aly Verjee, Tetiana Kyselova, Chris M A Kwaja, Suda Perera. 2024. "Forum: Making Peace with Un-Certainty: Reflections on the Role of Digital Technology in Peace Processes beyond the Data Hype." *International Studies Perspectives* 25 (2): 185–225. https://doi.org/10.1093/isp/ekad004.
HM Government. 2021. *Northern Ireland Protocol: The Way Forward.* https://assets.publishing.service.gov.uk/government/uploads/system/uploads/attachment_data/file/1008451/CCS207_CCS0721914902-005_Northern_Ireland_Protocol_Web_Accessible__1_.pdf.
Hobsbawm, E. J., and T. O. Ranger, eds. 1983. *The Invention of Tradition.* Past and Present Publications. Cambridge, New York: Cambridge University Press.
Hoijtink, Marijn, and Matthias Leese, eds. 2019. *Technology and Agency in International Relations.* Emerging Technologies, Ethics and International Affairs. London, New York: Routledge Taylor & Francis Group.
Hoijtink, Marijn, and Matthias Leese. 2019a. "How (Not) to Talk about Technology. International Relations and the Question of Agency." In *Technology and Agency in International Relations*, edited by Marijn Hoijtink and Matthias Leese. Emerging Technologies, Ethics and International Affairs. Routledge Taylor & Francis Group.
Hoijtink, Marijn, and Matthias Leese. 2019b. *Technology and Agency in International Relations.* Emerging Technologies, Ethics and International Affairs. Routledge Taylor & Francis Group.
Höne, Katharina. 2019. *Mediation and Artificial Intelligence: Notes on the Future of International Conflict Resolution.* DiploFoundation. https://www.diplomacy.edu/sites/default/files/Mediation_and_AI.pdf.
Hu, Lianyu, Mudi Jiang, Junjie Dong, Xinying Liu, and Zengyou He. 2024. "Interpretable Clustering: A Survey." arXiv:2409.00743. Version 1. Preprint, arXiv, September 1. https://doi.org/10.48550/arXiv.2409.00743.
Hug, Alexander, and Simon Mason. 2022. "Ceasefire Monitoring and Verification Technology." CSS Policy Perspectives. Zurich: Center for Security Studies, ETH Zurich.
Human Rights Watch. 2020. *Sri Lanka: Increasing Suppression of Dissent.* August 8. https://www.hrw.org/news/2020/08/08/sri-lanka-increasing-suppression-dissent.
Human Rights Watch. 2021. *Open Wounds and Mounting Dangers. Blocking Accountability for Grave Abuses in Sri Lanka.* https://www.hrw.org/report/2021/02/01/open-wounds-and-mounting-dangers/blocking-accountability-grave-abuses-sri-lanka.
Hunt, Charles T. 2017. "Beyond the Binaries: Towards a Relational Approach to Peacebuilding." *Global Change, Peace & Security* 29 (3): 209–227. https://doi.org/10.1080/14781158.2017.1360855.
Hunt, C. T. 2023. How Many Turns Make a Revolution? Whither the 'Dialogue of the Deaf' Between Peacebuilding Scholars and Practitioners. *Journal of Intervention and Statebuilding*, 17(4), 333–350. https://doi.org/10.1080/17502977.2023.2197446.

IBM. 2024. "What Is Zero-Shot Learning? | IBM." https://www.ibm.com/think/topics/zero-shot-learning.
Ide, Tobias. 2017. "Space, Discourse and Environmental Peacebuilding." *Third World Quarterly* 38 (3): 544–562. https://doi.org/10.1080/01436597.2016.1199261.
Imbleau, Martin. 2011. "Denial of the Holocaust, Genocide, and Crimes Against Humanity." In *Genocide Denials and the Law*, edited by Ludovic Hennebel and Thomas Hochmann. Oxford: Oxford University Press. https://doi.org/10.1093/acprof:oso/9780199738922.003.0008.
Independent Expert Advisory Group. 2014. "A World That Counts: Mobilising the Data Revolution for Sustainable Development." New York: United Nations Secretary-General.
Institute for Participatory Interaction in Development. 2021. *Final Evaluation. Promoting Reconciliation in Sri Lanka.UNDP/UNICEF/WHO.* Dehiwala. https://www.un.org/peacebuilding/sites/www.un.org.peacebuilding/files/documents/pbf_lka_a-2_final_evaluation_report_promoting_reconciliation_in_sri_lanka.pdf.
International Centre for Ethnic Studies. n.d. "About MoMaC – Digital Museum of Memory and Coexistence." Accessed April 16, 2025. https://momac.lk/about-momac/momac-intro/.
IPI. 2013. "New Technology and the Prevention of Violence and Conflict." *Stability: International Journal of Security & Development* 2 (3): 55. https://doi.org/10.5334/sta.cp.
The Irish Times. 1999. "Belfast Community Centre Still Defying Sectarianism." The Irish Times. https://www.irishtimes.com/news/belfast-community-centre-still-defying-sectarianism-1.239864.
Irwin, Colin. n.d. *Using Artificial Intelligence in Peacemaking—The Libya Experience.*
ITU and WSIS. 2005. "WSIS-05/TUNIS/DOC/7-E". Tunis. https://www.itu.int/net/wsis/docs2/tunis/off/7.html.
ITU DataHub. 2023. "ICT Statistics." ITU. https://www.itu.int:443/en/ITU-D/Statistics/Pages/stat/default.aspx.
Jackson, Patrick Thaddeus. 2011. *The Conduct of Inquiry in International Relations: Philosophy of Science and Its Implications for the Study of World Politics.* The New International Relations. London, New York: Routledge.
Jacobsen, Katia Lindskov, and Linda Monsees. 2019. "Co-Production. The Study of Productive Processes at the Level of Materiality and Discourse." In *Technology and Agency in International Relations*, edited by Marijn Hoijtink and Matthias Leese. Emerging Technologies, Ethics and International Affairs. Routledge Taylor & Francis Group.
Jaidka, Kokil, Alvin Zhou, and Yphtach Lelkes. 2019. "Brevity Is the Soul of Twitter: The Constraint Affordance and Political Discussion." *Journal of Communication* 69 (4): 345–372. https://doi.org/10.1093/joc/jqz023.
Janardhanan, Arun. 2024. "Putting Economic Justice over Ethnic Nationalism, How a Once-Fringe Coalition Swept to Majority in Sri Lanka Parliament." *The Indian Express*, November 16. https://indianexpress.com/article/india/srilanka-fringe-coalition-majority-economic-justice-over-ethnic-nationalism-9672129/.
Jandrić, Petar, Jeremy Knox, Tina Besley, Thomas Ryberg, Juha Suoranta, and Sarah Hayes. 2018. "Postdigital Science and Education." *Educational Philosophy and Theory* 50 (10): 893–899. https://doi.org/10.1080/00131857.2018.1454000.
Jandrić, Petar, Thomas Ryberg, Jeremy Knox, Callum McGregor, Nataša Lacković, Sarah Hayes, Juha Suoranta, Mark Smith, Anne Steketee, Peter McLaren, Michael Peters, Derek R. Ford, Gordon Asher, Georgina Stewart, and Andrew Gibbons. 2019. "Postdigital Dialogue." *Postdigital Science and Education* 1 (1): 163–189. https://doi.org/10.1007/s42438-018-0011-x.
Janet, Salmons, and Wilson Lynn. 2008. *Handbook of Research on Electronic Collaboration and Organizational Synergy.* Hershey and London: IGI Global.

Jarman, Neil. 1999. *Drawing Back from the Edge: Community Based Response to Violence in North Belfast.* Community Development Centre. https://cain.ulster.ac.uk/cdc/back/jarman99.htm.

Jarman, Neil, and Chris O'Halloran. 2001. "Recreational Rioting: Young People, Interface Areas and Violence." *Child Care in Practice* 7 (1): 2–16. https://doi.org/10.1080/13575270108413230.

Jasanoff, Sheila. 2004. "The Idiom of Co-Production." In *States of Knowledge: The Co-Production of Science and Social Order*, edited by Sheila Jasanoff. International Library of Sociology. London and New York: Routledge.

Jasanoff, Sheila, and Hilton R. Simmet. 2017. "No Funeral Bells: Public Reason in a 'Post-Truth' Age:" *Social Studies of Science*, ahead of print. London: Sage UK. https://doi.org/10.1177/0306312717731936.

Jiang, Ray, Silvia Chiappa, Tor Lattimore, András György, and Pushmeet Kohli. 2019. "Degenerate Feedback Loops in Recommender Systems." *arXiv:1902.10730 [Cs, Stat]*, ahead of print, February 27. https://doi.org/10.1145/3306618.3314288.

Jobarteh, Sanjally, and Haller Natalina. 2020. "Editorial—Digital Technologies: A Gamer Changer for Peace?" *À Propos—The KOFF Peacebuilding Magazine* (168). https://varia.swisspeace.ch/apropos/november-2020/.

Johnson, Douglas. 2014. *Federalism in the History of South Sudanese Political Thought – Rift Valley Institute*. Vol. 1. Rift Valley Institute Research Paper. The Rift Valley Institute. https://riftvalley.net/publication/federalism-history-south-sudanese-political-thought/.

Jones, Marc Owen. 2022. *Digital Authoritarianism in the Middle East: Deception, Disinformation and Social Media.* London: Hurst Publishers.

Asia Justice and Rights, Suriya, and Transitional Justice Asia Network. 2022. *Transitional Justice Fact Sheet: Sri Lanka.* https://asia-ajar.org/wp-content/uploads/2022/06/Sri-Lanka-English-Transitional-Justice-Factsheet.pdf.

Kahl, Anne, and Helena Puig Larrauri. 2013. "Technology for Peacebuilding." *Stability: International Journal of Security and Development* 2 (3): 15. https://doi.org/10.5334/sta.cv.

Kahl, Anne, Christy McConnell, and William Tsuma. 2012. "Crowdsourcing as a Tool in Conflict Prevention." *Conflict Trends* 2012 (1): 27–34.

Karlsrud, John. 2014. "Peacekeeping 4.0: Harnessing the Potential of Big Data, Social Media, and Cyber Technologies." In *Cyberspace and International Relations*, edited by Jan-Frederik Kremer and Benedikt Müller, 141–160. Berlin, Heidelberg: Springer. https://doi.org/10.1007/978-3-642-37481-4_9.

Kelman, Herbert. 2004. *Reconciliation as Identity Change: A Social-Psychological Perspective.* https://doi.org/10.1093/acprof:oso/9780195166439.003.0006.

Kenny, Mark. 2022. "Coarse and Effect: Normalised Anger Online as an Essential Precondition to Violence." In *Rethinking Social Media and Extremism*, edited by Paul Pickering and Shirley Leitch. Canberra: ANU Press. https://doi.org/10.22459/RSME.2022.

Kern, Stephen. 2003. *The Culture of Time and Space, 1880–1918: With a New Preface.* 2nd ed. Cambridge, Mass: Harvard University Press.

Klausen, Jytte. 2015. "Tweeting the *Jihad*: Social Media Networks of Western Foreign Fighters in Syria and Iraq." *Studies in Conflict & Terrorism* 38 (1): 1–22. https://doi.org/10.1080/1057610X.2014.974948.

Knox, Jeremy. 2019. "Postdigital Thinking." In *Postdigital Dialogue*, vol. 1, edited by Jandrić, P., Ryberg, T., Knox, J. et al. Postdigital Dialogue. Postdigit Sci Educ 1, 163–189. Postdigital Science and Education. https://doi.org/10.1007/s42438-018-0011-x.

Koehler, Daniel. 2014. "The Radical Online: Individual Radicalization Processes and the Role of the Internet." *Journal for Deradicalization* Winter 2014/15 (1): 1.

Kopponen, Aleksi, Antti Hahto, Petri Kettunen, Tommi Mikkonen, Niko Mäkitalo, Jarkko Nurmi, and Matti Rossi. 2022. "Empowering Citizens With Digital Twins: A Blueprint." *IEEE Internet Computing* 26 (5): 7–16. https://doi.org/10.1109/MIC.2022.3159683.

Krause, Jana, Werner Krause, and Piia Bränfors. 2018. "Women's Participation in Peace Negotiations and the Durability of Peace." *International Interactions* 44 (6): 985–1016. https://doi.org/10.1080/03050629.2018.1492386.

Krummenacher, Heinz. 2006. "Computer Assisted Early Warning—the FAST Example." In *Programming for Peace: Computer-Aided Methods for International Conflict Resolution and Prevention*, edited by Robert Trappl. Dordrecht: Springer Netherlands. https://doi.org/10.1007/1-4020-4390-2_4.

Krummenacher, Heinz, and Susanne Schmeidl. 2001. *Practical Challenges in Predicting Violent Conflict: FAST an Example of a Comprehensive Early-Warning Methodology*. Swisspeace.

Krummenacher, Heinz, Daniel Schwarz, and Matthias Siegfried. 2002. *Local Information Networks: Practical Requirements and Considerations*, Paper Prepared for the Annual ISA Conference 2002 in New Orleans, LA.

Kühn, Florian P. 2016. "International Peace Practice: Ambiguity, Contradictions and Perpetual Violence." In *The Politics of International Intervention: The Tyranny of Peace*, edited by Mandy Turner and Florian P. Kühn. London; New York: Routledge.

Kuol, Luka Biong D. 2020. "South Sudan: The Elusive Quest for a Resilient Social Contract?" *Journal of Intervention and Statebuilding* 14 (1): 64–83. https://doi.org/10.1080/17502977.2019.1627692.

Kydd, Andrew. 2003. "Which Side Are You On? Bias, Credibility, and Mediation." *American Journal of Political Science* 47 (4): 597–611.

Laffey, M., and S. Nadarajah. 2012. "The Hybridity of Liberal Peace: States, Diasporas and Insecurity." *Security Dialogue* 43 (5): 403–420. https://doi.org/10.1177/0967010612457974.

Lamb, Harriet. 2017. "Can Robots Do A Better Job Of Building Peace?" HuffPost UK, September 20. https://www.huffingtonpost.co.uk/harriet-lamb/robots-bring-us-peace_b_18031158.html.

Lambach, Daniel, Markus Bayer, Felix S. Bethke, Matteo Dressler, and Véronique Dudouet. 2020. *Nonviolent Resistance and Democratic Consolidation*. London; New York: Springer International Publishing.

Lanz, David, and Ahmed Eleiba. 2018. *The Good, the Bad Andthe Ugly: Social Media and Peace Mediation*. Policy Brief. Swisspeace.

Larrauri, Helena Puig, and Yeonju Jung. 2017. "Reimagining Peacebuilding Through Innovation Stockholm Forum on Peace and Development. Conference Report." Session Report 14. Stockholm: Stockholm Forum on Peace and Development. https://www.sipri.org/sites/default/files/2017-09/session_report_no_14.pdf.

Larrauri, Helena Puig, and Anne Kahl. 2013. "Technology for Peacebuilding." *Stability: International Journal of Security and Development* 2 (3): Art. 61. https://doi.org/10.5334/sta.cv.

Latour, Bruno. 2005. *Reassembling the Social: An Introduction to Actor-Network-Theory*. Clarendon Lectures in Management Studies. Oxford: Oxford University Press.

Lawrence, Charles R., Mari J. Matsuda, Richard Delgado, and Kimberlé Williams Crenshaw. 1993. "Introduction." In *Words That Wound: Critical Race Theory, Assaultive Speech, and the First Amendment*. Press Release 31. Westview Press.

Lederach, John Paul. 1995. *Conflict Transformation in Protracted Internal Conflicts: A Case for a Comprehensive Framework*. https://www.johnpaullederach.com/2023/03/conflict-transformation-in-protracted-internal-conflicts-the-case-for-a-comprehensive-framework/

Lederach, John Paul. 2005. *The Moral Imagination: The Art and Soul of Building Peace*. Oxford: Oxford University Press.

Lederach, John Paul, and Preeti Thapa. 2018. "Staying True in Nepal: Understanding Community Mediation Through Action Research." In *Cultural Encounters and Emergent Practices in Conflict Resolution Capacity-Building*, edited by Tamra Pearson d'Estrée and Ruth J. Parsons. Cham: Springer International Publishing. https://doi.org/10.1007/978-3-319-71102-7_2.

Lehti, Marko. 2024. "Agonistic Peacebuilding." In *Routledge Handbook of Peacebuilding*, 2nd ed. Oxon and New York: Routledge.

Lehti, Marko, and Vadim Romashov. 2022. "Suspending the Antagonism: Situated Agonistic Peace in a Border Bazaar." *Third World Quarterly* 43 (6): 1288–1306. https://doi.org/10.1080/01436597.2021.1962274.

Leib, Julia. 2023. "Of Peacekeepers and Pandemics: How Covid-19 Changed Strategic Communication of the UN Mission in South Sudan." *Global Governance: A Review of Multilateralism and International Organizations* 29 (1): 11–36. https://doi.org/10.1163/19426720-02901004.

Leonardi, Cherry. 2013. *Dealing with Government in South Sudan: Histories of Chiefship, Community & State*. Eastern Africa Series. Woodbridge: James Currey.

Leonardsson, Hanna, and Gustav Rudd. 2015. "The 'Local Turn' in Peacebuilding: A Literature Review of Effective and Emancipatory Local Peacebuilding." *Third World Quarterly* 36 (5): 825–839. https://doi.org/10.1080/01436597.2015.1029905.

Levitsky, Steven, and Lucan A. Way. 2010. *Competitive Authoritarianism: Hybrid Regimes after the Cold War*. Problems of International Politics. Cambridge: Cambridge University Press. https://doi.org/10.1017/CBO9780511781353.

Lewis, David, John Heathershaw, and Nick Megoran. 2018. "Illiberal Peace? Authoritarian Modes of Conflict Management." *Cooperation and Conflict* 53 (4): 486–506. https://doi.org/10.1177/0010836718765902.

Lewis, David G. 2020. "Sri Lanka's Schmittian Peace: Sovereignty, Enmity and Illiberal Order." *Conflict, Security & Development* 20 (1): 15–37. https://doi.org/10.1080/14678802.2019.1705067.

Van der Lijn, Jaïr. 2022. "The Impact of the Covid-19 Pandemic on Multilateral Peace Operations." Stockholm International Peace Research Institute. https://doi.org/10.55163/ZYXI8245.

Little, Adrian, and Sarah Maddison. 2017. "Reconciliation, Transformation, Struggle: An Introduction." *International Political Science Review* 38 (2): 145–154. https://doi.org/10.1177/0192512116681808.

Livingston, Steven, and Gregor Walter-Drop. 2014. *Bits and Atoms: Information and Communication Technology in Areas of Limited Statehood*. Oxford: Oxford University Press.

Lombardo, Giorgia. 2019. "What Happens When AI Writes a Novel?" Medium, December 3. https://medium.com/demagsign/what-happens-when-ai-writes-a-novel-173d407af567.

López López, Aura Liliana, and Bert Ingelaere. 2021. "The Missing Link in Hybrid Peacebuilding: Localized Peace Trajectories and Endogenous Knowledge." In *A Requiem for Peacebuilding?*, edited by Jorg Kustermans, Tom Sauer, and Barbara Segaert. Rethinking Peace and Conflict Studies. Cham: Springer International Publishing. https://doi.org/10.1007/978-3-030-56477-3_5.

Lührmann, Anna, Marcus Tannenberg, and Staffan I. Lindberg. 2018. "Regimes of the World (RoW): Opening New Avenues for the Comparative Study of Political Regimes." *Politics and Governance* 6 (1): 60–77.

Ma, ZIlin, Yiyang Mei, Claude Bruderlein, Krzysztof Z. Gajos, and Weiwei Pan. 2024. "'ChatGPT, Don't Tell Me What to Do': Designing AI for Context Analysis in Humanitarian Frontline Negotiations." arXiv:2410.09139. Version 1. Preprint, arXiv, October 11. https://doi.org/10.48550/arXiv.2410.09139.

Mac Ginty, Roger. 2011. *International Peacebuilding and Local Resistance: Hybrid Forms of Peace*. New York: Palgrave Macmillan.

Mac Ginty, Roger. 2021. *Everyday Peace: How So-Called Ordinary People Can Disrupt Violent Conflict*. Oxford: Oxford University Press.

Mac Ginty, Roger, and Oliver Richmond. 2016. "The Fallacy of Constructing Hybrid Political Orders: A Reappraisal of the Hybrid Turn in Peacebuilding." *International Peacekeeping* 23 (2): 219–239. https://doi.org/10.1080/13533312.2015.1099440.

Mac Ginty, Roger, and Oliver Richmond. 2013. "The Local Turn in Peace Building: A Critical Agenda for Peace." *Third World Quarterly* 34 (5): 763–783. https://doi.org/10.1080/01436597.2013.800750.

Mack, Andrew. 2014. "Measuring Peacebuilding Performance: Why We Need a 'Data Revolution.'" *Accord* (25): 109–112.

Maddison, Sarah. 2015. "Relational Transformation and Agonistic Dialogue in Divided Societies." *Political Studies* 63 (5): 1014–1030. https://doi.org/10.1111/1467-9248.12149.

Mallempati, Samatha. 2024. *The Rise of the JVP-Led NPP: A Clear Political Shift in Sri Lanka*. Indian Council of World Affairs. https://www.icwa.in/show_content.php?lang=1&level=3&ls_id=12091&lid=7367.

Mamdani, Mahmood. 1996. *Citizen and Subject: Contemporary Africa and the Legacy of Late Colonialism*. Princeton, N.J.: Princeton University Press.

Mandelik, Florence, and Ayad Mohamed. 2020. *Can You Hear Me? Taking Mediation Online: Reflections from Practice*. NOREF Brief. NOREF. https://noref.no/Publications/Themes/Peacebuilding-and-mediation/Can-you-hear-me-Taking-mediation-online-reflections-from-practice, Accessed August 09, 2022.

Marichal, Jose, and Richard Neve. 2019. "Antagonistic Bias: Developing a Typology of Agonistic Talk on Twitter Using Gun Control Networks." *Online Information Review* 44 (2): 343–363. https://doi.org/10.1108/OIR-11-2018-0338.

Martin-Shields, Charles. 2015. "Proving (Digital) Peace." In *#Peacetech: Everything You Need to Know, from Social Media in Afghanistan to Humanitarian Drones in Syria*, edited by Jessica Berns, 33–34. Washington, D.C: Alliance for Peacebuilding.

Masarat. 2021. Sumoud—Resilience. Bringing resilience back to the core of Palestinian policy-making. Palestinian Centre for Policy Research and Strategic Studies.

Matsuda, Mari J. 1993. "Public Response to Racist Speech: Considering the Victim's Story." In *Words That Wound: Critical Race Theory, Assaultive Speech, and the First Amendment*. London: Westview Press.

McCormack, Jayne. 2024. *NI's Government Has Returned Stormont—What You Need to Know*. January 30. https://www.bbc.com/news/uk-northern-ireland-67726389.

McKinley, Alexander, and Merin Shobhana Xavier. 2023. "The Deconstruction of Dafther Jailani: Muslim and Buddhist Contests of Original History in Sri Lanka." *History of Religions* 62 (3): 254–283. https://doi.org/10.1086/723307.

Mead, Walter Russell. 2014. "The Return of Geopolitics: The Revenge of the Revisionist Powers." *Foreign Affairs* 93 (3): 69–79.

Meier, Claudia. 2021. "Four Observations from Running Consultation Processes via WhatsApp." Build Up—Medium, March 31. https://howtobuildup.medium.com/four-observations-from-running-consultation-processes-via-whatsapp-f954c6c29f91.

Meier, Patrick. 2008. "The Past and Future of Crisis Mapping." *iRevolutions*, October 18. https://irevolutions.org/2008/10/18/future-of-crisis-mapping/.

Mercy Corps. 2020. "Advancing Peace in a Changed World. Covid-19 Effects on Conflict and How to Respond." https://www.mercycorps.org/sites/default/files/2020-09/Advancing_Peace_COVID-19_and_Conflict_Sept-2020_0.pdf.

Meta. 2020. "How Facebook Uses Super-Efficient AI Models to Detect Hate Speech." https://ai.meta.com/blog/how-facebook-uses-super-efficient-ai-models-to-detect-hate-speech/, Accessed July 08, 2024.

Milan, Stefania. 2015. "From Social Movements to Cloud Protesting: The Evolution of Collective Identity." *Information, Communication & Society* 18 (8): 887–900. https://doi.org/10.1080/1369118X.2015.1043135.

Mitchell, David. 2023. "Rescuing Reconciliation: Finding Its Role in Peace Research and Practice." *Third World Quarterly* 44 (8): 1737–1753. https://doi.org/10.1080/01436597.2023.2205120.

Modin-Lundin, Amie. 2022. *Aragalaya—Stories about Unity and Conflicts in a Crisis-Hit Sri Lanka*. http://urn.kb.se/resolve?urn=urn:nbn:se:uu:diva-489987.

Moe, Louise Wiuff, and Finn Stepputat. 2018. "Introduction: Peacebuilding in an Era of Pragmatism." *International Affairs* 94 (2): 293–299. https://doi.org/10.1093/ia/iiy035.

Mosco, Vincent. 2017. *Becoming Digital: Toward a Post-Internet Society*. Bingley: Emerald Group Publishing.

Mouffe, C. 2002. "For an Agonistic Public Sphere." In *Democracy unrealized: documenta 11, platform 1*, edited by O. Enwezor, 87–97. Hatje Cantz. https://westminsterresearch.westminster.ac.uk/item/93xxz/for-an-agonistic-public-sphere.

Mouffe, Chantal. 2005. *The Return of the Political*. Radical Thinkers. London, New York: Verso.

Moulana, Amjad. 2022. "Owing to @Dailymirror_SL bungling on calling Kuragala something a nation has been waiting for, I thought I'd shed some light on the myths and truths surrounding Kuragala. A thread . . . /1." Tweet. Twitter, May 15. https://x.com/MoulanaAmjad/status/1525798809103585281, Accessed December 08, 2024.

Murel, Jacob, and Eda Kavlakoglu. 2024. "What Is Collaborative Filtering? | IBM." https://www.ibm.com/think/topics/collaborative-filtering.

Murphy, Mary C. 2021. "Northern Ireland and Brexit: Where Sovereignty and Stability Collide?" *Journal of Contemporary European Studies* 29 (3): 405–418. https://doi.org/10.1080/14782804.2021.1891027.

Murphy, Paul L. 1995. "Samuel Walker. Hate Speech: The History of an American Controversy. Lincoln: University of Nebraska Press. 1994. Pp. 217. Cloth $25.00, Paper $11.95". *The American Historical Review* 100 (3): 965. https://doi.org/10.1086/ahr/100.3.965.

Newman, Edward, Roland Paris, and Oliver P. Richmond. 2009. *New Perspectives on Liberal Peacebuilding*. Tokyo, New York and Paris: United Nations University Press.

Niblett, Robin. 2017. "Liberalism in Retreat: The Demise of a Dream Out of Order." *Foreign Affairs* 96: 17–24.

Nilsson, Desirée. 2012. "Anchoring the Peace: Civil Society Actors in Peace Accords and Durable Peace." *International Interactions* 38 (2): 243–266. https://doi.org/10.1080/03050629.2012.659139.

Njeru, Shastry. 2009. "Information and Communication Technology (ICT), Gender, and Peacebuilding in Africa: A Case of Missed Connections." *Peace and Conflict Review* 3 (2): 32–40.

Norris, Pippa. 2019. *Cultural Backlash: Trump, Brexit, and Authoritarian Populism*. Cambridge: Cambridge University Press.

Norval, Aletta. 2009. "Democracy, Pluralization, and Voice." *Ethics & Global Politics* 2 (4): 297–320. https://doi.org/10.3402/egp.v2i4.2118.

Nyheim, David. 2008. *Can Violence, War and State Collapse Be Prevented?* OECD.

Nyheim, David. 2015. *Early Warning and Response to Violent Conflict: Time for a Rethink*. London: Saferworld.

NYU Center on International Cooperation, dir. 2021. *Harnessing Data for Early Warning: UNDP's Experience*. With Ethan Harrison. 52:23. https://www.youtube.com/watch?v=GBgU1geL6Ds.

O'Brien, Matt, and Stephen Wright. 2019. "Sri Lanka Blocks Social Media After Easter Sunday Bombings." *The Diplomat*, April 22. https://thediplomat.com/2019/04/sri-lanka-blocks-social-media-after-easter-sunday-bombings/.

O'Leary, Brendan, and Jamie Pow. 2023. "The Turning Point? The Northern Ireland Local Government Elections of 18 May 2023." Royal Irish Academy. https://www.ria.ie/blog/the-turning-point-the-northern-ireland-local-government-elections-of-18-may-2023/.

O'Neill, Julian. 2021. "Belfast: Rioting 'Was Worst Seen in Northern Ireland in Years.'" *BBC News*, April 8. https://www.bbc.com/news/uk-northern-ireland-56664868.

OAS and UNDP. 2016. *Practical Guide: Early Warning and Response Systems Design for Social Conflicts*, https://www.undp.org/latin-america/publications/practical-guide-early-warning-and-response-systems-design-social-conflicts.

OECD. 2009. *Preventing Violence, War and State Collapse: The Future of Conflict Early Warning and Response*. Organisation for Economic Co-operation and Development. https://www.oecd.org/content/dam/oecd/en/publications/reports/2009/02/preventing-violence-war-and-state-collapse_g1gha84f/9789264059818-en.pdf.

OHCHR. 2016. "South Sudan: Dangerous Rise in Ethnic Hate Speech Must Be Reined in – Zeid." *OHCHR*, October 25. https://www.ohchr.org/en/2016/10/south-sudan-dangerous-rise-ethnic-hate-speech-must-be-reined-zeid.

Okello, Ojok. 2022. *An Overview of the Media Landscape in South Sudan*. Konrad Adenauer Stiftung.

Onishi, Akira. 1987. "Global Early Warning System for Displaced Persons: Interlinkages of Environment, Development, Peace and Human Rights." *Technological Forecasting and Social Change* 31 (3): 269–299. https://doi.org/10.1016/0040-1625(87)90015-1.

Orjuela, Camilla. 2020. "Countering Buddhist Radicalisation: Emerging Peace Movements in Myanmar and Sri Lanka." *Third World Quarterly* 41 (1): 133–150. https://doi.org/10.1080/01436597.2019.1660631.

Panic, Branka. 2020. "Data for Peacebuilding and Prevention. The State of Play and the Path to Creating a Community of Practice." New York: NYU Center on International Cooperation.

Panic, Branka, and Paige Arthur. 2024. *AI for Peace*. AI for Everything. Boka Raton and Oxon: CRC Press, https://doi.org/10.1201/9781003359982.

Parekh, Bhikhu. 2012. "Is There a Case for Banning Hate Speech?" In *The Content and Context of Hate Speech: Rethinking Regulation and Responses*, edited by Michael Herz and Peter Molnar. Cambridge: Cambridge University Press. https://doi.org/10.1017/CBO9781139042871.006.

Paris, Roland. 2010. "Saving Liberal Peacebuilding." *Review of International Studies* 36 (02): 337.

Passoth, Jan-Hendrik, and Nicholas J. Rowland. 2015. "Who Is Acting in International Relations?" In *Human Beings in International Relations*, edited by Daniel Jacobi and Annette Freyberg-Inan. Cambridge: Cambridge University Press. https://doi.org/10.1017/CBO9781316337042.015.

Pepperell, R., and M. Punt, 2000. *The Postdigital Membrane: Imagination, Technology and Desire*. Intellect Books.

UN PBF. 2023. *Countering Hate Speech through Education and Advocacy for Improving Social Cohesion in Sri Lanka*. https://mptf.undp.org/sites/default/files/documents/2023-07/sri_lanka_00129535_irf_427_hate_speech_prevention_nce49_signed_gw.pdf.

Peters, Michael A., and Tina Besley. 2019. "Critical Philosophy of the Postdigital." *Postdigital Science and Education* 1 (1): 29–42. https://doi.org/10.1007/s42438-018-0004-9.

Piccolino, Giulia. 2015. "Winning Wars, Building (Illiberal) Peace? The Rise (and Possible Fall) of a Victor's Peace in Rwanda and Sri Lanka." *Third World Quarterly* 36 (9): 1770–1785. https://doi.org/10.1080/01436597.2015.1058150.

Pinaud, Clemence. 2014. "South Sudan: Civil War, Predation and the Making of a Military Aristocracy." *African Affairs* 113 (451): 192–211. https://doi.org/10.1093/afraf/adu019.

Price, Monroe, and Nicole Stremlau. 2012. "Media and Transitional Justice: Toward a Systematic Approach." *International Journal of Communication* 6 (2012), 1077–1099.

Radio Tamazuj. 2024. "Cybercrimes and Computer Misuse Bill 2024 Tabled in Parliament." *Radio Tamazuj*, September 2. https://www.radiotamazuj.org/en/news/article/cybercrimes-and-computer-misuse-bill-2024-tabled-in-parliament.

Ragnedda, Massimo. 2016. *The Third Digital Divide: A Weberian Approach to Digital Inequalities.* London: Routledge. https://doi.org/10.4324/9781315606002.

Rahmaty, Masooma, and Jasmine Jaghab. 2020. "Peacebuilding during a Pandemic: Keeping the Focus on Women's Inclusion." International Peace Institute, https://www.ipinst.org/2020/09/peacebuilding-during-a-pandemic-keeping-the-focus-on-womens-inclusion.

Rainie, Janna Anderson and Lee. 2023. "Themes: The Best and Most Beneficial Changes in Digital Life That Are Likely by 2035." *Pew Research Center*, June 21. https://www.pewresearch.org/internet/2023/06/21/themes-the-best-and-most-beneficial-changes-in-digital-life-that-are-likely-by-2035/.

Rammert, Werner. 2012. "Distributed Agency and Advanced Technology: Or: How to Analyze Constellations of Collective Inter-Agency." In *Agency without Actors? New Approaches to Collective Action*, edited by Jan-Hendrik Passoth, Birgit Peuker, and Michael Schillmeier. London: Routledge.

Ramsbotham, Oliver. 2010. *Transforming Violent Conflict.* 1st ed. London; New York: Routledge.

Randazzo, Elisa. 2016. "The Paradoxes of the 'Everyday': Scrutinising the Local Turn in Peace Building." *Third World Quarterly* 37 (8): 1351–1370. https://doi.org/10.1080/01436597.2015.1120154.

Reeve, Whitham D. 2003. "Voiceband Data Communications." In *Encyclopedia of Physical Science and Technology (Third Edition)*, edited by Robert A. Meyers, 529–558. New York: Academic Press. https://doi.org/10.1016/B0-12-227410-5/00814-0.

Reilly, Paul. 2022. "Parallel Lives Essay on Social Media and Divisions in Northern Ireland." *Dr Paul Reilly*, December 12. https://drpauljreilly.com/2022/12/12/parallel-lives-essay-on-social-media-and-divisions-in-northern-ireland/.

Remesh. 2024. "Percent Agree: Identify the Most and Least Agreed upon Open-End Responses." https://help.remesh.ai/percent-agree.

Remesh. n.d. *How Remesh Works.* Accessed January 24, 2025. https://4256459.fs1.hubspotusercontent-na1.net/hubfs/4256459/One%20Pagers/How%20Remesh%20Works.pdf.

Verité Research. 2013. *Devolving Land Powers. A Guide for Decision Makers.* https://www.veriteresearch.org/wp-content/uploads/2018/06/Devolving-Land-Powers_A-Guide-for-Decision-makers.pdf.

Reuters. 2021. "Internet Disrupted, Streets Quiet in South Sudan after Call for Protests." *The Standard*, August 30. http://www.standardmedia.co.ke/africa/article/2001422182/internet-disrupted-streets-quiet-in-south-sudan-after-call-for-protests.

Reynolds, Louis, and Lucie Parker. 2018. *Digital Resilience: Stronger Citizens Online.* Institute for Strategic Dialogue (ISD).

Richards, Imogen, and Mark Wood. 2020. "Legal and Security Frameworks for Responding to Online Violent Extremism: A Comparison of Far-Right and Jihadist Contexts." In *The Handbook of Collective Violence.* London: Routledge.

Richmond, Oliver P. 2010. "A Genealogy of Peace and Conflict Theory." In *Peacebuilding. Critical Develpments and Approaches.*, edited by Oliver P. Richmond. Palgrave Advances in Peacebuilding. Basingstoke: Palgrave Macmillan. https://link.springer.com/book/10.1057/9780230282681.

Richmond, Oliver P., and Ioannis Tellidis. 2020. "Analogue Crisis, Digital Renewal? Current Dilemmas of Peacebuilding." *Globalizations* 17 (6): 935–952. https://doi.org/10.1080/14747731.2020.1712169.

Richmond, Oliver P., and Gëzim Visoka, eds. 2021. *The Oxford Handbook of Peacebuilding, Statebuilding, and Peace Formation*. Oxford: Oxford University Press.

Richmond, Oliver P., Gëzim Visoka, and Ioannis Tellidis. 2023. *Peace in Digital International Relations: Prospects and Limitations*. 1st ed. Cambridge: Cambridge University Press. https://doi.org/10.1017/9781009396752.

Riebe, Thea, and Christian Reuter. 2019. "Dual-Use and Dilemmas for Cybersecurity, Peace and Technology Assessment." In *Information Technology for Peace and Security: IT Applications and Infrastructures in Conflicts, Crises, War, and Peace*, edited by Christian Reuter, 165–183. Wiesbaden: Springer Fachmedien. https://doi.org/10.1007/978-3-658-25652-4_8.

Roche, Stephane, Eliane Propeck-Zimmermann, and Boris Mericskay. 2013. "GeoWeb and Crisis Management: Issues and Perspectives of Volunteered Geographic Information." *GeoJournal* 78 (1): 21–40. https://doi.org/10.1007/s10708-011-9423-9.

Roessler, P., and M. Howard. 2009. "Post-Cold War Political Regimes: When Do Elections Matter?" In *Democratization by Elections: A New Mode of Transition*, edited by Staffan Lindberg. Baltimore: Johns Hopkins University Press.

Rohwerder, B. 2015. *Conflict early warning and early response (GSDRC Helpdesk Research Report 1195)*. Birmingham, UK: GSDRC, University of Birmingham.

Rueden, Laura von, Sebastian Mayer, Katharina Beckh, Bogdan Georgiev, Sven Giesselbach, Raoul Heese, Birgit Kirsch, Julius Pfrommer, Annika Pick, Rajkumar Ramamurthy, Michal Walczak, Jochen Garcke, Christian Bauckhage, Jannis Schuecker. 2020. "Informed Machine Learning—A Taxonomy and Survey of Integrating Knowledge into Learning Systems." *arXiv:1903.12394 [Cs, Stat]*, February 12. http://arxiv.org/abs/1903.12394.

Rupesinghe, K., and M. Kuroda, (Eds.). 1992. Early warning and conflict resolution (1. publ). St. Martin's Press.

Rupesinghe, Kumar. 1987. "The Quest for a Disaster Early Warning System: Giving a Voice to the Vulnerable." *Bulletin of Peace Proposals* 18 (2): 217–227.

Rupesinghe, Kumar. 1989. "Early Warnings: Some Conceptual Problems." *Bulletin of Peace Proposals* 20 (2): 183–191.

Rustad, Siri Aas. 2024. "Conflict Trends: A Global Overview." Oslo: Peace Research Institute Oslo (PRIO). https://cdn.cloud.prio.org/files/92a7aad5-3572-4886-9e9c-8aa155f1d0f4/Conflict_Trends-2024_DIGITAL.pdf?inline=true.

Rutherford, Alex. 2021. "What Machines Can Teach Us about How to Write Constitutions." Medium, January 18. https://medium.com/futuring-peace/what-machines-can-teach-us-about-how-to-write-constitutions-3d675e33533b.

Ryberg, Thomas. 2019. "Dragging the Digital Into the Mud." In *Postdigital Dialogue*, vol. 1. Postdigital Science and Education. https://doi.org/10.1007/s42438-018-0011-x.

Sabaratnam, Meera. 2013. "Avatars of Eurocentrism in the Critique of the Liberal Peace." *Security Dialogue* 44 (3): 259–278. https://doi.org/10.1177/0967010613485870.

Sæbø, Øystein, Jeremy Rose, and Leif Skiftenes Flak. 2008. "The Shape of eParticipation: Characterizing an Emerging Research Area." *Government Information Quarterly* 25 (3): 400–428. https://doi.org/10.1016/j.giq.2007.04.007.

Samaraweera, Buddhika. 2024. "Ethnic Issue: New Constitutional Solution Mooted Post-P'ment Poll." *The Morning*, October 2. https://themorning.lk//articles/Nm8T1ocq2C7jS hgFSO8z, Accessed October 2, 2024.

Sammut, Claude, and Geoffrey I. Webb, eds. 2010. "TF–IDF." In *Encyclopedia of Machine Learning*. Boston, MA: Springer US. https://doi.org/10.1007/978-0-387-30164-8_832.

Sanjana. 2005. "Beyond ICT: Designing Cutting Edge ICT Solutions for South Asia." *Sanjana Hattotuwa* (blog). July 28. https://sanjanah.wordpress.com/2005/07/28/beyond-ict-designing-cutting-edge-ict-solutions-for-south-asia/.

Savun, Burcu. 2008. "Information, Bias, and Mediation Success." *International Studies Quarterly* 52 (1): 25–47. https://doi.org/10.1111/j.1468-2478.2007.00490.x.

Sayes, Edwin. 2014. "Actor-Network Theory and Methodology: Just What Does It Mean to Say That Nonhumans Have Agency?" *Social Studies of Science* 44 (1): 134–149.

Scheible, Kristin. 2016. *Reading the Mahāvamsa: The Literary Aims of a Theravada Buddhist History*. New York: Columbia University Press.

Schierenbeck, Isabell. 2015. "Beyond the Local Turn Divide: Lessons Learnt, Relearnt and Unlearnt." *Third World Quarterly* 36 (5): 1023–1032. https://doi.org/10.1080/01436597.2015.1043991.

Schirch, Lisa. 2020. "25 Spheres of Digital Peacebuilding and PeaceTech." no. 93, 28. Toda Peace Institute and Alliance for Peacebuilding.

Schirch, Lisa. 2022. *The Digital Space and Peace Processes A Thought Piece*. ICT4Peace Foundation. https://ict4peace.org/wp-content/uploads/2022/06/P4P-Digital-Space-and-Peace-Processes-v2.pdf.

Schirch, Lisa. 2024. *Policy Brief. Defending Democracy with Deliberative Technology*. Report. University of Notre Dame. https://doi.org/10.7274/25338103.v1.

Schwab, Klaus, and Saadia Zahidi. 2020. *How Countries Are Performing on the Road to Recovery*. The Global Competitiveness Report/World Economic Forum, Special edition 2020. World Economic Forum.

Schweiger, Elisabeth, and Kerstin Tomiak. 2022. "Researching Silence: A Methodological Inquiry." *Millennium* 50 (3): 623–646. https://doi.org/10.1177/03058298221083999.

Scognamillo, Corrado, and Jonas Gutschke. 2021. "Understanding the World through Data—UNDP Crisis Risk Dashboard." United Nations World Data Forum. https://unstats.un.org/unsd/undataforum/blog/undp-crisis-risk-dashboards/.

UN Secretary-General. 2001. *Prevention of Armed Conflict: Report of the Secretary-General*. June 7. https://digitallibrary.un.org/record/442540.

UN Secretary-General. 2019. *United Nations Strategy and Plan of Action on Hate Speech*. United Nations. https://www.un.org/en/genocideprevention/documents/advising-and-mobilizing/Action_plan_on_hate_speech_EN.pdf.

UN Secretary-General. 2020. *Detailed Guidance on Implementation of the UN Strategy and Plan of Action for United Nations Field Presences*. United Nations. https://www.un.org/en/genocideprevention/documents/UN%20Strategy%20and%20PoA%20on%20Hate%20Speech_Guidance%20on%20Addressing%20in%20field.pdf.

UN Secretary-General. 2022. "#NoToHate Fact Sheet." https://www.un.org/sites/un2.un.org/files/notohate_fact_sheets_en.pdf.

Sellars, Andrew. 2016. "Defining Hate Speech." SSRN Scholarly Paper 2882244. Rochester, NY, December 1. https://doi.org/10.2139/ssrn.2882244.

Sending, Ole Jacob, and Iver B. Neumann. 2006. "Governance to Governmentality: Analyzing NGOs, States, and Power." *International Studies Quarterly* 50 (3): 651–672.

SentiOne. 2024. "Conversational AI Platform." SentiOne. https://sentione.com/.

Seoighe, Rachel. 2016. "Inscribing the Victor's Land: Nationalistic Authorship in Sri Lanka's Post-War Northeast." *Conflict, Security & Development* 16 (5): 443–471. https://doi.org/10.1080/14678802.2016.1219507.

SEUPB. 2021. *PEACEPLUS Programme 2021 – 2027 Programme Overview*. Special EU Programmes Body.

Shepherd, Jason. 2023. "Digital Twins and Public Policy." https://www.techuk.org/resource/digital-twins-and-public-policy.html.

Shevlin, Henry, Karina Vold, Matthew Crosby, and Marta Halina. 2019. "The Limits of Machine Intelligence." *EMBO Reports* 20 (10): e49177. https://doi.org/10.15252/embr.201949177.

Shiel, Gerry, and Lorraine Gilleece. 2015. "Literacy and Numeracy in Northern Ireland and the Republic of Ireland in International Studies." *The Irish Journal of Education/Iris Eireannach an Oideachais* 40: 3–28.

Shinko, Rosemary E. 2008. "Agonistic Peace: A Postmodern Reading." *Millennium: Journal of International Studies* 36 (3): 473–491. https://doi.org/10.1177/03058298080360030501.

Shinko, Rosemary E. 2022. "A Critical (Re)Reading of the Analytical Significance of Agonistic Peace." *Third World Quarterly* 43 (6): 1399–1407. https://doi.org/10.1080/01436597.2022.2042681.

Shumpert, Nirmeen. 2018. "Summary: 'The Danger of a Single Story' | Department of English." https://english.umd.edu/research-innovation/journals/interpolations/fall-2018/summary-danger-single-story.

Silva, Raquel da. 2018. "Memory Politics, Statebuilding and Social Movement Identity." *Journal of Intervention and Statebuilding* 12 (3): 435–441. https://doi.org/10.1080/17502977.2017.1416771.

De Simone, Sara. 2014. "Post-Conflict Decentralization: Dynamics of Land and Power in Unity State—South Sudan." *UNISCI Discussion Papers* UNISCI Discussion Papers, Nº 33: 35–55. https://doi.org/10.5209/rev_UNIS.2013.n33.44814.

Sinclair, Christine, and Sarah Hayes. 2019. "Between the Post and the Com-Post: Examining the Postdigital 'Work' of a Prefix." *Postdigital Science and Education* 1 (1): 119–131. https://doi.org/10.1007/s42438-018-0017-4.

Singer, Joel David, and Melvin Small. 1972. *The Wages of War, 1816–1965: A Statistical Handbook*. New York: Wiley.

Skaar, Elin. 2012. "Reconciliation in a Transitional Justice Perspective." SSRN Scholarly Paper No. 4168965. Social Science Research Network. https://papers.ssrn.com/abstract=4168965.

Small Arms Survey. n.d. "Context." Accessed January 15, 2025. https://www.smallarmssurvey.org/upper-nile-prepares-return-war/context.

Smith, Claire Q. 2014. "Illiberal Peace-Building in Hybrid Political Orders: Managing Violence during Indonesia's Contested Political Transition." *Third World Quarterly* 35 (8): 1509–1528. https://doi.org/10.1080/01436597.2014.946277.

Snodderly, Dan, ed. 2011. *Peace Terms. Glossary of Terms for Conflict Management and Peacebuilding*. Washington D.C. https://www.academia.edu/34176684/Peace_terms.

SparkBlue, dir. 2021. *Monthly CRD Webinar Sri Lanka*. 1:04:06. https://www.youtube.com/watch?v=K04OvAPXkOQ, Accessed May 21, 2024.

Special Rapporteur on the promotion and protection of the right to freedom of opinion and expression. 2022. *A/77/288: Disinformation and Freedom of Opinion and Expression during Armed Conflicts—Report of the Special Rapporteur on the Promotion and Protection of the Right to Freedom of Opinion and Expression*. United Nations. https://www.ohchr.org/en/documents/thematic-reports/a77288-disinformation-and-freedom-opinion-and-expression-during-armed.

Spiller, Neil. 2009. "Plectic Architecture: Towards a Theory of the Post-Digital in Architecture." *Technoetic Arts: A Journal of Speculative Research* 7 (2): 95–104. https://doi.org/10.1386/tear.7.2.95/1.

Standfield, Catriona. 2020. "Caught between Art and Science: The Women, Peace and Security Agenda in United Nations Mediation Narratives." *International Feminist Journal of Politics* 22 (5): 629–651. https://doi.org/10.1080/14616742.2020.1716632.

Stauffacher, Daniel, William J. Drake, Paul Currion, and Julia Steinberger, UN Information and Communication Technologies Task Force, and ICT4Peace Initiative. 2005. "Information and Communication Technology for Peace: The Role of ICT in Preventing, Responding to and Recovering from Conflict /: By Daniel Stauffacher, William Drake, Paul Currion and Julia Steinberger; Preface by Kofi Annan; Foreword by Micheline Calmy-Rey." In . UN; https://digitallibrary.un.org/record/561572.

Steibel, Fabro, and Elsa Estevez. 2015. "Designing Web 2.0 Tools for Online Public Consultation." In *Impact of Information Society Research in the Global South*, edited by Arul Chib, Julian May, and Roxana Barrantes, 243–263. Singapore: Springer. https://doi.org/10.1007/978-981-287-381-1_13.

Strömbom, Lisa, and Isabel Bramsen. 2022. "Agonistic Peace: Advancing Knowledge on Institutional Dynamics and Relational Transformation." *Third World Quarterly* 43 (6): 1237–1250. https://doi.org/10.1080/01436597.2022.2054797.

Strömbom, Lisa, Isabel Bramsen, and Anne Lene Stein. 2022. "Agonistic Peace Agreements? Analytical Tools and Dilemmas." *Review of International Studies* 48 (4): 689–704. https://doi.org/10.1017/S0260210522000055.

Sudan Tribune. 2013. "Kiir Urges Traditional Leaders to Put National Interest above All Else." *Sudan Tribune*, June 16. https://sudantribune.com/article46073/, Accessed July 13, 2024.

Suransky, A. C., and H. A. Alma. 2017. "An Agonistic Model of Dialogue." *Journal of Constructivist Psychology*, ahead of print, March 31. https://doi.org/10.1080/10720537.2017.1298487.

Taffel, Sy. 2016. "Perspectives on the Postdigital: Beyond Rhetorics of Progress and Novelty." *Convergence* 22 (3): 1–15. https://doi.org/10.1177/1354856514567827.

Tellidis, Ioannis. 2020. "Technology and Peace." In *The Palgrave Encyclopaedia of Peace and Conflict Studies*, edited by Oliver P. Richmond and Gezim Visoka. Cham: Palgrave Macmillan. https://doi.org/10.1007/978-3-030-11795-5_26-1.

Tellidis, Ioannis, and Stefanie Kappler. 2016. "Information and Communication Technologies in Peacebuilding: Implications, Opportunities and Challenges." *Cooperation and Conflict* 51 (1): 75–93. https://doi.org/10.1177/0010836715603752.

Thompson, Mark, and Geoff Walsham. 2010. "ICT Research in Africa: Need for a Strategic Developmental Focus." *Information Technology for Development* 16 (2): 112–127. https://doi.org/10.1080/02681101003737390.

Timmermann, Wibke. 2008. "Counteracting Hate Speech as a Way of Preventing Genocidal Violence." *Genocide Studies and Prevention* 3 (3): 353–374. https://doi.org/10.3138/gsp.3.3.353.

Tontodimamma, Alice, Eugenia Nissi, Annalina Sarra, and Lara Fontanella. 2021. "Thirty Years of Research into Hate Speech: Topics of Interest and Their Evolution." *Scientometrics* 126 (1): 157–179. https://doi.org/10.1007/s11192-020-03737-6.

Townsend, Emma. 2014. "Hate Speech or Genocidal Discourse? An Examination of Anti-Roma Sentiment in Contemporary Europe." *PORTAL Journal of Multidisciplinary International Studies* 11 (1): 1. https://doi.org/10.5130/portal.v11i1.3287.

Trappl, Robert. 2006. *Programming for Peace Computer-Aided Methods for International Conflict Resolution and Prevention*. Advances in Group Decision and Negotiation. Dordrecht: Springer.

Trappl, Robert. 2021. "Civil Society Coalition Demand for Change in South Sudan." *Sudan Tribune*, August 1. https://sudantribune.com/article184482/, Accessed April 15, 2024.

Tzavaras, Johann. 1989. "Heideggers Unterscheidung Zwischen Zuhandenheit Und Vorhandenheit." *Philosophisches Jahrbuch* 96 (2): 367.

UN News. 2014. *UN Mission in South Sudan Condemns Ethnic Killings in Bentiu | UN News*. April 21. https://news.un.org/en/story/2014/04/466522.

UN Office on Genocide Prevention. 2023. *Countering and Addressing Online Hate Speech: A Guide for Policy Makers and Practitioners*. United Nations. https://www.un.org/en/genocideprevention/documents/publications-and-resources/Countering_Online_Hate_Speech_Guide_policy_makers_practitioners_July_2023.pdf.

UNDP. 2022. "UNDP - Digital Strategy 2022–2025." https://digitalstrategy.undp.org/.

UNDP. 2022. *From Pilots to Practice: Risk-Informed Utilization of Online Data for Preventing Violent Extremism and Addressing Hate Speech*. UNDP. https://www.undp.org/sites/g/files/zskgke326/files/2022-05/UNDP-Guidance-Note-Risk-Informed-Utilization-of-Online-Data-for-PVE-and-Addressing-Hate-Speech.pdf.

UNDP. 2022. "Ideating on Conflict Early Warning and Response Systems." https://www.undp.org/south-sudan/ideating-conflict-early-warning-and-response-systems.

UNDP and SfCG. 2021. *IDENTIFYING, COUNTERING & REPORTING ONLINE HATE SPEECH*. https://www.sfcg.org/identifying-countering-reporting-online-hate-speech/.

UNDP South Sudan. 2021. *AlertMe Mobile App Wins Youth Innovation Competition to Strengthen Conflict Early Warning and Response in South Sudan*. https://www.undp.org/south-sudan/press-releases/alertme-mobile-app-wins-youth-innovation-competition-strengthen-conflict-early-warning-and-response-south-sudan, Accessed May 08, 2024.

UNDPPA and HD Centre. 2019. "Digital Technologies and Mediation in Armed Conflict." Geneva: United Nations Department of Political and Peacebuilding Affairs and Centre for Humanitarian Dialogue.

UNGA. 2022. "Countering Disinformation for the Promotion and Protection of Human Rights and Fundamental Freedoms A/77/287."

United Nations. 2012. *UN Guidance for Effective Mediation*. http://peacemaker.un.org/sites/peacemaker.un.org/files/GuidanceEffectiveMediation_UNDPA2012%28english%29_0.pdf.

United Nations. 2020a. "Amid COVID-19 Pandemic, Coordinated, Conflict-Sensitive Responses Crucial to Sustaining Peace, Secretary-General Tells Security Council." 2020. https://press.un.org/en/2020/sc14275.doc.htm.

United Nations. 2020b. *Security Council Endorses Conclusions of Berlin Conference on Libya, Adopting Resolution 2510 (2020) by 14 Votes in Favour, 1 Abstention | Meetings Coverage and Press Releases*. February 12. https://press.un.org/en/2020/sc14108.doc.htm.

United Nations. 2023. What is hate speech? United Nations. United Nations. https://www.un.org/en/hate-speech/understanding-hate-speech/what-is-hate-speech, Accessed April 26, 2026.

United Nations Peacekeeping. 2011. "Strategy for the Digital Transformation of UN Peacekeeping." New York: United Nations. https://peacekeeping.un.org/sites/default/files/strategy-for-the-digital-transformation-of-un-peacekeeping_en_final-01_15-08-2021_final.pdf.

University of Minnesota Law School Human Rights Clinic. n.d. *Responses to Questionnaire for Report on Anti-Muslim Hatred and Discrimination*. Accessed April 15, 2025. https://www.ohchr.org/sites/default/files/Documents/Issues/Religion/Islamophobia-AntiMuslim/Civil%20Society%20or%20Individuals/HRLitigationandInternationalLegalAdvocacyClinicMinnesota.pdf.

UNMISS. 2022. *South Sudan: UN Report Says 173 Civilians Killed in Four Months in Southern Unity State amid Continued Fighting*. Press Release. UMISS.

UNMISS. 2023. "Inter-School Cultural Exchanges Facilitated by UNMISS Seeks to Strengthen Social Cohesion." UNMISS, March 22. https://unmiss.unmissions.org/inter-school-cultural-exchanges-facilitated-unmiss-seeks-strengthen-social-cohesion.

UNMISS and OHCHR. 2022. "Attacks against Civilians in Southern Unity State, South Sudan (February–May 2022)." September 6. https://reliefweb.int/report/south-sudan/attacks-against-civilians-southern-unity-state-south-sudan-february-may-2022.

UNSC. 2020. *Letter Dated 28 April 2020 from the Panel of Experts on South Sudan Addressed to the President of the Security Council (S/2020/342)*. United Nations. https://www.securitycouncilreport.org/atf/cf/%7B65BFCF9B-6D27-4E9C-8CD3-CF6E4FF96FF9%7D/S_2020_342.pdf.

UNSC. 2022. "Technology and Security: Briefing: What's In Blue: Security Council Report." 2022. https://www.securitycouncilreport.org/whatsinblue/2022/05/technology-and-security-briefing.php.

UNSC. 2023. *S/RES/2686 (2023)*. https://press.un.org/en/2023/sc15321.doc.htm.

UNSCOL. 2020. "Transcript of UN Secretary-General's Virtual Press Encounter on the Appeal for Global Ceasefire." UNSCOL. 2020. https://unscol.unmissions.org/transcript-un-secretary-generals-virtual-press-encounter-appeal-global-ceasefire, Accessed March 14, 2025.

UNSG. 1992. *An Agenda for Peace: Preventive Diplomacy, Peacemaking and Peace-Keeping.* https://unesdoc.unesco.org/ark:/48223/pf0000092330.

UNSMIL. 2020, October 10. UNSMIL Statement on the resumption of intra-Libyan political and military talks. UNSMIL. https://unsmil.unmissions.org/unsmil-statement-resumption-intra-libyan-political-and-military-talks, Accessed November 29, 2022.

UNSMIL. 2021. "Libyan Political Dialogue Forum." UNSMIL. https://unsmil.unmissions.org/libyan-political-dialogue-forum.

UNU CPR. 2024. *2024 PBF Thematic Review: Synergies between Human Rights and Peacebuilding in PBF-Supported Programming.* https://www.un.org/peacebuilding/sites/www.un.org.peacebuilding/files/documents/2024_pbf_thematic_review_hires_0.pdf.

USIP. 2022. "From Factionalism to Foreign Interference: Libya's Conflict Remains Frozen." United States Institute of Peace. https://www.usip.org/publications/2022/11/factionalism-foreign-interference-libyas-conflict-remains-frozen, access date: 29.11.2022.

Varela, Miguel. 2021. "Technology and Peace Mediation: A Shift in the Mediator's Profession." In *Rethinking Peace Mediation: Challenges of Contemporary Peacemaking Practice*, edited by Catherine Turner. Bristol: Bristol University Press.

Verjee, Aly. 2022. "Ceasefire Monitoring under Fire: The OSCE, Technology, and the 2022 War in Ukraine." *Global Policy* 13 (5): 808–817. https://doi.org/10.1111/1758-5899.13123.

Von Einsiedel, Sebastian, Louise Bosetti, James Cockayne, Cale Salih, and Wilfred Wan. 2017. *Civil War Trends and the Changing Nature of Armed Conflict.* 10. Occasional Paper. United Nations University Centre for Policy Research. https://i.unu.edu/media/cpr.unu.edu/attachment/1558/OC_01-MajorRecentTrendsinViolentConflict.pdf.

VRA. 2018. "About Virtual Research Associates." https://www.vranet.com/Home/About.

Wachowiak, Mark. 2022. *Text Clustering Based on Tf-Idf Features.* February 28. https://ecampusontario.pressbooks.pub/nudh3/chapter/text-clustering-based-on-tf-idf-features/.

Wählisch, Martin. 2020. "Big Data, New Technologies, and Sustainable Peace: Challenges and Opportunities for the UN." *Journal of Peacebuilding & Development* 15 (1): 122–126. https://doi.org/10.1177/1542316619868984.

Wählisch, Martin. 2024. "AI and the Future of Mediation." In *Still Time to Talk*, edited by Teresa Whitfield. Accord. Conciliation Resources.

Waldron, Jeremy. 2014. *The Harm in Hate Speech.* Cambridge, MA: Harvard University Press.

Walker, Samuel. 1994. *Hate Speech: The History of an American Controversy.* Lincoln: University of Nebraska Press.

Wang, Ge. 2019. "Humans in the Loop: The Design of Interactive AI Systems." Human-Centered Artificial Intelligence. Stanford University. https://hai.stanford.edu/news/humans-loop-design-interactive-ai-systems.

Warakapitiya, Kasun. 2022. "May 9 Violence and Protests: 4000 Arrested." *The Sunday Times, Sri Lanka*, September 11. http://www.sundaytimes.lk/220911/news/may-9-violence-and-protests-4000-arrested-494822.html.

Welch, Jennifer R., Susan Halford, and Mark Weal. 2015. "Information and Communication Technologies (ICTs) and Peacebuilding: A Conceptual Framework." In *Proceedings of the ACM Web Science Conference*, 1–9. WebSci '15. Oxford: Association for Computing Machinery. https://doi.org/10.1145/2786451.2786479.

Whittaker, Joshua, Blythe McLennan, and John Handmer. 2015. "A Review of Informal Volunteerism in Emergencies and Disasters: Definition, Opportunities and Challenges." *International Journal of Disaster Risk Reduction* 13 (September): 358–368. https://doi.org/10.1016/j.ijdrr.2015.07.010.

Wild, Hannah, Jok Madut Jok, and Ronak Patel. 2018. "The Militarization of Cattle Raiding in South Sudan: How a Traditional Practice Became a Tool for Political Violence." *Journal of International Humanitarian Action* 3 (1): 2. https://doi.org/10.1186/s41018-018-0030-y.

Williams, Paul D. 2018. "Strategic Communications for Peace Operations: The African Union's Information War Against al-Shabaab." *Stability: International Journal of Security and Development* 7 (1): 3. https://doi.org/10.5334/sta.606.

Wilson, Jacqueline. 2019. *The Religious Landscape in South Sudan: Challenges and Opportunities for Engagement.* Washington D.C: United States Institute of Peace.

Wirasinha, Pranith. 2022. "The Disappeared and the Struggle of Love." *Groundviews*, June 5. https://groundviews.org/2022/06/05/the-disappeared-and-the-struggle-of-love/.

Wolff, Jonas. 2009. "De-Idealizing the Democratic Civil Peace: On the Political Economy of Democratic Stabilization and Pacification in Argentina and Ecuador." *Democratization* 16 (5): 998–1026. https://doi.org/10.1080/13510340903162143.

Wong, Julia Carrie, and Luke Harding. 2021. "'Facebook Isn't Interested in Countries like Ours': Azerbaijan Troll Network Returns Months after Ban." Technology. *The Guardian*, April 13. https://www.theguardian.com/technology/2021/apr/13/facebook-azerbaijan-ilham-aliyev.

World Bank. 2024. "World Development Indicators, Individuals Using the Internet (% of Population)." https://data.worldbank.org/indicator/IT.NET.USER.ZS.

Wright, Dawn J. 2016. "Toward a Digital Resilience." *Elementa: Science of the Anthropocene* 4: 000082. https://doi.org/10.12952/journal.elementa.000082

Wulf, Herbert, and Tobias Debiel. 2010. "Systemic Disconnects: Why Regional Organizations Fail to Use Early Warning and Response Mechanisms." Global Governance: A Review of Multilateralism and International Organizations. *Global Governance: A Review of Multilateralism and International Organizations* 16 (4): 525–547. https://doi.org/10.1163/19426720-01604007.

Yadav, Pavinder, Nidhi Gupta, and Pawan Kumar Sharma. 2023. "A Comprehensive Study towards High-Level Approaches for Weapon Detection Using Classical Machine Learning and Deep Learning Methods." *Expert Systems with Applications* 212 (February): 118698. https://doi.org/10.1016/j.eswa.2022.118698.

Yilmaz, Ihsan, Raja M. Ali Saleem, Mahmoud Pargoo, Syaza Shukri, Idznursham Ismail, and Kainat Shakil. 2022. *Religious Populism, Cyberspace and Digital Authoritarianism in Asia*: India, Indonesia, Malaysia, Pakistan, and Turkey. Report. Deakin University. https://dro.deakin.edu.au/articles/report/Religious_Populism_Cyberspace_and_Digital_Authoritarianism_in_Asia_India_Indonesia_Malaysia_Pakistan_and_Turkey/20621349?file=36822879.

Zelin, Aaron Y. 2015. "Picture Or It Didn't Happen: A Snapshot of the Islamic State's Official Media Output." *Perspectives on Terrorism*, 9 (4): 85–97.

Zuboff, Shoshana. 2015. "Big Other: Surveillance Capitalism and the Prospects of an Information Civilization." *Journal of Information Technology* 30 (1): 75–89. https://doi.org/10.1057/jit.2015.5.

Index

For the benefit of digital users, indexed terms that span two pages (e.g., 52–53) may, on occasion, appear on only one of those pages.

Figures are indicated by an italic *f*.

Activism, 8–9
digital, 36–37, 221–222
left-wing, 44–45
peace, 10–11, 219, 227
political, 8–9, 77, 215
social media, 77, 213, 219, 225
Addis Ababa, 121, 127–128
African National Congress (ANC), 34
African Union (AU), 44–45, 118–119, 121, 123, 127–129
African Peace and Security Architecture (APSA), 121
Continental Early Warning System (CEWS), 121–123, 127–129
Security Council (AUSC), 6–7
Agency, 51–52, 57–59
human, 57–59, 64, 69, 155, 158–160, 170–171, 186–187, 192
machine, 57, 158–160, 170, 186–187
relational, 51–52
Agonistic peace (theory), 12–13, 177, 189, 225–226, 230, 232
AI *see* Artificial Intelligence
Al-Gaddafi, Muammar, 165–166
Al-Qaeda, 87
AlertMe Application, 131, 133–134, 141–143, 151
Amazon, 38, 175
Analogue *vs.* digital, 9, 26–27, 52, 54, 78–79, 233
ANC *see* African National Congress
Annan, Kofi, 31, 119–120
Apartheid, 34
APSA *see* African Union, African Peace and Security Architecture
Apomediaries, 14–15, 62–63
integrative, 191, 192–193, 194–202, 208, 212–213, 216, 218–219, 220–223
programmatic, 191, 193, 216
reformative, 191, 194, 209–210, 216
Apomediation, 14–18, 42, 51–52, 62–69
challenges intermediaries, 65–66
digressive, 199–202
meaning-making, 64–65
regressive, 202–210
relation-making, 64–65
transformative, 191–192, 210–219
Armed conflict, 5–6, 9–10, 13, 14–17, 23–24, 26, 33–34, 42–43, 63, 66, 69–70, 79, 84, 126–127, 158–160, 162–163, 180–181, 186–187, 189, 190–191, 227–228
Artificial Intelligence (AI), 5–6, 26, 40, 57, 94. 118–119, 153
chatbots, 25, 159, 162, 232
generative, 157, 231–232, 233–235
generative pre-trained transformers (GPTs), 44, 162, 231–232
recommender systems, 175–176
training data, 24, 158, 162–163, 232
transparency, 161–162, 186–187, 232, 235
AU *see* African Union
AUSC *see* African Union, Security Council

Belfast, 132–134, 135–136, 201–202
Bidirectional encoder representations from transformers (BERT), 161–162
Bismarck, Otto von, 2–3
Brexit, 79, 199–202
Bruderlein, Claude, 162
Build Up, 35–37
Bureaucracy, 5, 30–31, 41, 46, 62, 125, 154–155, 165

Capitalism, 3–4, 36
CDC *see* Community Development Centre
Censorship, 17–18, 21–22, 72, 77–79, 101, 198–199, 202–203, 206*f*, 213–214
self-censorship, 205–206, 206*f*, 213–214, 215–216
Center for Policy Alternatives (CPA), 32–33
Cerf, Vinton, 36–38
CEWARN *see* Intergovernmental Authority for Development, Conflict Early Warning and Response Mechanism
CEWERU *see* South Sudan, Conflict Early Warning and Early Response Unit

CEWS *see* Continental Early Warning System
ChatGPT, 44, 158, 162
Civil society organizations (CSOs), 15–16, 19, 29, 34–35, 60–61, 63, 79–80, 91–92, 100–102, 110, 112–113, 159, 205–208, 212–213
Civil society, 5, 22, 26, 54–55, 71, 97–98, 101, 109–110, 126, 146, 153, 193–194, 196, 200, 203, 212–213
  activism, 34, 77–78, 80–81, 132–133, 135, 145–146, 148–149, 156, 208, 213–216, 218
  groups, 23–24, 75–76, 146, 213–214
  initiatives, 132, 148–149, 151, 190, 225
  networks, 192
Clientelism, 103–104, 203
  ethnic, 208
CMI-Martti Ahtisaari Peace Foundation, 180–186
Co-production, 59–60
Colombo, 219
Colonial legacies, 6–7, 15, 33–34, 39–40, 53, 60, 107–108, 121, 199–200
Colonialism, 6–7, 33–34, 53, 60, 81, 107–108, 121, 181, 199–200
Conflict transformation, 8–13, 48–49, 51–52, 80–81, 84–85, 172–173, 178, 180–181, 185–186, 188, 189–190, 192–193, 195–196, 198–200, 211–212, 220–222, 224–225, 229–230, 232–233, 235
COVID-19 pandemic, 1, 23–24, 29, 45–46, 98–99, 103, 153–155, 156–157, 160, 165–166, 182
CPA *see* Center for Policy Alternative
Critical mass, 195–199, 215–216
Critical theory (following Robert Cox), 17, 46–47, 231–232
Critical-reflexivity, 14, 17, 51, 60, 85–86, 154–155, 227–228
Crowdsourcing, 5, 14, 22, 35, 37, 80–81, 116–117, 129–133, 225–226
CrowdTangle, 95–96
CSOs *see* Civil society organizations
Cybersecurity, 31–32, 76–78
Cyberspace, 31–32, 56

Data-driven approaches, 4–5, 113, 126, 135, 151, 155, 158, 170–171*
*Data, Big Data, 25, 35, 122–123
*Data, dashboards, 10–11, 81, 113–117, 229–230
*Data, synthetic, 174–178, 232–235
*Data, visualization, 10–11, 67, 94, 112–117, 156–157, 184
DD&R *see* Digital divide and rule
Decolonial thought, 6–7, 10–11, 39–40, 48, 117–122
Defy-HateNow, 93–95, 98–99, 102–103
Diaspora, 75–76, 89, 103–104, 109–110, 194, 196, 203, 207, 218
Digi-political playing field, 18, 23–24, 69–75, 73*f*, 149–150, 188, 195–196, 197*f*, 200, 210, 213, 224
  adaptability, 18, 69–70, 71–72, 78–79, 98, 191–192
  slantedness of the, 22, 69–70, 70*f*, 85–86, 110, 118–119, 151–152, 196–197, 222, 225
Digital applications, 4–6, 7–8, 14, 28, 42–44, 45–46, 131, 158–160, 178, 232
Digital dialogues, 4, 13, 24, 25–26, 42–43, 47–48
Digital divide (or gap), 32–33, 54–55, 71–72
Digital divide and rule (DD&R), 23–24, 82–83, 190–192, 194–195, 196–198, 220–223, 227
Digital escapism, 3–4
Digital innovation, 4, 16–17, 20, 25–26, 30, 32, 40, 48–49, 53, 55, 118–119, 163–164, 186
Digital literacy, 17–18, 21–22, 54, 72–74, 75–76, 78, 93–94, 97–99, 110–111, 179, 203–204, 210, 214–215
Digital peacebuilding, 3–6, 26–28, 31–32, 35–39, 60
  Long durée, 1–2, 21–22, 51–53
Digital revolution, 5–6, 8, 25–27, 45–46, 51–53, 55
Digital technologies, 1–3, 4–5, 7–9, 10–11, 14, 16–18, 25–27, 28–29, 39–40, 43, 46–49, 52–53, 61, 77, 90–91, 116–117, 129, 155, 186, 224–225
Digital transformation, 26, 43
Digital twins, 234–235
Digitalization of peacebuilding, 26–27, 28–30
Disinformation, 2–3, 22, 56–57, 83–84, 88–91, 97–98, 100–102, 103–104, 110–111, 124, 211, 233
DPPA *see* United Nations, Department of Political and Peacebuilding Affairs

Early warning, 4–5, 13, 15–16, 22, 28, 31, 35, 45, 47–48, 54–55, 60–61, 62, 64–65, 76, 80–81, 112, 159–160, 192, 224–230, 234–235
  response gap, 128–130, 132, 139–140, 226–227
  field monitors, 126–128
Empiricism, 27, 171–173
Empowerment, 5, 164, 178, 180–181, 201–202, 212, 225–226
Enframing (concept), 17, 21, 27–28, 42–45, 116–117, 151, 228–230
EU *see* European Union
Eurocentrism, 10, 33–34, 55, 138–139
European Union (EU), 79, 199–201
Evidence-based approaches, 110, 114, 116, 134–135, 171

Facebook, 61–62, 66–67, 86, 89, 93–95, 96–97, 122–123, 215–216, 233
Fact-checking (debunking), 97–104, 108–110, 201, 233
FAST *see* Früh-Analyse von Spannungen und Tatsachenermittlung
FEWER *see* Forum for Early Warning and Early Action
Finland, 30, 32–33
Forum for Early Warning and Early Action (FEWER), 119–120, 122, 124–125, 128–129
France, 2
Früh-Analyse von Spannungen und Tatsachenermittlung (FAST) system, 118–119, 123–129, 133–134, 138–139

Gaza, 180–181, 186
Geneva, 29, 31–32, 155–156, 167, 169
Geographic information systems (GIS), 33–34, 35, 44–45, 53, 62
GIS *see* Geographic information systems
Global North, 4–5, 19–20, 32–34, 54, 120–121, 230
Global politics, 4–6, 9, 14, 80, 189–190, 230
Global South, 30, 55, 97, 120, 121, 127, 199–200
GNC *see* Libya, General National Congress
GNU *see* Libya, Government of National Unity
GPTs *see* Artificial Intelligence, generative pre-trained transformers
Grievances, 1, 7–8, 28–29, 52–53, 77–78, 82–83, 94–95, 105, 146–148, 190–194, 198, 208–210, 212, 216, 218, 220, 227, 233
Guterres, António, 1

Hamas, 180–181, 186
Hardware, 3, 55–56, 67
Harmful Speech, 22, 47–48, 78–79, 82, 112–113, 202–203, 224–225, 228–229, 231
Hashtag Generation, 92, 94–95, 96–97, 108–109, 112–115
Hate speech, 71, 84, 87, 89–91, 95–96, 112–113
Hattotuwa, Sanjana, 32–37
Heidegger, Martin, 17, 21, 27–28, 41–48, 229–230
Himmelfarb, Sheldon, 36–39
Human-machine networks (peer-to-peer), 14–17, 21–22, 51–52, 58, 62, 66–69, 80–81, 84, 129–130, 134, 141–142, 151, 154–156, 159–160, 170–171, 186–187, 191, 224, 229
Humour, 81, 109
Hyperreality, 233–235

ICTs *see* Information and Communication Technologies
ICT4Peace Foundation, 31–32
ICT4Peace, 21, 27–28, 30–34, 36–37, 39–40, 50, 52
IGAD *see* Intergovernmental Authority for Development
Information and Communication Technologies (ICTs), 1–2, 31–34, 39–41, 42, 52–53, 55–56, 61–62, 70, 118–119, 121, 134–135
Instagram, 61–62, 68*f*
Intergovernmental Authority for Development (IGAD), 121, 123, 130, 145–146
  Conflict Early Warning and Response Mechanism (CEWARN), 121, 122–123, 125–127, 130–131, 138–139, 146
International Panel on the Information Environment (IPIE), 39
Internet, 14–15, 29–30, 63
  censorship, 17–18, 77–79, 198, 202–203, 206*f*
  filtering, 8–9, 199, 213–214
  shutdowns (blackouts), 70, 198, 202–203, 205–206, 211
IPIE *see* International Panel on the Information Environment
Iran nuclear deal, 29, 155–156
Israel, 179–181, 183

Jacobs, Irwin, 36–37

K-means clustering, 160–161
Kiir, Salva, 54–55, 102–103, 130–131, 147–148, 202–203, 204–206

Large language models (LLMs), 25, 45–46, 94, 124, 158, 162, 171–172, 232, 234
Larrauri, Helena Puig, 35–36
Lexicalism, 94–95
Liberal peacebuilding, 6, 10, 189–190
Liberation Tigers of Tamil Eelam (LTTE), 115, 210, 213–214
Libya, 167, 179–180
  Political Dialogue Forum (LPDF), 155, 165–166
  Berlin Conference, 167–168
  General National Congress (GNC), 166
  Government of National Unity (GNU), 167
LLMs *see* Large language models
LPDF *see* Libya, Political Dialogue Forum
LTTE *see* Liberation Tigers of Tamil Eelam

*Machine learning, 4, 26, 44, 52–53, 54, 62, 94, 114–115, 118–119, 155, 157–158, 234
Marginalization, 163–164, 180, 189–190, 196, 201–202, 218, 226–227
Martti Ahtisaari, 32
Masarat - the Palestinian Center for Policy Research & Strategic Studies, 180–181
Memory, 189

Messaging services (mobile), 2–3, 28–29, 61, 63, 66, 68*f*, 75–76, 148–151, 156, 176, 194–195, 208–209
Meta, 94–95
Misinformation, 2–3, 8–9, 28, 42–43, 71–72, 83–84, 98–99, 110, 124
Missing-from-hand, 17, 21, 41, 47–49, 58–59, 80, 85–86, 94–95, 110–111, 152, 186, 225–227, 228–230
Mobile applications, 3–4, 25, 54–55, 61, 131–132, 142–143, 149–150
Mobile phones, 3–4, 44, 55–56, 66, 89, 130, 132–135, 136–143
Morse, Samuel, 2–3

Nationalism, 8–9, 86–87, 105–106
  ethno-nationalist populism, 77
  Sinhala-Nationalist, 104–105, 107–108, 109–110, 210–211, 218
Natural Language Processing (NLP), 4, 15–16, 28, 94, 118–119, 122–124, 161–162, 171–172
NLP *see* Natural Language Processing
Nokia, 32
Northern Ireland, 55, 73–74, 78–79, 132–135, 196–197, 199–202
  Community Development Centre (CDC), 132–133, 134–135, 137
  Good Friday Agreement, 79, 132, 199–201
  Republicans, 199
  Unionists, 199–201
NPP *see* Sri Lanka, National People's Power Party
NSS *see* South Sudan, National Security Service

One Text Method, 32–33
Online conferencing, 28–29, 44, 45–46, 50–51, 68*f*, 156–157, 166–168
Online meetings, 1–2, 29, 66, 153–154, 169–170
Online consultations, 10–11, 16–17, 66–67, 157, 164–170, 226–227
Online polarization, 78–79, 190–191, 196–197, 199, 206–207, 230
OpenAI, 162
Oslo Accords, 181

PA *see* Palestine, Palestinian Authority
Palestine, 180–186
Palestinian Authority (PA), 181
PCCA *see* People's Coalition for Civic Action
Peace mediation *see* Peacemaking
Peace mediators, 25, 158–159, 161–162
  intermediary role, 156, 169–170
Peace Research Institute Oslo (PRIO), 118
Peacebuilding
  authoritarian, 7, 60
  everyday, 189–191, 198
  human activity, 2–3, 28–29
  international, 5–7, 26–27, 62, 65–66, 199–200
  liberal, 6–7, 60
  local, 1, 7, 10, 116–117, 127–130, 136
  organizations, 1–2, 15–16, 19, 36, 76, 77, 90, 92–93, 204
  sincere approach, 171–172
  subjunctive approach, 173, 178–180
Peacekeeping, 4–5, 31
Peacemaking (Peace mediation), 153
  hybrid intelligence, 154–163
  inclusive, 153–155, 163–165, 167–170
  multi-track model, 153, 154–155, 163–164
PeaceTech, 4, 18–19, 21, 27–28, 30–40, 43, 50, 52, 54–55, 80–81, 227–228, 229–232
  Accelerator, 39, 54–55, 131
  Lab, 36–39
  Pioneers, 50, 120, 122, 128–129, 229–230
PESTEL framework, 184
Political intimidation, 205–206, 213–214, 215
Political oppression, 17–18, 82–83, 202–203, 212, 216–217
Populism, 5–6, 77, 87, 210
Post-digital, 9, 21–22, 51, 55–58, 62, 228–229
Present-at-hand/Vorhandenheit, 42–47
PRIO *see* Peace Research Institute Oslo
PTA *see* Sri Lanka, Prevention of Terrorism Act

Qualcomm, 36–37

Racism, 84–85, 87–88, 90
Radio, 28–29, 36–37, 52–53, 75–76, 83, 86, 117, 131
Ready-to-hand/Zuhandenheit, 42–47
Recommender systems, 3–4, 64, 75, 175–176, 194–195
Reconciliation, 5, 189–190
Reconstituted Joint Monitoring and Evaluation Commission (R-JMEC), 204, 207
Remesh, 164–165, 166–171, 174–176
Remote Sensing, 25, 35, 41, 52–53, 117, 121, 130, 156–157, 234–235
R-JMEC *see* Reconstituted Joint Monitoring and Evaluation Commission
Rohingya, 86, 112

Satellite Technology, 35, 41, 52–53, 117, 121, 130, 131, 156–157, 234–235
Search for Common Ground (SfCG), 36–37, 43, 90–91, 136–141
SDGs *see* sustainable development, sustainable development goals
SentiOne, 95
SfCG *see* Search for Common Ground
Social embeddedness of technology, 4–5, 38–42, 228–229

Social media, 2–4, 15, 18–19, 43
activism, 203, 206–207, 208–210, 213–221
content moderation, 96–97
countering hate speech, 103–104, 112–113
fact checking, 98–104, 106–109, 201
memes, 195, 216–218
platforms, 15, 25, 61–62, 66–67, 71, 80, 82–83, 87, 90–94, 95–97, 202–203, 231
Social movements, 5, 36, 191, 194
Socio-political, 33, 82, 93–95, 149–150
Socio-technicality, 9–11, 13–14, 50–51, 59–60, 229–232
Software, 3, 25, 28, 32–33, 44–45, 67, 80–81, 112, 125, 137
Solomon, Richard, 36–37
Somoud, 182–184, 185–186
South Africa, 34, 97
South Sudan, 17–19, 22–24, 28–30, 53, 54–55, 69–73, 73*f*, 74–80, 82–83, 85–86, 88–89, 98–100, 102–103, 109–111, 118, 121, 126–128, 130–132, 133–134, 136–138, 141–142, 145–151, 162, 194–198, 201, 202–210, 211–215, 218, 220–222, 225–227
civil war, 136–137, 147–148, 202–203
Conflict Early Warning and Early Response Unit (CEWERU), 121, 126–127, 130–131, 141–142, 143–144
inter-communal conflict, 148–149, 202–203, 206–208
local government, 53, 102–103, 130–131, 148–149
National Security Service (NSS), 103, 147–148, 205–206, 208–209
Peace and Reconciliation Commission (SSPRC), 130–131, 141–142, 143–144, 145–148
People's Coalition for Civic Action (PCCA), 77, 208–210, 214
Sudan People's Liberation Movement (SPLM), 208
Sudan People's Liberation Movement-In Opposition (SPLM-IO), 147–148
traditional authorities, 53, 102–103, 141–142, 144, 148–149
SPLM *see* South Sudan, Sudan People's Liberation Movement
SPLM-IO *see* South Sudan, Sudan People's Liberation Movement-In Opposition
Sri Lanka, 29–30, 32–33, 71, 73–74, 77–78, 104–109, 210–223
Aragalaya, 78, 198, 210–221, 227
Bodu Bala Sena, 211
civil war, 77, 216, 218–219
cultural heritage, 105–106
Easter Sunday terror attacks, 211
Janatha Vimukthi Peramuna (JVP) Party, 101
Jathika Jana Balawegaya, 221
Muslim minority, 107, 114–115, 211, 213–214, 216–219
National People's Power (NPP) Party, 101, 221
Prevention of Terrorism Act (PTA), 213–214, 216–217
Rajapaksa regime, 77–78, 101, 210–212, 213–216, 218, 220–221
Sinhala majority, 104–105, 106–107, 210–212
Sinhala-Buddhist nationalism, 101, 104–105, 107–109, 211
Social Media Monks, 83, 109–110
Sri Lanka Unites, 212–213
Tamil minority, 77–78, 104–108, 210. 216–217, 218–219
SSPRC *see* South Sudan, Peace and Reconciliation Commission
Stauffacher, Daniel, 31, 33
Strategic communication, 5, 193–194, 203–204, 207, 232
Surveillance, 8–9, 35, 43, 70, 76–77, 101, 120, 179, 180, 213–214, 215–216
Sustainable development, 33, 90
Sustainable Development Goals, SDGs, 26
Swisspeace (Swiss Peace Foundation), 122
Switzerland, 30–32, 166

Tech-solutionism, 4, 46, 117–119, 122–123, 125–126, 131–132, 134–135, 154–155, 228–229
Technical innovation, 1–2, 8, 45–46
Telegraphy, 2–3
Television, 29, 52–53, 78–79, 86, 167, 190, 233
Term Frequency-Inverse Document Frequency (TF-IDF), 114–116, 160
Text mining, 66–67, 114–115, 123–124, 172, 229–230
TF-IDF *see* Term Frequency-Inverse Document Frequency
Tool-centrism, 3–5, 8, 14, 30–32, 36–37, 40–41, 231–232
Transitional justice, 10, 28–29, 189–190, 211–212, 221
Truth seeking, 189–190, 211–212
Tunis, 31
Tut Makuach (Nuer Prophet), 147–148
Twitter/X, 66–67, 122–123, 193–194, 201, 207–208

UN *see* United Nations
Uncertainty, 110–111, 123–124, 156, 172–173
UNDP *see* United Nations, Development Programme
United Kingdom, 79, 132–133, 199

United Nations (UN), 1, 15, 19, 28–29, 43–45, 84, 89–90
  Department of Political and Peacebuilding Affairs (DPPA), 39, 43, 60–61, 113–114
  Development Programme (UNDP), 35, 43, 54–55, 113–115, 131, 142, 145–146, 149–150
  General Assembly, 91
  Global Pulse, 35
  Human Rights Council, 211–212
  Innovation Cell, 35, 39, 158–159
  peace operations, 26, 28–29, 35, 43, 44–45, 55–56, 101–102
  Peacebuilding Fund, 60–61, 90–92
  Secretary-General, 31, 90–91, 119–120
  Security Council (UNSC), 6–7, 15–16, 90, 121, 127–128, 146–147, 159
  Special Representatives of the Secretary-General, 153, 166
  United Nations Mission in South Sudan (UNMISS), 213
  United Nations Support Mission in Libya (UNSMIL), 154–155, 165–170
United States (US), 30, 79, 87–88, 118–119, 199–200
United States Institute of Peace (USIP), 36–37, 39–40
UNMISS *see* United Nations, United Nations Mission in South Sudan
UNSC *see* United Nations, Security Council
UNSMIL *see* United Nations, United Nations Support Mission in Libya
US *see* United States
USIP *see* United States Institute of Peace

Voice, 164, 174, 177–178, 179–180, 187–188, 210, 226–227
Volunteers, 15–16, 19, 35, 56–57, 76, 80–81, 96, 98, 132–134, 137–139, 151, 193

West Bank, 181
WhatsApp, 98–99, 148–149, 206–207
  groups, 89–90, 137, 206–208, 215–216
White Army, 147–148
Williams, Stephanie (Acting Special Representative of the UN), 166, 167–168, 169–170
WSIS *see* World Summit on the Information Society
World Summit on the Information Society (WSIS), 31–33, 34–35

X *see* Twitter

Zoom, 154, 166, 168